Management of Lakes and Reservoirs during Global Climate Change

NATO ASI Series

Advanced Science Institutes Series

A Series presenting the results of activities sponsored by the NATO Science Committee, which aims at the dissemination of advanced scientific and technological knowledge, with a view to strengthening links between scientific communities.

The Series is published by an international board of publishers in conjunction with the NATO Scientific Affairs Division

A	**Life Sciences**	Plenum Publishing Corporation
B	**Physics**	London and New York
C	**Mathematical and Physical Sciences**	Kluwer Academic Publishers
D	**Behavioural and Social Sciences**	Dordrecht, Boston and London
E	**Applied Sciences**	
F	**Computer and Systems Sciences**	Springer-Verlag
G	**Ecological Sciences**	Berlin, Heidelberg, New York, London,
H	**Cell Biology**	Paris and Tokyo
I	**Global Environmental Change**	

PARTNERSHIP SUB-SERIES

1.	**Disarmament Technologies**	Kluwer Academic Publishers
2.	**Environment**	Springer-Verlag / Kluwer Academic Publishers
3.	**High Technology**	Kluwer Academic Publishers
4.	**Science and Technology Policy**	Kluwer Academic Publishers
5.	**Computer Networking**	Kluwer Academic Publishers

The Partnership Sub-Series incorporates activities undertaken in collaboration with NATO's Cooperation Partners, the countries of the CIS and Central and Eastern Europe, in Priority Areas of concern to those countries.

NATO-PCO-DATA BASE

The electronic index to the NATO ASI Series provides full bibliographical references (with keywords and/or abstracts) to more than 50000 contributions from international scientists published in all sections of the NATO ASI Series.
Access to the NATO-PCO-DATA BASE is possible in two ways:

– via CD-ROM "NATO-PCO-DATA BASE" with user-friendly retrieval software in English, French and German (© WTV GmbH and DATAWARE Technologies Inc. 1989).

The CD-ROM can be ordered through any member of the Board of Publishers or through NATO-PCO, Overijse, Belgium.

Series 2: Environment – Vol. 42

Management of Lakes and Reservoirs during Global Climate Change

edited by

D. Glen George
NERC Institute of Freshwater Ecology,
Windermere Laboratory,
Far Sawrey, Ambleside, Cumbria, U.K.

J. Gwunfryn Jones
Freshwater Biological Association,
The Ferry House,
Far Sawrey, Ambleside, Cumbria, U.K.

Pavel Punčochář
T.G. Masaryk Water Research Institute,
Prague, Czech Republic

Colin S. Reynolds
Freshwater Biological Association,
The Ferry House,
Far Sawrey, Ambleside, Cumbria, U.K.

and

David W. Sutcliffe
Freshwater Biological Association,
The Ferry House,
Far Sawrey, Ambleside, Cumbria, U.K.

Springer-Science+Business Media, B.V.

Proceedings of the NATO Advanced Research Workshop on
Management of Lakes and Reservoirs during Global Climate Change
Prague, Czech Republic
11–15 November 1995

A C.I.P. Catalogue record for this book is available from the Library of Congress.

ISBN 978-94-010-6085-1 ISBN 978-94-011-4966-2 (eBook)
DOI 10.1007/978-94-011-4966-2

Printed on acid-free paper

CONTENTS

Working Group Topic

Detecting and predicting the responses of lakes to global climate change

Working Group Topic

Working Group Topic

PREFACE

If present trends continue, most climatologists agree that the concentration of carbon dioxide in the atmosphere will have doubled by the year 2050. This increase in CO_2 will have a major effect on the global climate and substantially alter the physical, chemical and biological characteristics of lakes throughout the world. In recent years, it has become clear that year-to-year changes in the weather have a major effect on the seasonal dynamics of lakes. Many water quality problems that were once regarded as "local" phenomena are now known to be influenced by changes in the weather that operate on a regional or even global scale. For example, blooms of toxic blue-green algae can be induced by prolonged reductions in the intensity of wind-mixing as well as increased supplies of nutrients. Long-term studies in the English Lake District have shown that many of these variations are quasi-cyclical in nature and can be related to long-term changes in the distribution of atmospheric pressure over the Atlantic Ocean. It is not yet clear what effect these changes have on the dynamics of European lakes but much of the historical data required to extend these analyses to continental Europe is already available. In the early 1970s the International Biological Programme served as a particularly effective focus for comparative limnological research in eastern as well as western Europe. Since then, there has been a tendency to underestimate the value of such integrated studies but several European laboratories still maintain long-term records of the changes recorded in a variety of lakes and reservoirs.

This NATO advanced research workshop was planned as an introductory meeting where limnologists from a number of European countries would be able to renew earlier contacts and plan integrated studies that would consider the impact of regional as well as local changes in the weather. The workshop was organised by Dr Pavel Punčochář at the T. G. Masaryk Water Research Institute in Prague, and Professor Gwynfryn Jones at the Freshwater Biological Association, Ambleside, UK. It included special contributions from Professor T. Davies at the Climate Research Unit in East Anglia, and Dr T. K. Kratz from the Long Term Ecological Research Group in Wisconsin, USA. The organisers also take this opportunity to thank the NATO Science Committee for their generous support of the workshop, held in Prague during November 1995.

This volume of invited contributions, on the management of lakes and reservoirs during global climate change, has been edited by Dr Glen George at the Institute of Freshwater Ecology, Windermere, UK, Professor Colin Reynolds at the Freshwater Biological Association, Ambleside, UK, and Dr David Sutcliffe at the FBA, who also prepared the camera-ready copy for the publisher. The editors thank all of the contributors to the volume and participants at the workshop: a list of these is given below. Thanks are also due to Dr Jack Talling for commenting on several manuscripts, Hayley Jump and Rosalie Sutcliffe for retyping some manuscripts, and Trevor Furnass for redrawing and adapting numerous text-figures.

LIST OF PARTICIPANTS

Adrian, R. (Mrs)
Institut für Gewässerökologie und Binnenfischerei, Müggelseedamm 260, D-12587 Berlin, Germany.

Bojanovsky, B.
Dept. Ecology & Environmental Protection, Sofia University, Dragan Tzancov Street 8, 1421 Sofia, Bulgaria.

Buchtele, J.
Institute of Hydrodynamics, CAS, Podbabská 13, 166 12 Prague 6, Czech Republic.

Bucka, H.
K. Starmach Institute of Freshwater Biology, PAS, SŁawkowska 17, 31-016 Kraków, Poland.

Constantinescu, T. L.
Romanian Water Authority, Edgar Quinet Str. 6, 70106 Bucharest, Romania.

Cruz-Pizarro, L.
Institute del Agua, University of Granada, Granada, Spain.

Davies, T. D.
Climate Research Unit, University of East Anglia, Norwich, Norfolk NR4 7TJ, UK.

Desortová, B. (Mrs)
T. G. Masaryk Water Research Institute, Podbabská 30, 160 62 Prague, Czech Republic.

Drabkova, V. G. (Mrs)
Institute for Lake Research RAS, Sevastyanov 9, 196199 St. Petersburg, Russia.

Duras, J.
Vltava River Board Corp., Denisovo Nabrezi 14, Plzeň, Czech Republic.

Gaedke, U. (Mrs)
Limnologisches Institute, Universität Konstanz, D-78434 Konstanz, Germany.

George, G. D.
Institute of Freshwater Ecology, Windermere Laboratory, Far Sawrey, Ambleside, Cumbria LA22 OLP, UK.

Gerdeaux, D.
Laboratoire d'Hydrobiologie Lacustre INRA, 75 Av. de Corzent, BP 511, F74203 Thonon Cedex, France.

Havel, L.
T. G. Masaryk Water Research Institute, Podbabská 30, 160 62 Prague 6, Czech Republic.

Holobrada, M. (Mrs)
Institute of Water Management, Nabr. gen. Svobodu; 81249 Bratislava, Slovak Republic.

Horicka, Z. (Mrs)
Dept. of Hydrobiology, Faculty of Life Sciences, Charles University, Vinicna 7, 120 00 Prague 2, Czech Republic.

Hudec, I.
Zoological Institute, SAS, Lofferova 10, 040 00 Kosice, Slovak Republic.

Jones, J. G.
Freshwater Biological Association, The Ferry House, Far Sawrey, Ambleside, Cumbria LA22 OLP, UK.

Keskitalo, J.
Lammi Biological Station, University of Helsinki, Paajarventie, FIN-16900 Lammi, Finland.

Kratz, T.
Center for Limnology, University of Wisconsin-Madison, 10810 County Highway, Boulder Junction, Wisconsin 54512, U.S.A.

Mineeva, N. (Mrs)
Institute for Biology of Inland Waters, Borok, 152742 Yaroslavl, Russia.

Ozimek, T. (Mrs)
Dept. of Hydrobiology, University of Warsaw, Banacha 2, 02 097 Warsaw, Poland.

Padisák, J.
Balaton Limnological Institute, Hungarian Academy of Science, H-8237 Tihany, Hungary.

Punčochář, P.
T. G. Masaryk Water Research Institute, Podbabská 30, 160 62 Prague 6, Czech Republic.

Reynolds, C. S.
Freshwater Biological Association, The Ferry House, Far Sawrey, Ambleside, Cumbria LA22 OLP, UK.

Rosendorf, P.
T. G. Masaryk Water Research Institute, Podbabská 30, 160 62 Prague 6, Czech Republic.

Straškrabová, V. (Mrs)
Hydrobiological Institute ASCR, Na Sadkach 7, 370 05 Ceske Budejovice, Czech Republic.

Straškraba, M.
Biomat. Laboratory, Entomological Institute, CAS, Branisovska 31, 370 05 Česke Budejovice, Czech Republic.

Winfield, I. J.
Institute of Freshwater Ecology, Windermere Laboratory, Far Sawrey, Ambleside, Cumbria LA22 OLP, UK.

CHANGES IN ATMOSPHERIC CIRCULATION AND CLIMATE OVER THE NORTH ATLANTIC AND EUROPE

T. D. DAVIES, D. VINER AND P. D. JONES
Climatic Research Unit
University of East Anglia
Norwich NR4 7TJ, UK

1. Introduction

This paper is a survey of some of the recent changes in features of the atmospheric circulation that influence some aspects of European climate. It is written with its role as a scene-setter for a volume on the management of lakes and reservoirs in Europe during environmental changes very much in mind. In particular, since some following contributions look to the possible links between lake behaviour and conditions well to the west of Europe, there is emphasis on the partial control of European climate by atmospheric and oceanic circulations in the North Atlantic. The focus is also on the recent observed climatic history, rather than on projections of possible future climates. Information on plausible future climate scenarios is available from sources such as [1] or, in more regional detail, via the Climate Impacts LINK Project at the Climatic Research Unit, University of East Anglia [2]. The period of the last 150 years of climate history which is considered here has received much attention because it is the period with best records and because part of its character is almost certainly a result of the response of the atmosphere and oceans to global warming caused by anthropogenic emissions of greenhouse gases and sulphate aerosol.

2. The last century in context

Europe has the most detailed climate history of any part of the world [3]. Instrumental observations were pioneered in Europe, and many climate reconstruction techniques were tested in the continent. Consequently, the past climate record is known in some considerable detail over the last few hundred years. In spite of this, there is considerable uncertainty over the spatial cohesiveness – over the continent – of decadal-scale changes [3, 4]. This emphasises the need to consider subcontinental-scale climate responses to

D.G. George et al.(eds.), Management of Lakes and Reservoirs during Global Climate Change, 1–13.

large-scale changes in the atmospheric circulation.

Since 1500 AD, Europe has seen noticeably warm and noticeably cool decades, but relatively few have been synchronous over the whole of the European continent. Jones & Bradley [3] assert that the evidence for the assumed period of protracted Europe-wide coolness embracing the 16th to the 19th centuries is actually rather weak, but that there do appear to have been marked and widespread cool periods in the 17th and 19th centuries. Since the 1920s, unusually warm conditions have prevailed, possibly related to the relative absence of large volcanic eruptions and, in all probability, to increasing concentrations of greenhouse gases due to human activities [5]. Nevertheless, sub-regional differences have been pronounced even in the climate record of the most recent decades; these need to be borne in mind in any assessement of European climate changes over the last century-or-so. When interactions between the overlying atmosphere and lakes are to be considered, it is also necessary to take into account meso-scale and local-scale circulations and effects. This is especially so when the lake locale is mountainous. So, the precise character of the changing climate in a lake-basin will depend – to a greater or lesser extent – on sub-regional and local influences. Nevertheless, there are elements of the changing circulation and climate on a larger-scale which should be considered as a starting point for the development of an approach to the management of lakes and reservoirs during global change.

3. An Atlantic atmospheric circulation control on the European climate

Two "centres of action" dominate the surface pressure pattern over the eastern Atlantic Ocean; the Iceland Low and the Azores High. A useful index to describe atmospheric circulation conditions "upstream" of Europe is the difference in pressure over the Azores region and pressure over Iceland, which characterises a feature known as the North Atlantic Oscillation (NAO) (see, for example, [6]). The oscillation is the link between the two "centres of action"; when the Azores High is more intense (higher pressure), the Iceland Low also tends to be be more intense (lower pressure). This NAO "signal" over years is present for all seasons, although it changes its precise character with the seasons. Such linked behaviour, over geographically widely-separated regions, is known as a teleconnection. The NAO is an important component of the interannual variability of the whole Northern Hemisphere circulation [7]. In the region of the Northern Hemisphere north of 20°N, it is more important than variations in the Southern Oscillation (see, for example, [8, 9]), the most globally-important pressure oscillation.

The winter season is the period of greatest interest, since this is when the strongest pressure gradients and greatest interannual variability occurs [10]. There is a "seesaw" in winter temperature severity which has been associated with extreme behaviour of the NAO [11]. When the NAO index has high values, there is a strong south-to-north pressure gradient, producing strong westerly flow over the North Atlantic (during high NAO index winters, the westerlies onto Europe are over 8 m s^{-1} stronger than during low NAO winters [6]), leading to warm conditions over much of Europe (except for the eastern Mediterranean where greater cold is experienced), and cold conditions in the Labrador and Greenland region (Fig. 1). The opposite phase condition occurs when there is a reversal in the pressure gradient between the two "centres of action" (so the

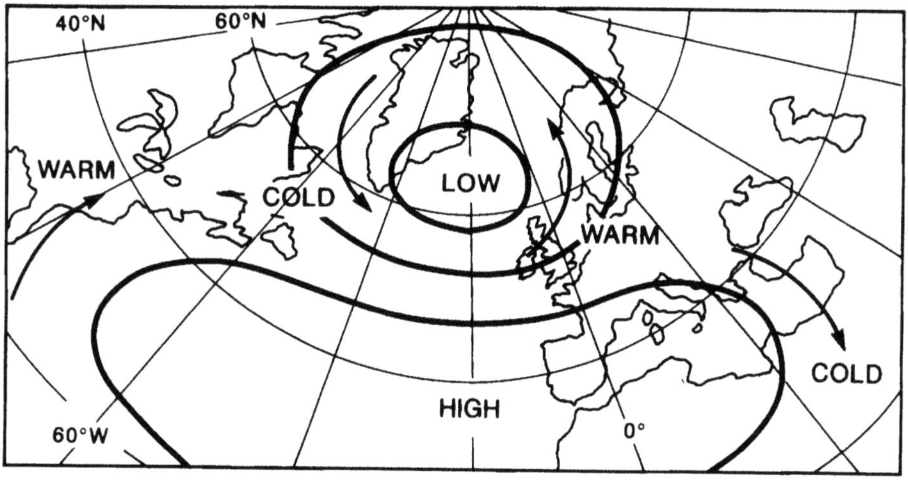

FIGURE 1. Schematic map showing the relationship between high values of the NAO (strong westerly flow) and warm and cold conditions in Europe and the Labrador region, respectively.

flow portrayed in Fig. 1 would be reversed), usually accompanied by persistent blocking highs in the region between Greenland and Scandinavia [10]. During such events, there are unusually cold winters in Europe. A reversal, or pronounced weakening, of the flow in summer is often associated with drought conditions over many parts of Europe.

Fig. 2 shows the record of the winter NAO index since 1861. The lowest winter values in recent decades occurred during much of the 1960s and in 1995–96, when blocking highs were persistent over Scandinavia or the eastern North Atlantic. Both winters were extremely cold in parts of Europe. The run of high NAO index values in most recent winters, on the other hand, has been associated with relatively mild winters in Europe. The contrast between 1995–96 and the previous 15 years made this recent winter appear very unusual. From the turn of the 20th century to ca. 1930, the NAO index values were high, leading to strong westerly winds across the Atlantic, and warmer than normal winter temperatures over much of Europe [12]. From the 1940s to the early 1970s, the winter NAO index exhibited a decline, and winter temperatures in Europe were often anomalously low [13]. The past three decades have experienced increasing values of the winter NAO index. Hurrell [6] points out that this recent behaviour of the NAO index is linked with the anomalous coldness of the last decade-or-so in the eastern Mediterranean, and the very warm conditions over Scandinavia and the rest of northern Europe. The summer NAO index values are much less variable, but similar decade-scale oscillations are also apparent in this record.

The links between the NAO and European climate are more complex than those implied in Fig. 1; there is more to the relationship than the stronger advection of warm maritime air under winter high NAO index conditions. Changes in atmospheric circulation patterns over the North Atlantic, which the NAO index is reflecting, are

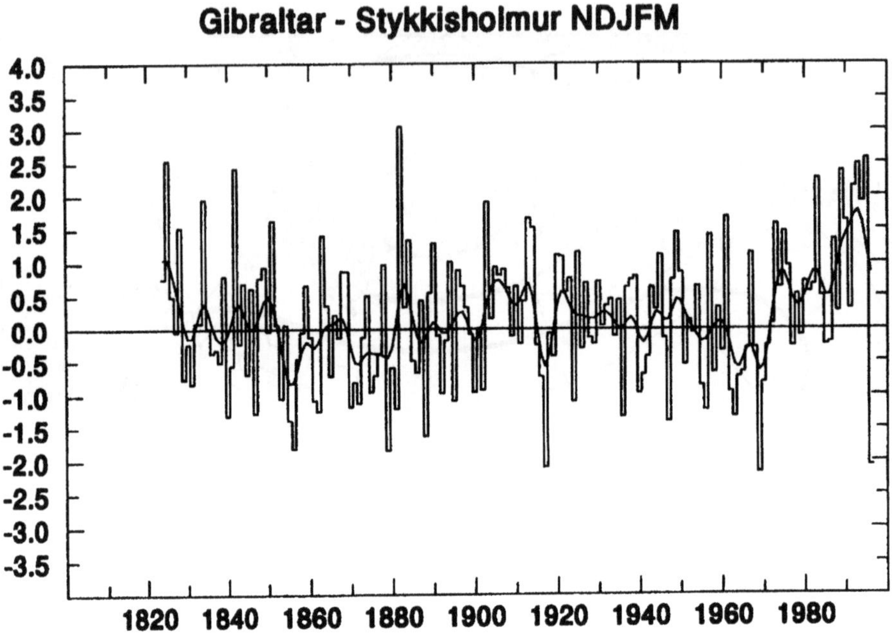

FIGURE 2. The winter (in this case, November to March) North Atlantic Oscillation Index. The value represents the difference between the normalised sea-level pressure at Gibralter and the normalised sea-level pressure at Stykkisholmur, Iceland. The heavier line represents a 10-point Gaussian smoothing.

associated with shifts in storm tracks and depression activity [14, 15]. Such changes influence atmospheric moisture transport. When the NAO index has a high value, in line with storms pushing further into northern Europe, the zone of maximum moisture transport extends further to the east and north into northern Europe and Scandinavia [6]. There is an associated reduction of moisture transport into southern Europe and the Mediterranean. These relationships are reflected in precipitation amounts across Europe, with winter precipitation amounts being greater across northern Europe and Scandinavia, and lower across southern Europe and the Mediterranean, at times of high NAO index value [6]. The reverse situation occurs during negative or low NAO index phase years, as in 1995–96.

Briffa et al. [16] examined the frequency of different weather types over the British Isles since 1861, indicating that there had been significant changes in their frequency. The most obvious change was a decline in westerly types since the early decades in the 20th century, and a corresponding increase in the frequency of anticyclonic and cyclonic types since the 1940s. Wanner [17] described changes in the main categories of European-scale weather-types since the 1880s; these changes were comparable to the changes over the British Isles. Since the 1970s there has been a marked increase in the frequency of south-westerly weather types over the British Isles (Fig. 3). This is

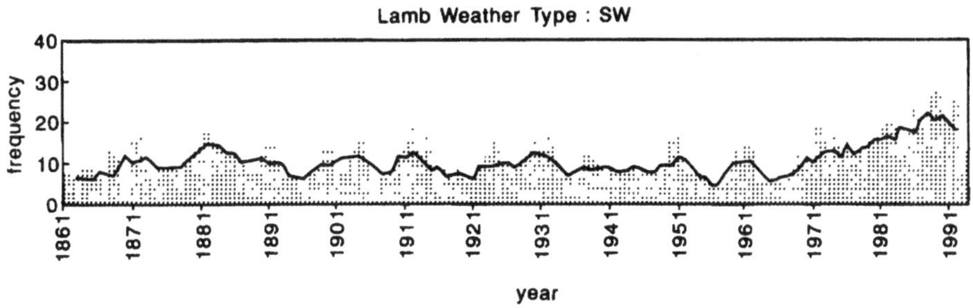

FIGURE 3. Annual frequency of the south-westerly Lamb Weather Type over the British Isles.

consistent with an increase in the frequency of European-scale daily weather types with a southerly component [18, 19], and is regarded as relating directly to the recent positive trend in the NAO index [20]. Inevitably, the records of weather type frequency over many decades are not in perfect correspondence with the NAO index record, indicating that the index does not capture all the relevant information about the nature of the circulation "upstream" of Europe which is pertinent to the observed climate of Europe, or that the relationship is not stationary as other factors come into play. Nevertheless, it does encapsulate sufficient information for it to be a very useful indicator of the behaviour of a geographical region of the atmosphere which is known to influence the surface climate of Europe – although there is much spatial variation in that influence (e.g. [6]).

4. Links with the ocean

The NAO has been linked to sea-surface temperatures (SST) in the North Atlantic. From year to year the SST patterns are probably forced by the atmospheric circulation – the winds changing the air–sea energy fluxes [21]. The interactions are, however, complicated and there is evidence that SST patterns in the western part of the North Atlantic influence the weather of north-western Europe on time-scales of months [22]. Warm SST anomalies in this part of the ocean tend to precede a greater incidence of cyclonic weather types over north-west Europe in the following months, whereas a cold SST anomaly is frequently followed by months that are more anticyclonic in character. The precise linking mechanism between the ocean and the atmosphere appears to be related to the shift in the position of the maximum surface temperature gradient, affecting the formation and path of depressions. The links have been shown to be stronger in some decades of this century compared to others, giving rise to periods with good forecast potential and others with poor potential.

SST patterns in parts of the North Atlantic are intimately linked with sea-ice distributions, and there are clearly established relationships between sea-ice extent around Iceland and European climate [23]. Periods of persistent northerly winds over the northern Atlantic and western Europe, for example, bring cooling to the continental landmass and ice to the northern shores of Iceland. The high NAO index values of the early decades of the 1900s denoted strong winds and greater warm air advection, and ice was a relatively infrequent visitor to Iceland. The long recorded history in Iceland and the changing periods of ice prevalence, particularly on the northern coasts, provide potential for a much longer record of NAO variability [24].

Over time-scales of several years to decades, although two-way interactions between the atmosphere and ocean still operate, there are indications that the SST anomalies (this time over a larger area of the ocean) are playing an important role in forcing the circulation of the overlying atmosphere [25, 26]. There is evidence for rapid and pronounced climate fluctuations in the Atlantic from historical and proxy records [6, 24]. Some may be related to fluctuations in the NAO [21, 25] and, in this regard, there has been considerable interest in the deep and vigorous overturning in the Atlantic Ocean which is due to the formation of North Atlantic Deep Water (NADW). The sinking of this watermass in the northern-most parts of the ocean causes the Gulf Stream and its North Atlantic Drift extension to turn more northward to replace the sinking water. This does not happen in the Pacific Ocean which has no deep water formation. The sinking of the NADW is a consequence of its high density, due to its high salinity as well as its low temperature. It is self-sustaining, to a degree, since the high salinity is due to the northward transport of saline water from tropical latitudes; its warmth is also a factor, since the induced evaporation further increases salinity. The sinking is sensitive to changes in the input of fresh water to the North Atlantic, and so it is possible that the overturning in the Atlantic Ocean could vary in strength, stop, or even reverse. Such switches could occur over periods of hundreds of years, with important repercussions for the atmospheric circulation over the North Atlantic and the climate of Europe, although the linkages are likely to be complex [26]. Such reversals have clearly occurred in the distant past during Ice Ages, and in the rapid deglaciation phases (e.g. the Younger Dryas; [27]).

There are some tantalising indications from computer models that the sinking in the North Atlantic may fluctuate over shorter time-scales; possible oscillations of around 40 to 60 years have been reported [28]. The reason for such oscillatory behaviour is uncertain, but some aspects may be triggered by a short-term change in the input of fresh water in the sinking region (north of 60°N), associated with ice input from the Arctic and its melt, or even precipitation); other aspects might be more self-sustaining. The implication for European climate is that these oscillations may be reflected in changes in the SST patterns of the North Atlantic. The reason why these computer model results are tantalising is because climate reconstructions from parts of Europe also exhibit oscillatory-type behaviour on the same time-scale [28].

The teleconnection character of the NAO has been mentioned. Globally, the most pronounced teleconnections are those associated with the El Nino Southern Oscillation (ENSO), which is a surface pressure oscillation across the tropical Pacific. Although strong ENSO signals have not been detected in the atmospheric circulation over the North Atlantic or Europe, there is some evidence of possible linkages with European-scale weather patterns. Fraedrich [29, 30] identified associations between warm and cold

ENSO events and European synoptic types, and between ENSO extremes and European-scale climate anomalies. Wilby [31] has also noted some weak links between ENSO and the frequency of cyclonic and anticyclonic weather types and precipitation over the British Isles.

5. Observed surface climate variables in Europe

The last 100-or-so years have been characterised by global warming, the patterns of which are generally consistent with simulations from General Circulation Models (GCMs) forced by increasing greenhouse gas concentrations and atmospheric sulphate aerosol concentrations [31]. Hurrell points out that changes in the NAO have contributed to the pattern of temperature change across Europe [6]. The changes in the NAO may be part of the global response to an enhanced greenhouse gas effect, or they may be partly- or un-related to such a response.

The observed annual temperature records for sectors of Europe are shown in Fig. 4. Since these are annual means, they are not directly comparable with the NAO index values displayed in Fig. 2, which are for the period November–March. However, there are some related features. All sectors experienced a rise in temperatures in the first half of the 20th century, although the timing varied by sector (in fact, a pronounced warming occurred in the last decade of the 19th century in "middle" Europe). There then followed a flattening off in the curves; to be supplanted by a temperature rise in recent decades, which is most pronounced in "northern" and "middle" Europe (and the western Mediterranean; the eastern Mediterranean has experienced a recent temperature decline, in line with the NAO changes in winter).

Annual precipitation amounts in certain parts of Europe have exhibited pronounced trends over time. Fig. 5 shows an increase between ca. 1910 and 1930 for "northern" Europe, and a pronounced increase since the 1970s. In the rest of Europe the recent decrease in the "southern" sector is most marked, and has led to great concern about "desertification" in the Mediterranean Basin. These observations are broadly consistent with the links between changes in the NAO and precipitation outlined in the previous section (remembering that the precipitation data are annual totals).

Changes in climate variables such as mean annual temperature and precipitation, described above, reflect changes in atmospheric circulation patterns, conveniently encapsulated by weather type classifications, such as the Lamb or Grosswetterlagen categories. Weather types encapsulate much broader weather conditions, including cloud-cover, wind speed, and even air pollution concentrations. For example, Palutikof et al. [32] linked wind speed variations over the British Isles to changes in the frequency of synoptic types, and Davies et al. [33] explained much of the variation in precipitation composition at sites in the UK by variations in weather types. Ohmura et al. [34] observed a marked decrease in global radiation at stations in central Europe, by as much as 20%, from 1959 to the end of the 1970s, with a recovery thereafter. A substantial part of these changes result from cloud cover variations, and the radiation fluctuations were considered to be the main reason for changes in the rate of evapotranspiration at a hydrological station near Zurich [35].

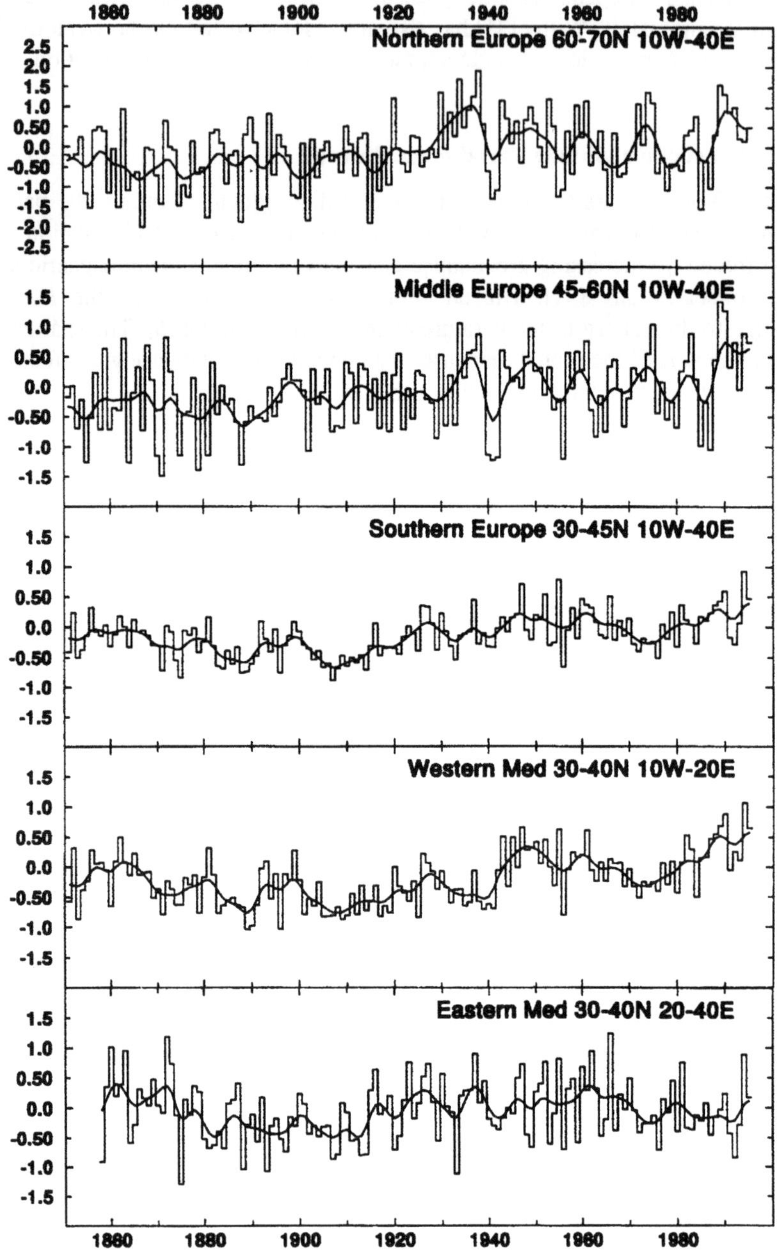

FIGURE 4. Annual mean areally-averaged temperatures for (land-areas in) sectors of "Europe". The values are annual departures (°C) from the mean for 1961 to 1990. The heavier line represents a 10-point Gaussian smoothing.

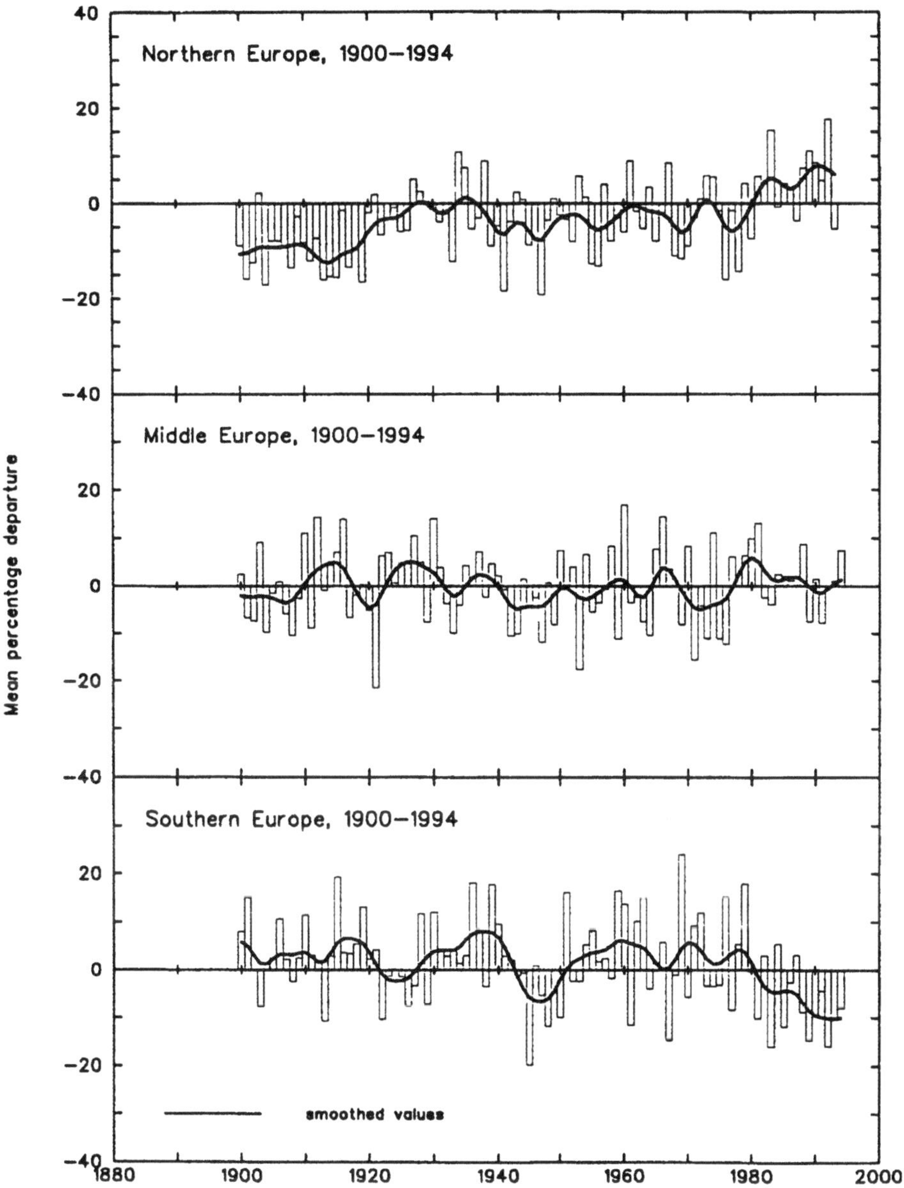

FIGURE 5. Annual mean areally-averaged precipitation for sectors in Europe. The values are percentage departures from the mean for 1961 to 1990. The heavier line represents a 10-point Gaussian smoothing (figure provided by Mike Hulme, Climatic Research Unit, University of East Anglia).

It should be borne in mind that not all of the observed changes in a particular climate variable may be explained by changes in weather types. With global warming, overall tropospheric temperature and moisture content have increased. Although the evidence is not unequivocal, there are indications that storms in the North Atlantic have recently become more intense [36], so the relationships between weather types and individual climate variables will not be stationary.

6. Conclusions

There are links between certain aspects of the European climate and the atmospheric circulation over the North Atlantic. Interactions with the Atlantic Ocean circulation are important, although they are complex and vary with time-scale. The interactions may have changed as a result of Man's activities, and they may change in the future. Some recent observed changes are consistent with the climate system responding to the continuing enhanced greenhouse gas effect, modified by the effects of atmospheric aerosols and dust. However, there is little doubt that "natural variability" has played a role, and will continue to play a part in the future. There is an overwhelming consensus that the European climate will continue to change because of global warming. A challenge will be to separate out the "background effects" (possible increased cloudiness/precipitation due to a greater atmospheric moisture content) from those resulting from changes in the frequency of weather types (with their individual suites of weather variables), also caused by global warming. A related and immense challenge is to "downscale" the projections of atmosphere/ocean computer models to produce future climate scenarios for individual lake basins. This is problematic at present, but needs to be pursued vigorously over the coming years if sensible strategies are to be developed for lake and basin management.

References

[1]. Carter, T. R., Parry, M. L., Harascuva, H. & Nushioku S. (1994). *IPCC technical guidelines for assessing climate change impacts and adaptations.* University College, London. 59 pp.

[2]. Viner, D. & Hulme, M. (1993). *The climate impacts LINK project. Providing climate change scenarios for impacts assessment in the UK.* Climatic Research Unit, University of East Anglia, UK.

[3]. Jones, P. D & Bradley, R. S. (1995). Climatic variations over the last 500 years. In: *Climate since A.D. 1500* (eds. R. S. Bradley & P. D. Jones). Routledge, London, pp. 649-665.

[4]. Bradley, R. S. & Jones, P. D. (1993). "Little Ice Age" summer temperature variations; their nature and relevance to recent global warming trends. *The Holocene* **3**, 367-376.

[5]. Schimel, D. et al. (1996). Radiative forcing of climate change. In: *Climate change 1995, the science of climate change* (eds. J. T. Houghton, L. G. Meira Filho, B. A. Callander, N. et al.). Cambridge University Press, 572 pp.

[6]. Hurrell, J. W. (1995). Decadal trends in the North Atlantic Oscillation: regional temperatures and precipitation. *Science* **269**, 676-679.

[7]. Hurrell, J. W. (1996). Influence of variations in extratropical wintertime teleconnections on Northern Hemisphere temperature. *Geophys. Res. Lttrs*, **23**, 665-668.

[8]. Ropelewski, C. F. & Jones, P. D. (1987). An extension of the Tahiti-Darwin Southern Oscillation Index. *Monthly Weather Review* **115**, 2161-2165.

[9]. Trenberth, K. E. & Hoar, T. J. (1996). The 1990-1995 El Nino-Southern Event: longest on record. *Geophys. Res. Lttrs* **23**, 57-60.

[10]. Moses, T., Kiladis, G. N., Diaz H. F. & Barry, R. G. (1987). Characteristics and frequency of reversals in mean sea-level pressure in the North Atlantic sector and their relationship to long-term temperature trends. *Journ. Climatology* **7**, 13-30.

[11]. Van Loon, H. & Rogers, J. C. (1978). The seesaw in winter temperatures between Greenland and Northern Europe: Part I: General Description. *Monthly Weather Review* **106**, 296-310.

[12]. Parker, D. E. & Folland, C. K. (1988). The nature of climatic variability. *Meteorol. Magazine* **117**, 201-210.

[13]. Van Loon, H. & Williams, J. (1976). The connection between trends of mean temperature and circulation at the surface, Part 1. Winter. *Monthly Weather Review* **104**, 365-380.

[14]. Rogers, J. C. (1990). Patterns of low-frequency monthly sea-level pressure variability (1899-1986) and associated wave cyclone frequencies. *Journ. Climate* **3**, 1364-1379.

[15]. Hurrell, J. W. (1995). Turbulent eddy forcing of the rotational flow during the Northern Winter. *Journ. Atmospheric Science* **52**, 2286-2301.

[16]. Briffa, K. R., Jones, P. D. & Kelly, P. M. (1990). Principal component analysis of the Lamb catalogue of daily weather types: Part 2, seasonal frequencies and update to 1987. *Internat. Journ. Climatology* **10**, 549-563.

[17]. Wanner, H. (1994). *The Atlantic-European circulation patterns and their significance for climate change in the Alps.* Report 1/94 to the National Science Foundation. Geography Institute, University Berne, Switzerland. 10 pp.

[18]. Bardossy, A. & Caspary, H. J. (1990). Detection of climate change in Europe by analysing European atmospheric circulation patterns from 1881-1989. *Theoretical Applied Climatology* **42**, 155-167.

[19]. Gerstengarbe, F.-W. & Werner, P. C. (1993). Katalog der Grosswetterlagen Europas nach Paul Hess und Helmuth Brezowski 1881-1992. Ber. Dt. Wetterd. 113, Offenbach a.M. 249 pp.

[20]. Schär, C., Davies, T. D., Frei, C., Wanner, H., Widmann, M., Wild, M. & Davies, H. C. (1996). Current Alpine Climate; Chapter 2 in SPP/CLEAR Book.

[21]. Barlow, L. K., White, J. W. C., Barry, R. G., Rogers, J. C. & Grootes P. M. (1993). The North Atlantic Oscillation signal in deuterium and deuterium excess signals in the Greenland Ice Sheet Project 2 Ice Core 1840-1970. *Geophys. Research Letters* **20**, 2901-2904.

[22]. Ratcliffe, R. A. S. & Murray, R. (1970). New lag associations between North Atlantic sea temperature and European pressure applied to long-range weather forecasting. *Quarterly Journal Royal Meteorological Society* **96**, 226-246.

[23]. Kelly, P. M., Goodess, C. M. & Cherry B. S. G. (1987). The interpretation of the Icelandic sea-ice record. *Journ. Geophysical Research* **92**, 10835-10843.

[24]. Ogilvie, A. E. J. (1995). Documentary evidence for changes in the climate of Iceland. In: *Climate Since AD 1500* (eds. R. S. Bradley & P. D. Jones). Routledge, London, pp. 92-117.

[25]. Deser, C. & Blackmon M. L. (1993). Surface climate variations over the North Atlantic Ocean during winter; 1900-1989. *Journ. Climate* **6**, 1743.

[26]. Kushnir, Y. (1994). Interdecadal variations in North Atlantic sea surface temperature and associated atmospheric conditions. *Journ. Climate* **7**, 142-157.

[27]. Dansgaard, W., White, J. W. C. & Johnsen, S. J. (1989). The abrupt termination of the Younger Dryas climate event. *Nature* **339**, 532-534.

[28]. Held, I. M. (1993). Large-scale dynamics and global warming. *Bulletin American Meteorological Society* **74**, 228-241.

[29]. Broecker, W. S. & Denton, G. H. (1989). The role of ocean-atmosphere reorganizations in glacial cycles. *Geochimica et Cosmochimica Acta* **53**, 2465-2502.

[30]. Delworth, T., Manabe, S. & Stouffer R. J. (1993). Interdecadal variations of the thermohaline circulation in a coupled ocean-atmosphere model. *Journ. Climate* **6**, 1993-2011.

[31]. Stocker, T. F. (1994). The variable ocean. *Nature* **367**, 221-222.

[32]. Fraedrich, K. (1990). European Grosswetter during the warm and cold extremes of the El Nino/Southern Oscillation. *Internat. Journ. Climatology* **10**, 21-31.

[33]. Fraedrich, K. & Muller, K. (1992). Climate anomalies in Europe associated with ENSO extremes. *Internat. Journ. Climatology* **12**, 25-31.

[34]. Wilby, R. (1993). Evidence of ENSO in the synoptic climate of the British Isles. *Weather* **48**, 234-239.

[35]. Santer, B. D., Taylor, K. E., Wigley, T. M. L., Johns, T. C., Jones P. D. et al. (1996). A search for human influences on the thermal structure of the atmosphere. *Nature* **381**, 39-46.

[36]. Palutikof, J. P., Kelly, P. M. & Davies, T. D. (1986). Windspeed variations and climatic change. *Wind Engineering* **10**, 182-189.

[37]. Davies, T. D., Dorling, S. R., Pierce, C. E., Barthelmie, R. J. & Farmer, G. (1991). The meteorological control on the anthropogenic ion content of precipitation at three sites in the UK: the utility of Lamb Weather Types. *Internat. Journ. Climatology* **11**, 795-807.

[38]. Ohmura, A., Gilgen, H. & Wild, M. (1989). *Global energy balance archive GEBA. Report 1; Introduction.* Zurcher Geografische Schriften Nr. 34, Zurich. 62pp.

[39]. Ohmura, A. & Lang, H. (1989). Secular variation of global radiation in Europe. In: *IRS '88: Current problems in atmospheric radiation* (eds. J. Lenoble & J.-F. Geleyn). DEEPAK Publishing Company, Hampton, Virginia, pp. 298-301.

[40]. Stein, O. & Hense, A. (1994). A reconstructed time series of the number of extreme low pressure events since 1880. *Meteorol. Zeitschrift, N. F.* **3**, 43-46.

LINKAGES BETWEEN ATMOSPHERIC WEATHER AND THE DYNAMICS OF LIMNETIC PHYTOPLANKTON

COLIN S. REYNOLDS
Freshwater Biological Association
NERC Institute of Freshwater Ecology
Far Sawrey, Ambleside
Cumbria LA22 0LP, UK

Abstract

The paper re-examines several of the datasets that contributed to the formulation of models of phytoplankton growth promoted by weather-influenced properties of the water column. These factors include temperature, period of stratification and depth of the thermocline. The paper gives examples of phytoplankton populations that variously developed in limnetic enclosures in direct response to stabilisation, to mixing, or to regular alternation between the two. The thesis is backed by more general presentations of the species-specific responses to changing physical conditions in two eutrophic lakes. The significance of these responses to the interpretation and detection of longer-term responses to climate are discussed.

1. Introduction

For as long as attention has been focussed on the variety of phytoplankton to be found in lakes, there have been attempts to establish general relationships governing the association of particular species or species-associations with the physicochemical conditions provided by certain types of lake habitat (small or large, shallow or deep, nutrient-rich or nutrient-poor, with high or low N:P, and so on) or with particular hydrographic conditions (mixed or stratified, well-flushed or otherwise). Among the more robust of these, in the sense of matching observation to theory, are those which recognise the linkages among the morphology, physiology and ecology of individual species and the physical conditions under which their best growth and replication performances are realised [1]. These deductions have led to the delimitation of distinct and explicable associations of species, often quite remote phylogenetically, but sharing

D.G. George et al.(eds.), Management of Lakes and Reservoirs during Global Climate Change, 15–38.

convergent adaptations to maximise some aspect of their performance in situ [1–3]. At the time of writing, twenty-six definable associations had been characterised, each comprising species with common functional adaptations, either (say) to rapid growth or to operation at low light intensities or at low nutrient concentrations. Significantly, the species in any given association share similar sizes, morphologies and ecologies, though they are not necessarily closely related in a phylogenetic sense.

Moreover, the species-associations have been shown to be allied themselves to one or other of three broad strategic categories, according to whether they are primarily invasive contenders (C), acquisitive (S) or attuning (R), or they intermediately blend some combination of two or even all three sets of properties (C-S, C-R, S-R or C-S-R: [4]). The primary separation of these groupings was proposed by Reynolds [1] but by the terms I, II and III. The revised separation was heavily influenced by Grime's [5] deliberations on terrestrial plant strategies and his C-S-R model carried conspicuous attractions to Reynolds [6]. Subsequent consideration of the applicability of Grime's terminology [5], implicit in the wholesale adoption of Grime's scheme (strategies are primarily Competitive, Stress-tolerant or distubance-tolerant, i.e., Ruderal), has led Reynolds to some retraction from applying quite such a literal translation of Grime's ideas to plankton [4]. Nevertheless, a similar conceptual framework around the clear linkages that exist between the morphological differentiation of phytoplankton and the particular life-cycles of individual species is not invalidated. Ascribing more appropriate adjectives to the diverging strategies among planktonic plants has overcome most of these difficulties anyway [4]; Grime's concept of primary adaptive strategies nevertheless remains at the heart of the planktonic analogue.

It must be stated that neither Grime's original strictures nor their application to the lives of phytoplankton have won universal acceptance by peers working in the respective fields. Some of Grime's views have been fiercely debated [7–10]. Among students of the temporal and spatial variability in the distribution of phytoplankton, thinking is still oriented towards the phylogenetic subdivision of the photoautotrophic plankton and to the importance of interspecific competition for nutrient resources on the basis of what are actually quite small interspecific differences in the absolute nutrient requirements of individual species [11]. The supposition that representatives of given phylogenetic groups of photoautotrophs will be sequentially, or even differently, turned "on" or "off" by particular concentrations of essential nutrients, or even by the ratios of concentrations to one another, continues to be the preferred paradigm. The abilities of planktonic algae to take up nutrients and carbon dioxide, even at diminished concentrations, yet at rates still sufficient to saturate the temperature-dependent growth capacity, have been shown [12], whereas, in many instances, the rate of in-situ cell replication is often likely to be as fast as the temperature and aggregate day-length allow [13]. Moreover, the fastest rate at which individual phytoplankton will grow and reproduce is well-predicted by the size and shape of the organism [13]. Of the species present, those which are likely to become dominant will be those that maintain the fastest rate of replication under the conditions obtaining, or through initially fielding the most recruits, or, yet again, by having the means to overcome rapid rates of mortality due to sedimentation, grazers, parasites and pathogens. This alternative paradigm invokes an alternative principle, that rather than be tuned to a ratio of resource requirements, phytoplankton species will grow as well as they can, wherever and whenever they have the opportunity.

Evidence against which to test the hypothesis that the in-situ replication rates of phytoplankton are regulated principally through fluctuations in the physical environment and not, save at chronically-low or recently-exhausted nutrient availabilities, by the capacities to take up nutrients, is not readily amassed. The problem is the familiar one of establishing just how fast new cells are being recruited to the population in the field, when the direct or analogue measurements of altered abundance do not anticipate the rates of simultaneous removal by grazers, sinking or advection. The measurement of photosynthetic capacity is not equivalent to growth, except at low average insolation [14], so the measured rate of photosynthetic carbon-fixation exceeds cell replication by an unknown extent. Methods have been developed to approximate cell replication rates in situ but, so far, these are confined to species having distinctive forms and a recognisable process of cell division (as in some dinoflagellates, colonial chlorophytes and desmids: [15–19]). The use of stains to detect nuclear division and its rate [20–22] offers improved prospects for successful address of the problem of in-situ growth measurement, though it has not yet enjoyed widespread application to ecological problems (but see [23]). Pending such resolution being brought to the problem of in situ growth, the most compelling evidence of replication-rate responses necessarily comes from reconstructions based upon the sums of net rates of change in populations maintained under realistic experimental conditions, and measured contemporaneous rates of loss to sedimentation and grazing. This was the experimental design applied to the large limnetic enclosures employed for this purpose by Lund & Reynolds [24]. The results have been worked up in various ways and presented in a series of publications [14, 25–28]. The combined dataset, which has been the subject of separate review [29], underpins the quantification advanced to justify the paradigm of a morphological–functional basis of phytoplankton ecology, involving not just net growth rates but dynamics-based ideas of population development, community assembly, succession and dominance, as well as its susceptibility to externally-imposed disturbance [4]. Doing this work in large limnetic enclosures as opposed to whole lakes obviates a major source of advection: the populations are captive. The limnological conditions in the enclosures were intensively monitored and the opportunities for environmental manipulation of (e.g.) nutrient loadings and dosing frequencies, to regulate the consumer food-chain and to mechanically alter the hydrography, were readily exploited [29]. At no time, however, were results compared against each other, owing to the inability to replicate treatments with any degree of statistical probity.

The purpose of this paper is not to make another thoretical abstract of the data, neither is it a review of the original data. Rather, it presents anew a small compendium of responses by specific populations to deliberate experimental manipulations of the immediate environment, which were designed to imitate sharp, weather-generated fluctuations of the physical environment. So far as is possible, the representations consider both net and reconstructed replication rates, but the cases are selected to demonstrate qualitatively just how sensitive in-situ growth rates are to the kinds of hydrographic variability invoked by weather events. In order to support the stand that the results are reproducible elsewhere, the observations are concluded with a series of summaries of the physical conditions under which many of the same species grew in two well-studied eutrophic natural lakes in the English northwest Midlands, Crose Mere and Rostherne Mere.

2. Methods

The design and maintenance of the enclosures in Blelham Tarn, English Lake District, the manipulative biasses exerted and the sampling methods, protocols and the analyses for the various determinands applied were those in contemporary use at the Windermere Laboratory and are fully described in [25], [27] and [30]. Three enclosures, A, B, and C, comprised butylite cylinders, supported by buoyancy collars and held in place by a heavy chain ring anchored in the bottom sediment, each being just over 45 m in diameter and isolating, respectively, ca. 18530, 18370 and 16200 m^3 of water. Lowering sections of the collar permitted enclosure waters to equilibrate with those of the surrounding lake; raising them isolated the enclosure and initiated the experimental operation, usually early in the growth season.

The fertility of an enclosure could be allowed to run down, or be maintained or enhanced through the dispersion and dissolution of inorganic chemicals at the water surface, in amounts and at frequencies determined in the experimental design. Except where deliberately stocked, the enclosures supported low fish populations or were netted to the point of being fishless. Early isolation of Enclosures A and B, which were anchored in uniformly deep water (mean depths ca. 11.4 and 11.2 m), also ensured low populations of the planktivorous larvae of *Chaoborus* [24]. Phytoplankton development in these enclosures was thus generally exposed to grazing by a zooplankton biomass that was almost unconstrained by predators. However, the siting of Enclosure C on the side of the Blelham Tarn basin, where it straddled a depth gradient (from 2.5 to 13 m; mean depth, 9.9 m), did not exclude organisms overwintering in shallow water. Enclosure C was also equipped with an air-lift pump, capable of pumping water from ca. 8.5 m depth and discharging it at the surface. Whereas the enclosures stratified thermally, with similar frequency and structure as the outside tarn waters, pump operation would increase the depth of the surface mixed layer and draw down the thermocline to the depth of abstraction at a rate of 1.6 m d^{-1}.

Enclosures were sampled twice or thee times per week, sometimes more frequently, in order to determine the dynamic reactions of populations to altered conditions. Routine sampling was carried out with a 5-m polyethylene hose [31], supplemented in summer months with a 1-m Friedinger bottle, constructed in the workshops of the Freshwater Biological Association [32]. All algal results given here were obtained by direct counting of material sedimented in Lugol's Iodine and enumerated either directly under an inverted microscope [33] or, after concentration and subsampling to a calibrated glass-slide, counted at high-power [34]. Most samples were duplicated for chemical analysis of pH and concentations of dissolved reactive silicon (SRSi), soluble reactive phosphorus (SRP), total phosphorus (TP) and dissolved forms of combined nitrogen (ammonium, nitrate and nitrite), herein bulked as "DIN" (dissolved inorganic nitrogen); the contemporaneously extant standard methods were used throughout, as detailed in [35]. Other details of the experimental conditions relating to the particular observations are noted in Section 3.

More or less similar sampling and analytical methods were followed in the studies of Crose Mere and Rostherne Mere [36, 37], although sampling was generally more frequent at Crose Mere (3 to 7 days as opposed to 14 days).

3. Phytoplankton responses in the Blelham Tarn Enclosures

3.1. CASE 1: *ANKYRA* IN ENCLOSURE A, 1983 (Fig. 1)

Ankyra judayi (G.M.Sm.) Fott was a frequent respondent to warm, well-insolated and nutrient-rich conditions in the enclosures, despite its relatively poor representation in the plankton of the natural lake [29]. It is considered to represent the *X1* association of eutrophic nanoplankton, being a typical *r*-selected, C-species [2]. The most impressive

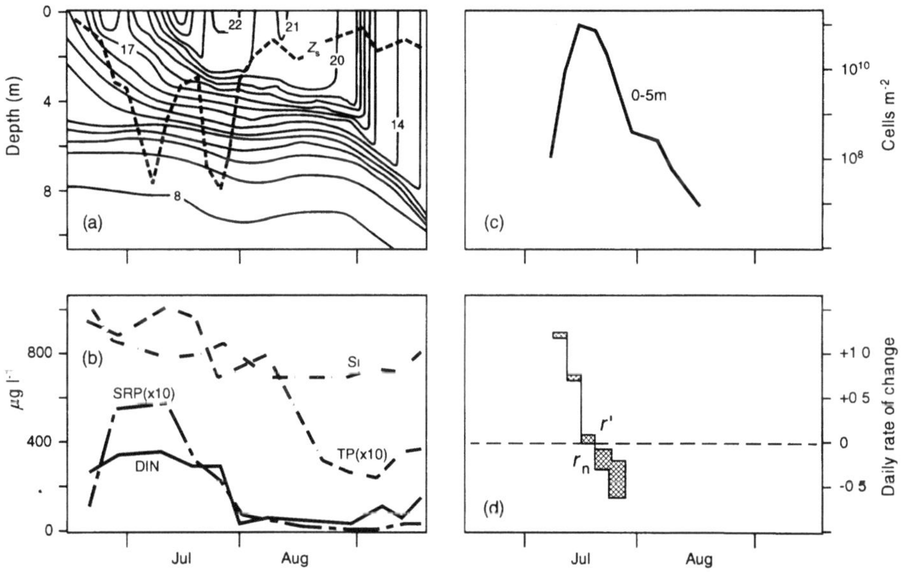

FIGURE 1. Features of the development of a population of *Ankyra judayi* in Blelham Enclosure A in 1983: (a) Vertical distribution of isotherms and depth of Secchi-disc extinction (broken line); (b) measured concentrations in integrated column samples (0–5 m: Si = dissolved reactive silicon in μg Si l^{-1}; DIN = dissolved nitrate-N + nitrite-N + ammonium-N, in μg N l^{-1}; SRP = soluble reactive phosphorus and TP = total phosphorus, in μg P l^{-1}, but on 10-fold exaggerated scale); (c) changes in the total population, in cells m^{-2}; (d) the rate of observed population change, r_n , and the reconstructed rate of cell replication, r' , over the full water depth, with the difference, the loss rate, here assumed to be due to grazing, shown by hatching. Original; data from [14].

performance of this alga previously observed in the Blelham enclosures was in July 1978, when a net in-situ rate of increase of 0.497 d^{-1} culminated in a maximum average population of 226,000 cells ml^{-1} over the upper 5 m of Enclosure A (equivalent to 1530 × 10^9 cells m^{-2}; [30]). In the case presented here (Fig 1), the net rates of population increase (r_n) in the layer extending from the surface to a depth of 5 m (0–5 m, in Fig 1c: up to 1.25 d^{-1}) were of greater magnitude, initially representing almost two doublings per day. This net rate slowed towards the middle of the month, although Fig. 1d reveals that the sharp increase in the rate of consumption by filter-feeding zooplankton (especially *Daphnia*; grazing loss rate represented by hatching), mainly in response to the increasing concentration of suitable food organisms, came too late to have been responsible for the slow-down in the reconstructed in-situ rate of cell replication, r^2. However, grazing was shown to be substantially implicated in the removal of the maximum biomass, equivalent to 150 × 10^9 cells m^{-2}.

The key question is what could have prompted the explosive increase of *Ankyra* in the first place. As pointed out elsewhere [14], the field experiments in 1983 sought to measure algal growth under the best possible field conditions and in adjacent systems that were challenged either chemically (Enclosure B, starved of phosphorus) or physically (Enclosure C, artficially mixed). As can be seen from Fig. 1, the response of *Ankyra* followed a single addition of phosphorus and nitrogen to the enclosure (Fig. 1b) and its coincidence with a phase of warm, still weather, which led to a marked increase in surface temperature (> 25°C), thermal stability and high water clarity (Fig. 1a; Secchi-disc extinction depth: 7.5 m). The coincidence of elevated temperature, high received irradiance and an abundant resource-base is precisely that considered to have stimulated many other enclosure populations of *Ankyra* or *Chlorella* or *Rhodomonas*; the observations are neither atypical nor unexpected for planktonic C-species.

3.2. CASE 2: *ASTERIONELLA* IN ENCLOSURE B, 1978 (Fig. 2)

Asterionella formosa Hass. is an extremely common planktonic diatom in small mesotrophic and mildly eutrophic lakes, especially but not exclusively in temperate regions. Its ecology has been very well studied [38–41]. Reynolds [2] considers it to be an *r*-selected, R-species, characteristic of at least two phytoplankton associations (*B, C*). It is able to maintain a relatively rapid rate of increase in cold water and on low aggregate daily light doses, provided that it has access to adequate phosphorus and, especially, silicon, that it is not subjected to excessive rates of removal by filter-feeding zooplankton and that, crucially, the depth of the surface mixed layer (in practice, > 2–3 m) is such to offset sinking loss rates to below the rate at which new cells can be recruited by growth [42]. The case illustrated here, the sensitivity of *Asterionella* dynamics to weather-driven changes in mixed-layer thickness, is demonstrated (Fig. 2).

When Enclosure B was isolated in March 1978, *Asterionella* was already increasing strongly, all the above criteria then being satisfied. Additions of assimilable phosphorus and silicon contributed to the prolongation of a chemically-favourable environment and the production of the remarkably pure population whose dynamics were eventually reported by [25]: the episode produced an estimated 335 (± 28) × 10^9 cells m^{-2}, with a maximum observed standing population of 273 × 10^9 cells m^{-2} on 15 March (Fig. 2c);

note that the text-figure presents the counts extrapolated for columns of 3, 5 and 11 m beneath one square metre, in order to show the uniformity of vertical distribution, until that point, and the divergences following the first traces of vertical stratification, shown in (Fig. 2a). An observed net rate of increase of 0.147 d^{-1}, compared with a true replication rate, calculated from silicon uptake, of 0.154 d^{-1}, was achieved to this stage but, thereafter, replication rate slowed, sinking loss rate increased (hatched area in Fig. 2d) and net rate of change quickly declined to negative values after the mixed layer shrank from the greater part of 11 m to < 2 m. Whereas not less than 12.5×10^9 cells m^{-2} are accountable to grazing losses, measurements showed that not less than 311×10^9 cells m^{-2} (> 81% of the cells produced) were recruited to the enclosure sediment.

Similar findings have been made on other populations in Enclosure B, confirming the sensitivity to mixed depth [29]. Note that it is not stratification *per se* that is avoided: several summer populations were reported in the same paper, but always after mixing episodes had extended the mixed-layer thickness to 4 or 5 m; a greater proportion of these summer-produced populations were consumed by grazers.

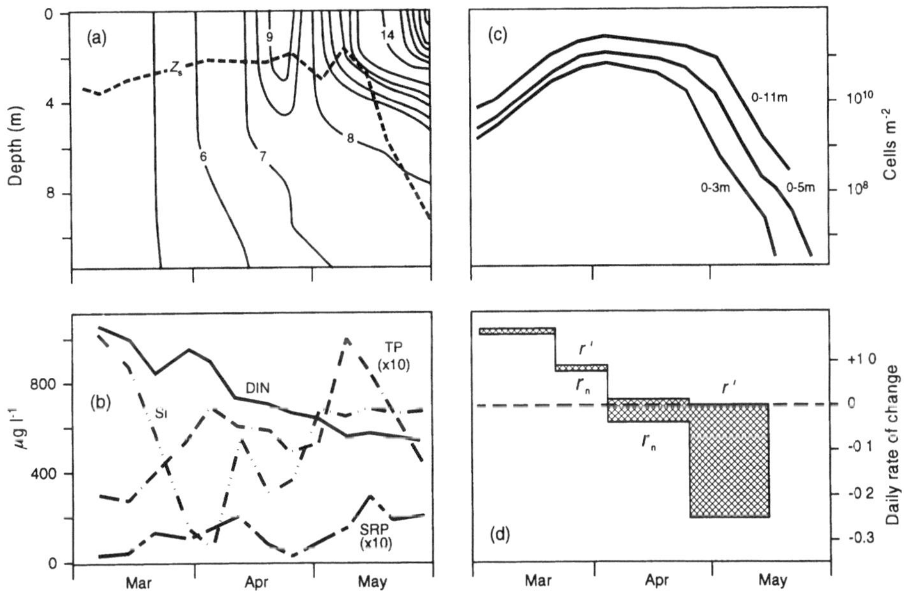

FIGURE 2. Features of the development of a population of *Asterionella formosa* in Blelham Enclosure B in 1978: (a) Vertical distribution of isotherms and depth of Secchi-disc extinction (broken line); (b) measured concentrations in integrated column samples (0–5 m: Si = dissolved reactive silicon in μg Si l^{-1}; DIN = dissolved nitrate-N + nitrite-N + ammonium-N, in μg N l^{-1}; SRP = soluble reactive phosphorus and TP = total phosphorus, in μg P l^{-1}, but on 10-fold exaggerated scale); (c) changes in the population, in cells m^{-2}, in water columnns extending from the surface to 3, 5 and 11 m; (d) the rate of observed population change, r_n , and the reconstructed rate of cell replication, r', over the full water depth, with the difference, the loss rate, here assumed to be due to sinking, shown by hatching. Original; data from [25].

3.3. CASE 3: *FRAGILARIA* IN ENCLOSURE C, 1981 (Fig. 3)

Fragilaria crotonensis Kitton is another common diatom, generally associated with more eutrophic lakes than *Asterionella*. Although phylogenetically close to *Asterionella* and resembling it in many ways, *Fragilaria* seems less well-adaptable to low light doses but its colonial habit (forming ribbons rather than spirals) confers on it a demonstrable resistance to planktonic filter-feeding crustaceans [27, 43]. These distinctions are evident in lakes where they both occur: *Asterionella* may be more prolific in the early season, when light doses are strongly regulating; the C and P associations in which the *r*-selected, R-species *Fragilaria crotonensis* often figures prominently [2], do so more in summer when grazers are relatively active. In addition to a sustained supply of soluble reactive silicon, both share the dependence upon wind and an adequate mixing depth.

The series of artificial mixing episodes applied to Enclosure C during 1981 (see [26]), when the nutrient levels were maintained so that they should not be allowed to fall to rate-limiting concentrations, is chosen to illustrate this point (Fig. 3): the

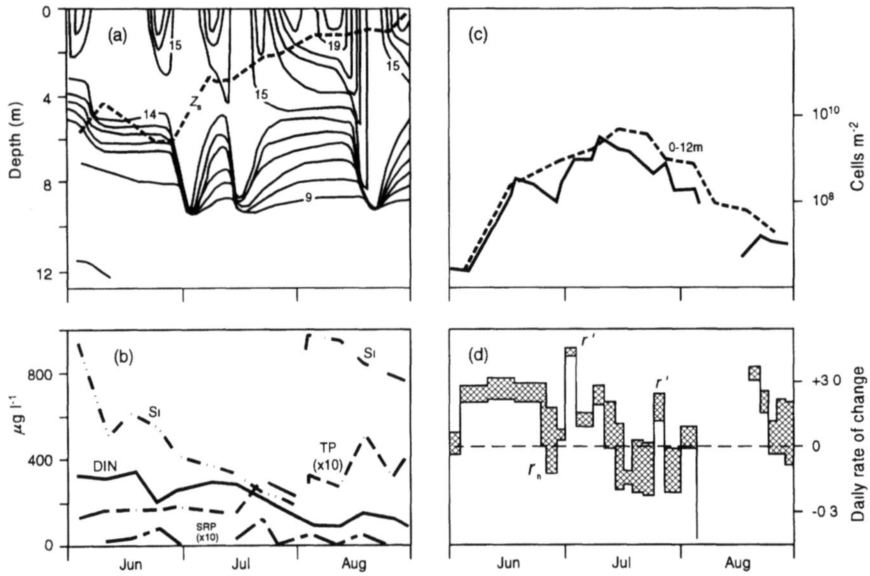

FIGURE 3. Features of the development of a population of *Fragilaria crotonensis* in Blelham Enclosure C in 1981: (a) Vertical distribution of isotherms and depth of Secchi-disc extinction (broken line); (b) measured concentrations in integrated column samples (0–5 m: Si = dissolved reactive silicon in µg Si l^{-1}; DIN = dissolved nitrate-N + nitrite-N + ammonium-N, in µg N l^{-1}; SRP = soluble reactive phosphorus and TP = total phosphorus, in µg P l^{-1}, but on 10-fold exaggerated scale); (c) changes in the total population, in cells m^{-2}, in water columns extending from the surface to 5 m (solid line) and 12 m (broken line); (d) the rate of observed population change, r_n , and the reconstructed rate of cell replication, r' , over the full water depth, with the difference, the loss rate, here assumed to be due to sinking, shown by hatching. Original; data from [27].

mixing episodes are shown by the sharp onsets of isothermy through 8.5 m of water; the subsequent quiescent periods occur when, almost as sharply, the enclosure resumed the temperature structure of the outside lake (Fig. 3a). It is interesting that the total abundance through the enclosure (Fig. 3c: broken line) appears as a single, flattish peak, although its inflections represent a much-damped representation of recruitment (mostly by growth) and depletion (almost entirely due to sinking) in the upper 5 m (Fig. 3c: solid line). The dynamics of the changes in the 5-m population were the subject of a detailed analysis [26], from which the traces of gross and net rates of change, shown in Fig. 3d, are derived. It is of interest to note that the disparity between r' and r_n is generally in inverse proportion to the depth of the first isotherm in the preceding period but that r' was simultaneously stimulated by the onset of mixing. From a later analysis [44], it appeared that the August episode may have been initiated by resuspension of *Fragilaria* from the small area of shallow bottom sediment of Enclosure C. The severe restriction of light penetration at that time is thought to have hampered a more sustained rate of increase.

3.4. CASE 4: *ANABAENA* IN ENCLOSURE C, 1982 (Fig. 4)

Almost reciprocal responses to intermittent artificial mixing were observed with the development of a population of *Anabaena flos-aquae* Bréb. ex Born. et Flah. in Enclosure C during 1982 (Fig. 4). *Anabaena* is a nitrogen-fixing, buoyancy-regulating, bloom-forming cyanobacterium, characterising association *H* of Reynolds [2] and exemplifying properties associated with S-species, rather more (*K*-) selected by its ability to self-replicate rapidly. Indeed, for three months the alga kept itself mainly in the upper part of the water column so that, only rarely, did the total number of cells m^{-2} through 13 m differ significantly from that in 5 m (Fig. 4b). Sinking and grazing losses are generally small for *K*-selected populations: the trace in Fig. 4d shows net rates of population change only but these clearly describe the alternating phases of increase and plateau in Fig. 4c. Both fit well to the alternations between mixing and quiescence, with the most powerful recruitment after the near-surface stratification had re-formed. Surface scum formation and akinete formation were responsible for the sharp drop in biomass during July (Fig. 4d; see also [45]).

3.5. CASE 5: *PLANKTOTHRIX (OSCILLATORIA)* IN ENCLOSURE C, 1982 (Figs 4 and 5)

Oscillatoria agardhii Gom. var. *isothrix* Skuja, now properly ascribed to *Planktothrix mougeotii* (Bory ex Gom.) Anag. et Kom., is a *K*-selected R-species [2] which has been commonly found in the English Lake District. Large populations have built up on occasions in the Blelham Enclosures, especially during winter isolation [24]. The species is able to operate for long periods on low daily light doses, so is pre-adapted to persistent mixing in light-deficient water columns, but it is not adapted to high frequency variability in mixing, quiescent episodes that often lead to photoinhibition and, in extremes, photolysis and death [45]. Case 5 was chosen to illustrate some of the reactions of *Planktothrix* to environmental variability during the same period in the

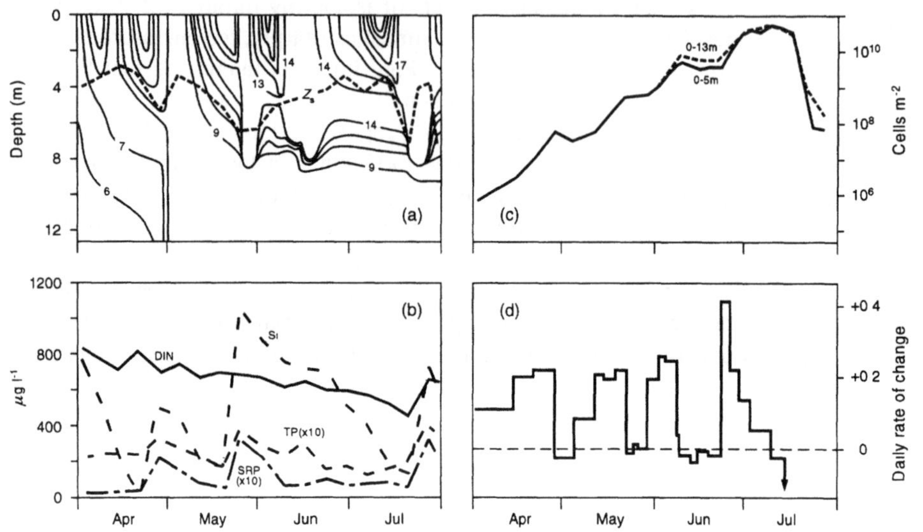

FIGURE 4. Features of the development of a population of *Anabaena flos-aquae* in Blelham Enclosure C in 1982: (a) Vertical distribution of isotherms and depth of Secchi-disc extinction (broken line); (b) measured concentrations in integrated column samples (0–5 m: Si = dissolved reactive silicon in µg Si l^{-1}; DIN = dissolved nitrate-N + nitrite-N + ammonium-N, in µg N l^{-1}; SRP = soluble reactive phosphorus and TP = total phosphorus, in µg P l^{-1}, but on 10-fold exaggerated scale); (c) changes in the total population, in cells m^{-2}, in water columns extending from the surface to 5 m (solid line) and to 13 m (broken line); (d) the rate of observed population change. Original; data from [27].

same enclosure as Case 4. Striking differences between the requirements of *Planktothrix* and those of (say) *Anabaena* serve to warn about the dangers of generalising from one species of cyanobacterium to the ecologies of other kinds of cyanobacteria. As depicted in Fig. 4, *Planktothrix* increased its biomass through the mixing episodes but could do no more than hold its numbers in the quiescent phases (Figs 5c and 5d). Comparison of Figs 4d and 5d reveals the almost precise reciprocity of the rapid-growth phases of the two species.

3.6. CASE 6: *CERATIUM* IN ENCLOSURE C, 1980 (Fig. 6)

Ceratium hirundinella O. F. Müll. is a dinoflagellate with strong characteristics of a *K*-selected S-species. It is common in the lakes of the Windermere catchment but its early reputation of not growing well in the Blelham enclosures was shown to be an inoculum effect [24], for its replication rate proved to be remarkably sustained in Enclosure C, the only one enclosing shallow sediments. The alga grew consistently through the cool

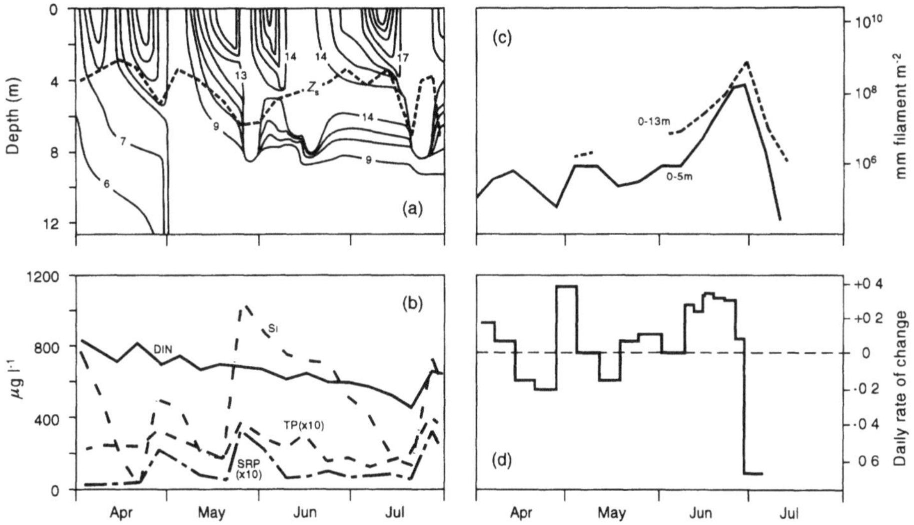

FIGURE 5. Features of the development of a population of *Planktothrix mougeotii* in Blelham Enclosure C in 1982: (a) Vertical distribution of isotherms and depth of Secchi-disc extinction (broken line); (b) measured concentrations in integrated column samples (0–5 m: Si = dissolved reactive silicon in µg Si l^{-1}; DIN = dissolved nitrate-N + nitrite-N + ammonium-N, in µg N l^{-1}; SRP = soluble reactive phosphorus and TP = total phosphorus, in µg P l^{-1}, but on 10-fold exaggerated scale); (c) changes in the total population, in cells m^{-2}, in water columns extending from the surface to 5 m (solid line) and to 13 m (broken line); (d) the rate of observed population change. Original; data from [27].

summer of 1980 in the fairly clear waters of the epilimnion (Fig. 6a), despite, in this instance, there having been almost indetectable amounts of soluble phosphorus through most of this period; yet the alga sustained a growth rate of 0.05 to 0.09 d^{-1}. It is significant that total phosphorus increased slowly over this period (Fig. 6b); it is not proven but it is possible that *Ceratium* increase was maintained by small quantities of phosphorus either dissolving into the epilimnion from the adjacent shallow sediments or by diffusion from the anoxic hypolimnion. This is not a primary issue here but is discussed by Reynolds [44].

Decline in the vegetative population occurred concomitantly with two phases of formation of cysts (Fig. 6c). The second of these was strongly related to natural autumnal expansion of the epilimnetic circulation. It is possible but not necessarily probable that phosphorus deficiency was implicated in the end of seasonal growth. However, it is entirely expected of a *K*-strategist alga that it might maintain its slow rate of growth for longer than many faster, smaller, so-called competitors for resources.

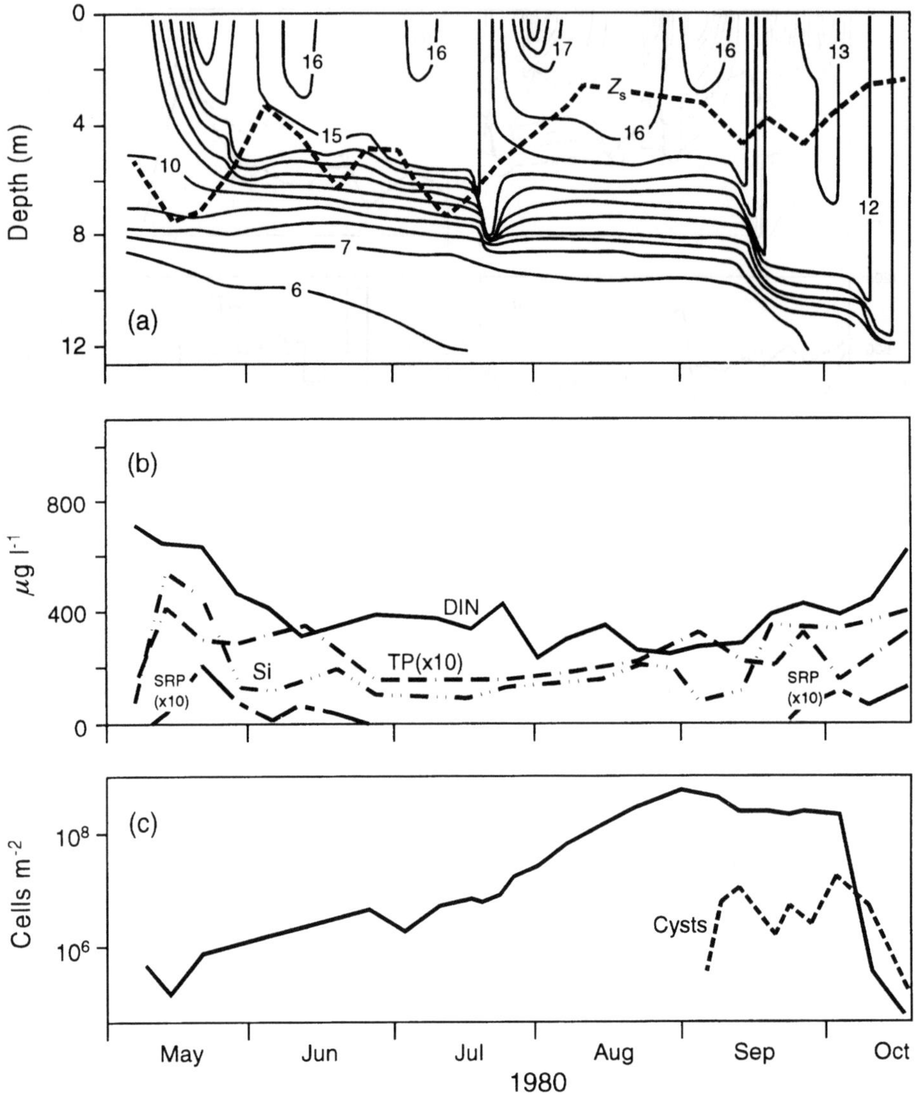

FIGURE 6. Features of the development of a population of *Ceratium hirundinella* in Blelham Enclosure C in 1980: (a) Vertical distribution of isotherms and depth of Secchi-disc extinction (broken line); (b) measured concentrations in integrated column samples (0–5 m: Si = dissolved reactive silicon in μg Si l^{-1}; DIN = dissolved nitrate-N + nitrite-N + ammonium-N, in μg N l^{-1}; SRP = soluble reactive phosphorus and TP = total phosphorus, in μg P l^{-1}, but on 10-fold exaggerated scale); (c) changes in the total population, in cells m^{-2} (solid line) and overwintering cysts m^{-2} (broken line), also numbers in the water column extending from the surface to 5 m. Original; data from [24].

4. Spatial and temporal patterns in recruitment of phytoplankton in eutrophic lakes

As pointed out in the Introduction, it is not intended that the six cases considered in Section 3 should be turned into "proofs" to support existing hypotheses. Neither is it appropriate for the theoretical abstractions to be wheeled out as a sort of appendix to justify the choice of studies (reference has been made to most of them in the text and the reader may determine how much validation is required). However, it is appropriate to invoke another set of the authors' observations dealing with a considerable body of quantitative data in a somewhat qualitative fashion, in order to suggest that the growth responses are consistent among the functional groups of species recognised. These pertain to Crose Mere and Rostherne Mere in the English northwest Midlands; both lakes are small but deep enough to stratify, both are fed by groundwater, have long retention times and are calcareous, and both are very nutrient-rich and the variability in dominant populations is due primarily and consistently to the incidence of weather-led events [36, 37, 46].

Each of the points plotted in Figs 7 to 10 represents a 1-week or 2-week period within the defined time range, during which the alga concerned increased its mass (how quickly or by what increment is not represented – the intention is to identify conditions conducive to net recruitment through growth. Believing that the relevant conditions centre around the physical conditions (temperature, aggregate daily photoperiod, etc., and their interaction with the depth and stability of the water column) at the START of the 1- or 2-week growth period, the axes selected are: the square of the Brunt-Väsälä frequency for the upper 6 m [$N^2 = (g/\rho)\ (\delta\rho/\delta z)$, where g is gravitational acceleration, ρ is the density of the water and $(\delta\rho/\delta z)$ is the average density gradient from the water surface to 6-m depth as calculated from the temperature profile; the higher is N, then the tighter and more stable is the stratification]; the depth (h_m, in metres) of the surface mixed layer (following [37], defined as the depth from the surface to the point where the gradient of density on depth first exceeds 0.02 kg m^{-3} m^{-1}); the temperature of that layer (θ); and the aggregate most probable daily photoperiod (ϕ, in h d^{-1}) analogised as $\phi = \Gamma\ (vol_s / vol_m)$, where Γ is the length of the solar day, sunrise to sunset, vol_m is the volume of the surface mixed layer and vol_s is the volume of the lake within the depth range of the Secchi-disc measurement (clearly, the limiting condition has to be applied that if $vol_s > vol_m$, then $\phi = \Gamma$).

The data plotted in Fig. 7 represent the net-increase performances of major phytoplankton in Crose Mere during two years (1972, 1973) distinguished by strong differences in weather-forcing and in the sequence of algal dominants. Yet, as was pointed out [36] and may be judged from the figures here, the between-year growth responses to the forcing were entirely consistent and indistinguishable. In a matrix of stability versus mixed depth, the affinity between four (R) species of diatom and mixed depth > 2 to 3 m is apparent, as is that of the (S) species of *Ceratium, Microcystis* and *Volvox* for high stability. In order of increasing "S-ness", the green algae *Pediastrum, Closterium* and *Eudorina* and the cyanobacterium *Anabaena*, fall intermediately between the R and S characters, while *Cryptomonas* alone seems to have had the flexibility to be able to grow under the full range of conditions represented by these dimensions.

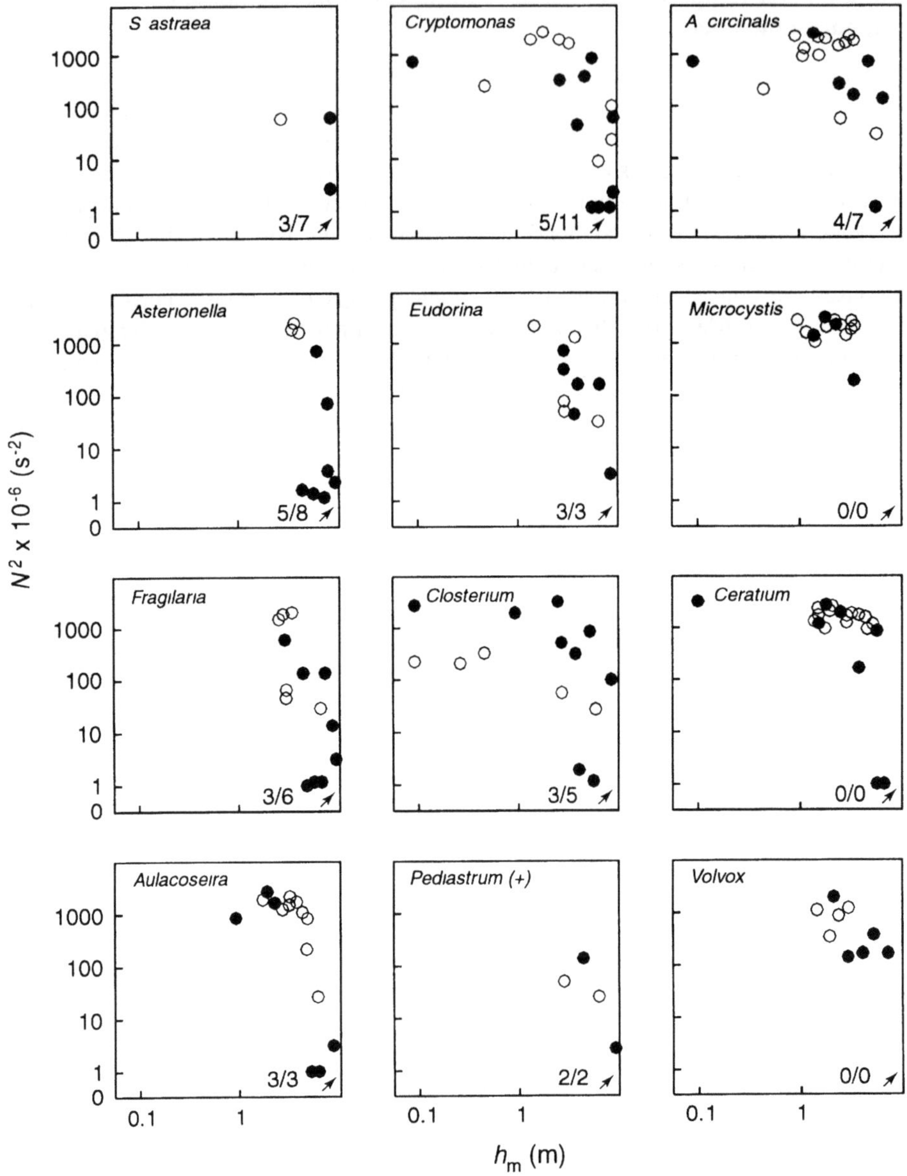

FIGURE 7. Stability (ordinate) and mixing (abscissa) conditions in Crose Mere preceding several (3 to 14) days of sustained net increase in standing populations of the named phytoplankton; after [36]. Data are distinguished according to whether the observation pertains to 1972 (solid) or 1973 (open) circles. Note that all values of $N^2 < 1$ are plotted as $N^2 = 1$.The figures in the lower right hand corner of each graph refer to the number of points falling at z = 9.3 m and $N^2 = 1$ in either year, that for 1972 being cited first in each instance.

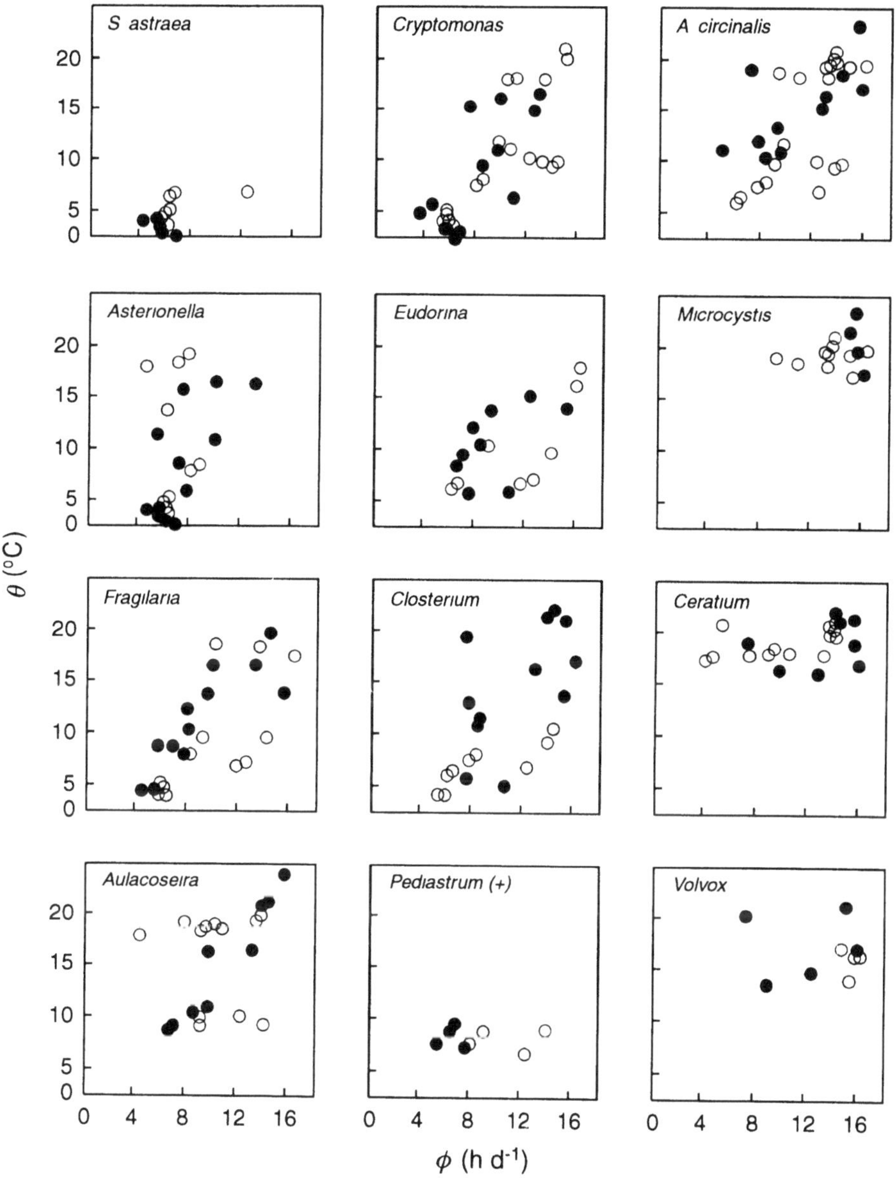

FIGURE 8. Temperature (ordinate) and light-dose (abscissa) conditions in Crose Mere preceding several (3 to 14) days of sustained net increase in standing populations of the named phytoplankton; after [36]. Data are distinguished according to whether the observation pertains to 1972 (solid) or 1973 (open) circles.

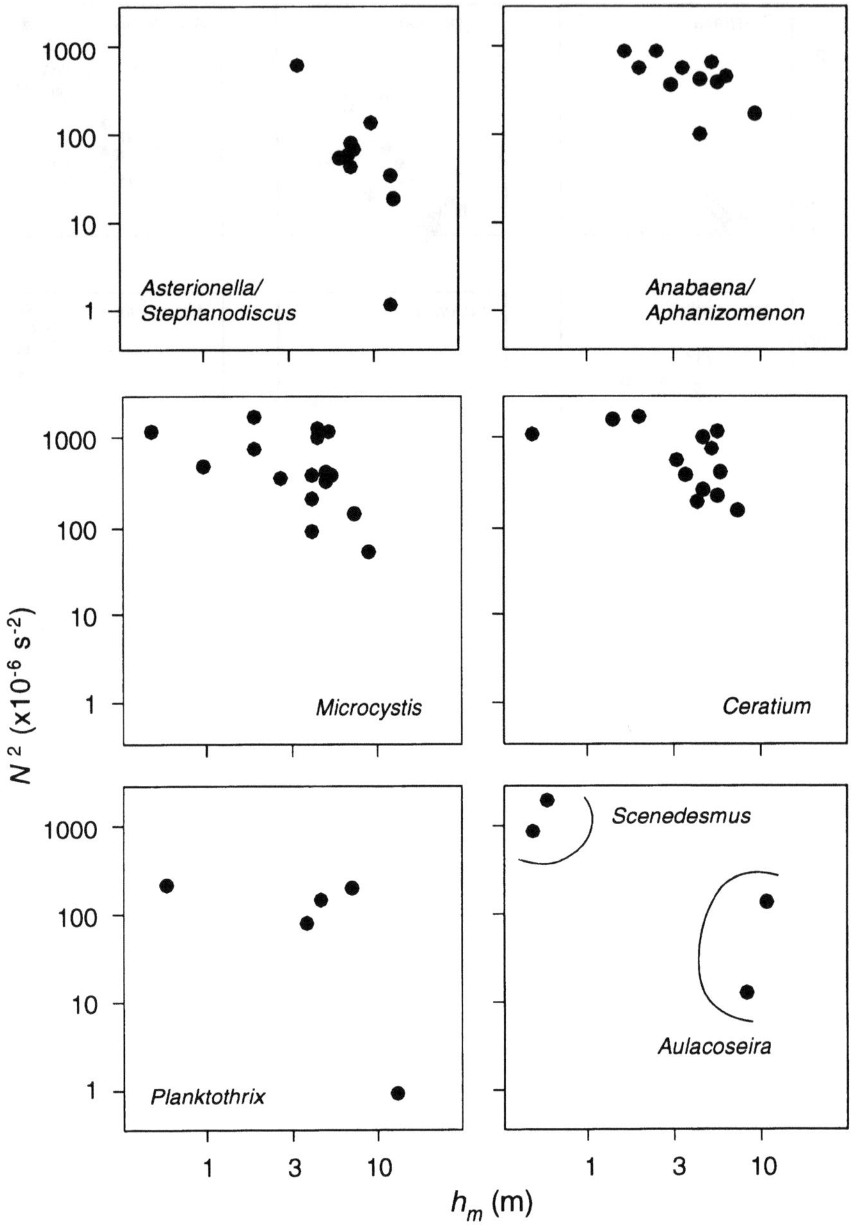

FIGURE 9. Stability (ordinate) and mixing (abscissa) conditions in Rostherne Mere preceding several (generally 14) days of sustained net increase in standing populations of the named phytoplankton; after [37]. Note that all values of $N^2 < 1$ are plotted as $N^2 = 1$.

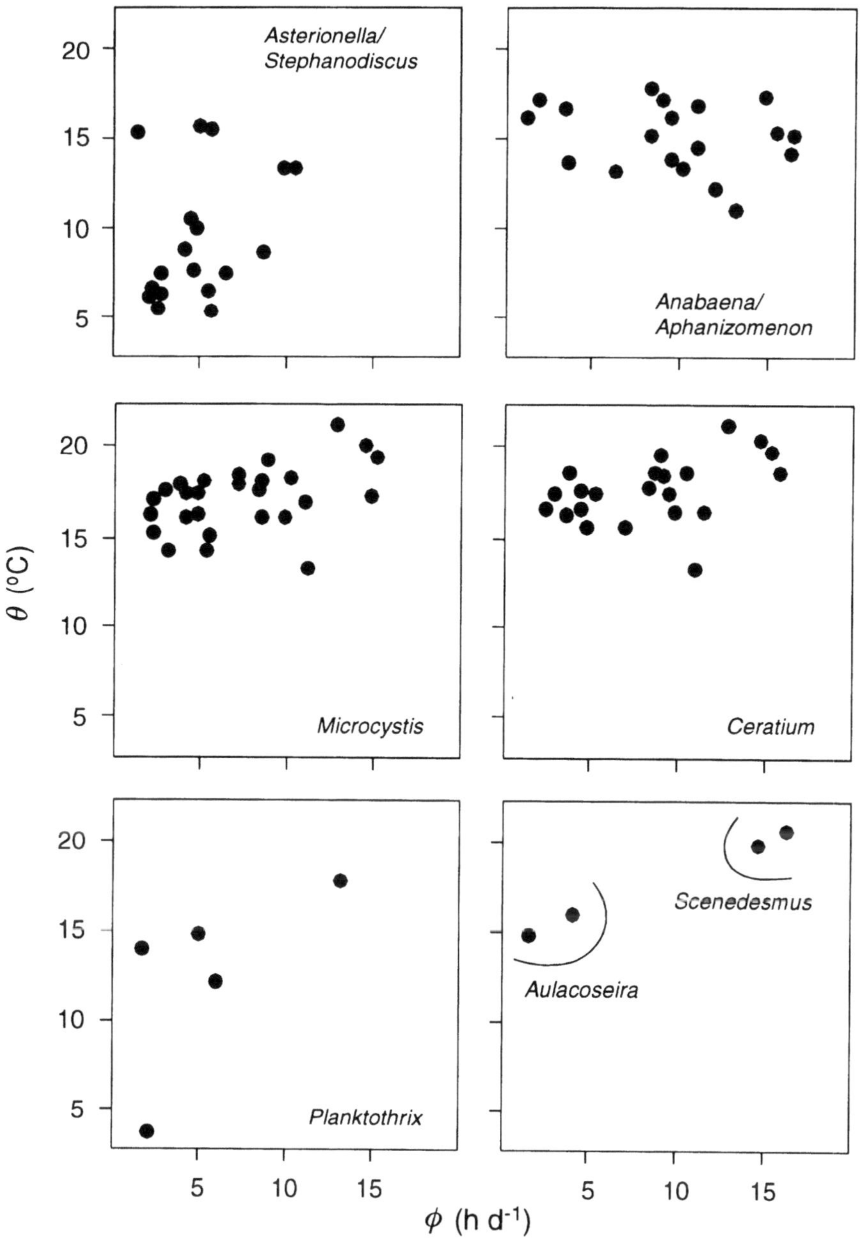

FIGURE 10. Temperature (ordinate) and light-dose (abscissa) conditions in Rostherne Mere preceding several (generally 14) days of sustained net increase in standing populations of the named phytoplankton; after [37].

When the same data are re-plotted in relation to water temperature and aggregate photoperiod (Fig. 8), analogous separations are detected; although the ***Stephanodiscus*** grew exclusively in vernal (cold-water) populations and ***Aulacoseira granulata*** was correspondingly excluded therefrom, all the diatoms considered showed low net light-thresholds for growth. The contrasting extreme is provided by ***Ceratium***, which has a rather higher temperature threshold, and ***Microcystis***, which has a higher light threshold as well. For the species showing intermediate responses, temperature thresholds tended to be more critical than any day-length criterion, with *Cryptomonas* sp(p). apparently able to grow at lower temperatures and light-thresholds than the others.

Analogous data pertaining to the growth of many of the same algae in Rostherne Mere, but over a different time band, are represented in Figs 9 and 10. In this relatively deep and opaque lake, the onset of stratification is quite crucial to the development of any plankton at all and, especially, of the summer S-species [47]. The distribution of *Anabaena* (and *Aphanizomenon*) spp. in the stability versus mixed-depth plot (Fig. 9) resembles that of *Microcystis* and *Ceratium* in apparently having its net increase confined to occasions of shallow-epilimnion formation. Development of prominent diatoms, including *Aulacoseira* on occasions, is, as ever, dependent upon an adequate depth of mixing but is by no means confined to unstratified periods: Reynolds & Bellinger [37] considered that a fully-mixed Rostherne Mere was incapable of supporting any vigorous phytoplankton growth, except in four to five months in summer, because the minimal aggregate exposure to daylight would be inadequate. The impact of summer stratification in extending the photoperiods of growing cells is shown by the distributions of points towards the top right-hand corners of the subdiagrams of Fig. 10. Those for *Anabaena, Microcystis* and *Ceratium* plainly spread further in this direction than do the three diatoms or *Planktothrix*.

The inclusion of *Scenedesmus* in these growth plots deserves explanation. This common alga of enriched ponds and lowland rivers is the nearest example of an *r*-selected C-species considered here, and it generally occurs at background levels in Rostherne Mere. It produced a large, dominant population in the mere in the summer of 1983, at a time of very strong near-surface stratification and shallow mixing. This performance was interpreted as an indication that, hydraulically as well as chemically, the near-surface layer had developed the conditions analogous more to a pond than a lake [37]. The reactions of systems and – in turn – their biota, to extreme events, can often illuminate the behaviour of the system in less extreme conditions and offer clues to the requirements of the organism that are more regularly satisfied elsewhere.

5. Discussion

One set of detailed observations of a few species in a contrived limnetic system scarcely constitutes the basis of a grand theory of phytoplankton ecology. Moreover, to support it with a further set of observations invoking only selected factors (however much others may be discounted safely) is not guaranteed to add a great deal of credibility to its tenure. On the other hand, there are developing theories which relate systematic differences among the temporal distribution and responses of phytoplankton populations to environmental circumstances, to the 3–4 orders of magnitude range in the linear dimensions of phytoplankton and the 2-plus orders of magnitude range in the ratio of surface-area-to-volume. Together, these morphological characteristics demonstrably influence the abilities of phytoplankton to intercept and harvest light, the rate of uptake of nutrients relative to mass, respiration rates, the rates of material transport and organisation within cells and, ultimately, the rates at which new cells can be replicated (e.g. [11]). The same differences help us to classify the adaptations and reactivities of species as being primarily more C-like, S-like or R-like and whether they are selected more through the ability to assemble mass quickly (*r*-selection) or conservatively (*K*-selection).

Following this reasoning, of course, it is very simple to point out that the current data comply with the theory, for indeed, they have been set up to do just this. However, it is no less valid to say that it was upon such observations that the theory began to be assembled in the first instance [6, 26]. Although less well-subscribed to than other contemporary theories in phytoplankton ecology, the present view of primary adaptive strategies, underpinned by consistent morphological traits, is more successful in predicting compositional trends on the basis of unique properties of the phylogenetic groups which may be represented, or of subtle activations and suppressions by the ratios in which certain nutrients are present in solution, even when they are not rate-limiting, or of the activities of organisms at higher trophic levels of organisms. If the ability to write accurate, deterministic models to simulate seasonal changes in the abundance and composition of real phytoplankton communities is any measure of the consistent and correct interpretation of the driving variability, then the morphological–strategic base of PROTECH models [2, 48, 49] lend sound credibility to the explanations advanced for each of the cases presented.

The relevance of this renewed deduction to the overall theme of the compendium is in being able to assert that the species composition of phytoplankton assemblages, subject as it is to a myriad of interactions working upon an available species stock, is powerfully influenced by atmospheric weather conditions, through the proximal transfer functions of temperature range and fluctuation, hydraulics and wind-forcing. To secure interannual biasses in composition, however, requires either the persistence of one kind of weather pattern (such as that which favoured *Scenedesmus* dominance in Rostherne Mere once in eighteen years) or a greater frequency with which certain kinds of weather are prevalent and which determinc, collectively, whether the year will be judged retrospectively to have been unusually warm or cool, or wet or dry. As some analyses have shown [37, 50], the relative abundances of main species are closely correlated with precisely those proximal drivers that have been highlighted: temperature, period of stratification, depth of thermocline, etc. (George et al. [50]; also see [51]), go one stage further back to detect interannual cyclicity and patterns in the "average weather" and find, convincingly, that the patterns in northwest Europe are strongly related to such global regulators as the jetstream and the circumpolar air currents and, significantly, by the position of the North Atlantic Drift Current, the so-called "Gulf Stream" [52].

The analysis of trends in the responses of biotic communities to subtle variations in "average weather" and over temporal appropriate scales is of great importance to extrapolating future scenarios of global change. Major changes in the terrestrial vegetation since the last glaciation have been related properly to gross fluctuations in climate. Individual generations of trees, however, may have always been subject to small interannual differences in the weather but, broadly, the fluctuations would need to have been within an envelope of normality perceived by that generation: they have not, by themselves, produced a population response (although the success of flowering, fruiting and recruitment may well have fluctuated interannually). Phytoplankton generations occupy periods of days, normally, and a string of population responses occupy a summer, while the lakes themselves integrate and average daily weather fluctuations to annual characteristics. The value of maintaining routine programmes of sampling and counting phytoplankton in selected lakes as sensitive, scaled respondents to ecospheric variability, can no longer be doubted.

Acknowledgements

I am grateful to the organisers and sponsors of the meeting in Prague for the opportunity to present this paper. All the field work was undertaken on behalf of the Freshwater Biological Association and, in large part, grant-aided by the Natural Environment Research Council. All the work on the Blelham Tarn Enclosures was jointly funded by the UK Department of the Environment. This support is gratefully acknowledged.

References

[1]. Reynolds, C. S. (1984). Phytoplankton periodicity: the interactions of form, function and environmental variability. *Freshwater Biology* **14**, 111-142.

[2]. Reynolds, C. S. (1996). Plant life of the pelagic. *Proceedings of the International Association for Theoretical and Applied Limnology* **26**, (in press).

[3]. Olrik, K. (1994). *Phytoplankton - ecology.* Miljøsministiert, København.

[4]. Reynolds, C. S. (1995). Successional change in the planktonic vegetation: species, structures, scales. In: *Molecular ecology of aquatic microbes* (ed. I. Joint). Springer-Verlag, Berlin, pp. 115-132.

[5]. Grime, J. P. (1979). *Plant strategies and vegetation processes.* Wiley-Interscience, Chichester.

[6]. Reynolds, C. S. (1988). Functional morphology and the adaptive strategies of freshwater phytoplankton. In: *Growth and reproductive strategies of freshwater phytoplankton* (ed. C. D. Sandgren). Cambridge University Press, New York, pp. 388-433.

[7]. Tilman, D. (1982). *Resource competition and community structure.* Princeton University Press, Princeton.

[8]. Tilman, D. (1988). *Plant strategies and the dynamics and structure of plant communities*. Princeton University Press, Princeton.

[9]. Loehle, C. (1988). Problems with the triangular model for representing plant strategies. *Ecology* **69**, 284-286.

[10]. Grace, J. B. (1991). A clarification of the debate between Grime and Tilman. *Functional Ecology* **5**, 583-587.

[11]. Reynolds, C. S. (1993). Swings and roundabouts: engineering the environment of algal growth. In: *Urban waterside regeneration, problems and prospects* (eds. K. N. White et al.). Ellis Horwood, Chichester, pp. 330-349.

[12]. Reynolds, C. S. (1990). Temporal scales of variability in pelagic environments and the response of phytoplankton. *Freshwater Biology* **23**, 25-53.

[13]. Reynolds, C. S. (1989). Physical determinants of phytoplankton succession. In: *Plankton ecology* (ed. U. Sommer). Brock-Springer, Madison, pp. 9-56..

[14]. Reynolds, C. S., Harris, G. P. & Gouldney, D. N. (1985). Comparison of carbon-specific growth rates and cellular increase of phytoplankton in large limnetic enclosures. *Journal of Plankton Research* **7**, 791-820.

[15]. Pollingher, U. & Serruya, C. (1976). Phased division of *Peridinium cinctum* f. *westii* (Dinophyceae) and development of the Lake Kinneret (Israel) bloom. *Journal of Phycology* **12**, 163-170.

[16]. Heller, M. D. (1977). The phased division of the freshwater dinoflagellate *Ceratium hirundinella* and its use as a method of assessing growth in a natural population. *Freshwater Biology* **7**, 527-533.

[17]. Frempong, E. (1984). A seasonal sequence of diel distribution patterns for the planktonic dinoflagellate Ceratium hirundinella in a eutrophic lake. *Freshwater Biology* **14**, 401-421.

[18]. Reynolds, C. S. (1983). Growth-rate responses of *Volvox aureus* Ehrenb. (Chlorophyta, Volvocales) to variability in the physical environment. *British Phycological Journal* **18**, 433-442.

[19]. Alvarez Cobelas, M., Velasco, J. L., Rubio, A. & Brook, A. J. (1988). Phased cell division in a field population of *Staurastrum longiradiatum* (Conjugatophyceae: Desmidaceae). *Archiv für Hydrobiologie* **112**, 1-20.

[20]. Coleman, W. A. (1980). Enhanced detection of bacteria in Al environments by fluorochrome staining of DNA. *Limnology and Oceanography* **25**, 948-951.

[21]. Porter, K. G. & Feig, Y. S. (1980). The use of DAPI for identifying and counting aquatic microflora. *Limnology and Oceanography* **25**, 943-948.

[22]. Braunwarth, C. & Sommer, U. (1985). Analyses of the in situ growth rates of Cryptophyceae by use of the mitotic index. *Limnology and Oceanography* **30**, 893-897.

[23]. Ojala, A. & Jones, R. I. (1993). Spring development and mitotic division pattern of a *Cryptomonas* sp. in an acidified lake. *European Journal of Phycology* **28**, 17-24.

[24]. Lund, J. W. G. & Reynolds, C. S. (1982). The development and operation of large limnetic enclosures in Blelham Tarn, English Lake District, and their contribution to phytoplankton ecology. In: *Progress in phycological research, Vol. 1* (eds. F. E. Round & D. J. Chapman). Elsevier, Amsterdam, pp. 1-65.

[25]. Reynolds, C. S., Thompson, J. M., Ferguson, A. J. D. & Wiseman, S. W. (1982). Loss processes in the population dynamics of phytoplankton maintained in closed systems. *Journal of Plankton Research* **4**, 561-600.

[26]. Reynolds, C. S., Wiseman, S. W., Godfrey, B. M. & Butterwick, C. (1983). Some effects of artificial mixing on the dynamics of phytoplankton in large limnetic enclosures. *Journal of Plankton Research* **5**, 203-234.

[27]. Reynolds, C. S., Wiseman, S. W. & Clarke, M. J. O. (1984). Growth- and loss-rate responses of phytoplankton to intermittent artificial mixing and their potential application to the control of planktonic algal biomass. *Journal of Applied Ecology* **21**, 11-39.

[28]. Reynolds, C. S. (1988). The concept of ecological succession applied to the seasonal periodicity of phytoplankton. *Verh. Internat. Verein. Limnol.* **23**, 683-691

[29]. Reynolds, C. S. (1986). Experimental manipulations of the phytoplankton periodicity in large limnetic enclosures in Blelham Tarn, English Lake District. *Hydrobiologia* **138**, 43-64.

[30]. Reynolds, C. S. & Wiseman, S. W. (1982). Sinking looses of phytoplankton maintained in closed limnetic systems. *Journal of Plankton Research* **4**, 489-522.

[31]. Lund, J. W. G. & Talling, J. F. (1957). Botanical limnological methods, with special reference to the algae. *Botanical Reviews* **23**, 489-583.

[32]. Irish, A. E. (1980). A modified 1-m Friedinger sampler - a description and some selected results. *Freshwater Biology* **10**, 135-139.

[33]. Lund, J. W. G., Kipling, C. & Le Cren, E. D. (1958). The inverted microscope method of estimating algal numbers and the statistical basis of estimates by counting. *Hydrobiologia* **11**, 143-170.

[34]. Lund, J. W .G. (1959). A simple counting chamber for nanoplankton. *Limnnology and Oceanography* **4**, 57-65

[35]. Mackereth, F. J. H., Heron, J. & Talling, J. F. (1978). Water analysis: some revised methods for limnologists. *Scientific Publications of the Freshwater Biological Association* **36**, 120 pp.

[36]. Reynolds, C. S. & Reynolds, J. B. (1985). The atypical seasonality of phytoplankton in Crose Mere in 1972: an independent test of the hypothesis that variability in the physical environment regulates community dynamics and structure. *British Phycological Journal* **20**, 227-242.

[37]. Reynolds, C. S. & Bellinger, E. D. (1992). Patterns of abundance and dominance of the phytoplankton of Rostherne Mere, England: evidence from an 18-year data set. *Aquatic Sciences* **54**, 10-36.

[38]. Lund, J. W. G. (1949). Studies on *Asterionella*. I. The origin and nature of the cells producing seasonal maxima. *Journal of Ecology* **37**, 389-419.

[39]. Lund, J. W. G. (1950). Studies on *Asterionella formosa* Hass. II. Nutrient depletion and the spring maximum. *Journal of Ecology* **38**, 1-35.

[40]. Sommer, U. (1987). Factors controlling the seasonal variation in phytoplankton species composition - a case study for a deep, nutrient-rich lake. In: *Progress in phycological research, Vol. 5* (eds. F. E. Round & D. J. Chapman). Biopress, Bristol, pp. 123-178.

[41]. Maberly, S. C., Hurley, M. A., Butterwick, C., Corry, J. E., Heaney, S. I., Irish, A. E., Jaworski, G. H. M., Lund, J. W. G., Reynolds, C. S. & Roscoe, J. V. (1994). The rise and fall of *Asterionella formosa* in the South Basin of Windermere: analysis of a 45-year series of data. *Freshwater Biology* **31**, 19-34.

[42]. Reynolds, C. S. (1984). *The ecology of freshwater phytoplankton.* Cambridge University Press, Cambridge.

[43]. Ferguson, A. J. D., Thompson, J. M. & Reynolds, C. S. (1982). Structure and dynamics of zooplankton communities maintained in closed systems, with special reference to the algal food supply. *Journal of Plankton Research* **4**, 523-543.

[44]. Reynolds, C. S. (1996). Phosphorus recycling in lakes; evidence from large limnetic enclosures for the importance of shallow sediments. *Freshwater Biology* **35**, 623-645.

[45]. Reynolds, C. S. (1984). Artificial induction of surface blooms of cyanobacteria. *Verh. Internat. Verein. Limnol.* **22**, 638-643.

[46]. Reynolds, C. S. (1979). The limnology of the eutrophic meres of the Shropshire-Cheshire Plain. *Field Studies* **5**, 93-173.

[47]. Reynolds, C. S. (1978). Notes on the phytoplankton periodicity of Rostherne Mere, Cheshire, 1967-1977. *British Phycological Bulletin* **13**, 329-335.

[48]. Hilton, J., Irish, A. E.. & Reynolds, C. S. (1992). Active reservoir management: a model solution. In: *Eutrophication, research and application to water supply* (eds. D. W. Sutcliffe & J. G. Jones). Freshwater Biological Association, Ambleside, pp. 185-196.

[49]. Reynolds, C. S. & Irish, A. E. (1996). Modelling phytoplankton dynamics in reservoirs: the problem of in situ growth rates. In: *Proceedings of a symposium on algal models, Reading* (in press).

[50]. George, D. G., Hewitt, D. P., Lund, J. W. G. & Smyly, W. J. P. (1990). The relative effects of enrichment and climate change on long-term dynamics of *Daphnia* in Esthwaite Water, Cumbria. *Freshwater Biology* **23**, 55-70.

[51]. George, D. G. & Harris, G. P. (1985). The effect of climate on long-term changes in crustacean zooplankton biomass of Lake Windermere, U.K. *Nature* **316**, 536-539.

[52]. George, D. G. & Taylor, A. H. (1995). UK lake plankton and the Gulf Stream. *Nature* **378**, 139.

THE INFLUENCE OF WEATHER CONDITIONS ON THE SEASONAL PLANKTON DEVELOPMENT IN A LARGE AND DEEP LAKE (L. CONSTANCE)

I. THE IMPACT OF IRRADIANCE, AIR TEMPERATURE AND WIND ON THE ALGAL SPRING DEVELOPMENT - AN ANALYSIS BASED ON LONG-TERM MEASUREMENTS

URSULA GAEDKE[1], ANGELIKA SEIFRIED[1]
AND REINER KÜMMERLIN[2]

[1]*Limnologisches Institut, Universität Konstanz*
D-78434 Konstanz, Germany

[2]*Institut für Seenforschung, Postfach 4146*
D-88081 Langenargen, Germany

Abstract

Long-term (ca. 1979–1995), weekly or biweekly measurements of algal biovolume and chlorophyll were related to daily records of irradiance, air temperature and wind speed in order to evaluate the impact of meteorological conditions on spring algal growth in large and deep Lake Constance. A significant influence could only be established by analysing the changes recorded over short periods of time, or by contrasting meteorological conditions in years characterised by an early onset of phytoplankton development (1989, 1990, 1991, 1993 and 1994) with the conditions recorded in more typical years (1979 to 1988, and 1992). The first-mentioned "early" years had significantly higher algal biovolumes and chlorophyll concentrations in March and lower ones in May, suggesting an earlier onset and termination of vernal algal development. Irradiance during "early" years reached average values or less whereas air temperatures around March were significantly higher (+ 3.5°C on average). The comparison of time-series enabled only limited insight into the synergistic effects of irradiance, air temperature and wind on plankton dynamics. These synergistic effects are further evaluated in subsequent studies using hydrodynamical models to compute water column stability and its dependence on various meteorological factors.

D.G. George et al.(eds.), Management of Lakes and Reservoirs during Global Climate Change, 39–55.

1. Introduction

Presumed changes of global circulation patterns and local weather conditions may have far reaching consequences for lake ecosystems during the next centuries (e.g. [1, 2]). A first step towards a better understanding of the potential response of pelagic food webs to climatic changes is to analyse the effect of actual year-to-year fluctuations in the weather on the seasonal dynamics of standing stocks of pelagic communities. The present study and three following papers in this volume summarise the results of such an analysis for large and deep Lake Constance, based on comprehensive long-term measurements of physical and biological parameters and an elaborate hydrodynamical model. In this first paper, we examine the response of the phytoplankton community in spring to different meteorological conditions, basing our analysis on data acquired between 1979 and 1995. Subsequent papers present a hydrodynamical model which enables a quantification of water column stability (Ollinger & Bäuerle, II in this volume, pp. 57–70), relate phytoplankton spring growth to water column stability (Gaedke et al., III in this volume, pp. 71–84), and evaluate reaction of zooplankton to fluctuations of spring water temperature and of the onset of spring algal growth (Straile & Geller, IV in this volume, pp. 85–92).

Before analysing time-series of biological and meteorological parameters of this kind, three key questions have to be considered: (1) how can seasonal and interannual fluctuations in weather conditions be quantified and compared in an ecologically meaningful way?; (2) which of the most commonly measured meteorological parameters or parameter combinations provide the best indication of "the weather" as perceived by "the lake ecosystem"?; (3) which limnological variables provide the best measure of a lake response to global environmental change?

(1) *Interannual comparison of meteorological conditions.* In Central Europe, weather conditions typically change on a time-scale of days rather than weeks. Consequently, it is very unlikely that weather conditions during one year or one season are consistently above or below the long-term average. Thus, we need to consider relatively short periods of time (ca. 1–3 weeks) separately, especially when investigating organisms like phytoplankton which have generation times in the order of days. The identification of unusual weather must also be related to the sensitivity of the biological community to various types of deviations from the mean (e.g. duration, magnitude).

(2) *Impact of various meteorological parameters on lake ecosystems.* Day-to-day changes in irradiance have a direct effect on the underwater light climate and indirectly influence the primary production, phytoplankton succession and ultimately the structure of the entire pelagic food web. Additionally, irradiance influences, in concert with wind and air temperature, the water temperature and vertical stability of the water column which, in turn, is of great importance for all process rates like production and respiration, but also for the light history and nutrient supply (Fig. 1). In spring the indirect effects of meteorological conditions via water column stabilisation are thought to be of great importance in deep unsheltered lakes like Constance (compare Gaedke et al. III). Lacking direct measures of water column stability and vertical mixing, we look for observed meteorological parameters that provide suitable indicators.

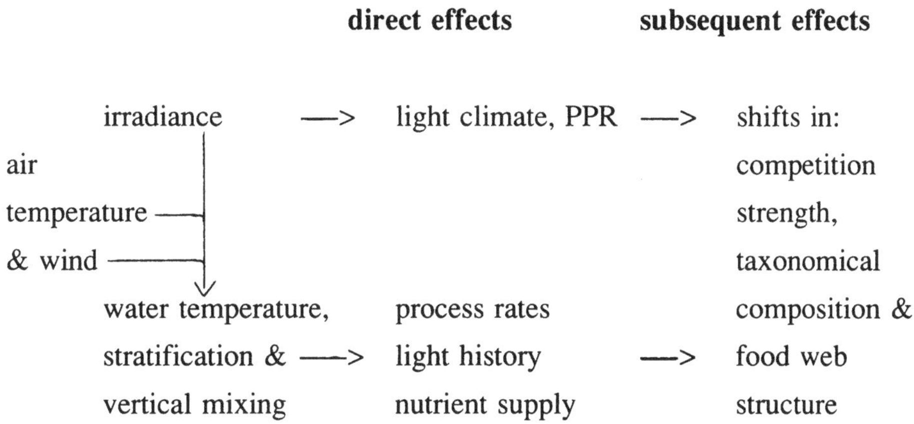

FIGURE 1. Scheme of the direct and indirect effects of weather conditions on pelagic food webs. (For details see the text and Gaedke et al., this volume pp. 71–84).

(3) *Organisms, processes and periods most sensitive to weather conditions.* Some knowledge of hydrodynamical and chemical conditions is required as essential background information for further analyses. Regarding the biological community, the large variety of individual species may be grouped along a gradient of trophic positions (roughly, e.g., phytoplankton > zooplankton > fish). This gradient corresponds to (a) a decreasing direct impact of the light and nutrient regime, and (b) approximately, to an increase of body size which, in turn, implies an increase in reaction time to disturbances. Thus, small organisms at the base of the food web are expected to have the most sensitive reactions to short-term weather fluctuations. A prominent exception from this generalisation is the vulnerability of fish stocks and their recruitment to weather conditions during early larval development (e.g. [3, 4]). The numerous species may be aggregated to varying extents into different hierarchical levels (e.g. populations, communities, the entire food web). The strength and reaction time of the response to seasonal and interannual fluctuations depend on the hierarchical level of aggregation. The sensitivity of individual species is on average greater than that of communities or the entire plankton [5], which suggests that one should focus on individual species. However, predictions related to global change will often concentrate on functional aspects of the ecosystem, i.e. higher levels of aggregation. The relative impact of physical and biological factors on plankton dynamics varies greatly throughout the year. We have strong indications that the plankton community is most responsive to changes in the physical conditions in Lake Constance in early spring (e.g. [6–9]).

To summarise, the present study will focus on the phytoplankton community as an entirety, and its reaction to changes in the irradiance, the air temperature, and the wind speed in spring. Water temperature, though of some importance, is not considered here because it is strongly influenced by internal seiche movements at our sampling site, which leads to pronounced fluctuations at a given depth with frequencies similar to those induced by weather changes [10].

2. Study site, materials and methods

Lake Constance (in German: Bodensee) is a large (476 km^2) and deep (z_{max}= 252 m) pre-alpine (47°50'N) lake of warm-monomictic character, at the northern fringe of the Alps. As expected from its large size, stratification in Constance is on average less pronounced and less stable than is typically found in small lakes. During the period of stratification, the well-mixed epilimnion is typically about 10 m deep and is followed by a thick metalimnion (ca. 15–30 m) which is characterised by a gradually decreasing temperature. Hypolimnetic temperatures are about 4–5°C throughout the year. The upper main basin (Obersee) of Lake Constance, considered here, was never covered by ice during the period of investigation. Constance is presently undergoing re-oligotrophication and it changed from meso-eutrophic to oligo-mesotrophic conditions during the study period (1979–1995). The onset of algal growth in spring is most probably not affected by the re-oligotrophication process because concentrations of all nutrients are sufficiently high to allow maximum growth rates during this time [8].

Irradiance was measured on the roof of the institute next to the lake and complemented by comparable measurements at a local weather station. The long-term mean irradiance was calculated as the moving average (5 days) of the mean irradiance of each day of the year during 1979 to 1994. Records of air temperature and wind speed were obtained from the German Weather Service.

Field samples were taken weekly during the season and ca. biweekly in winter at different depths at a 147-m deep sampling site in the fjordlike northwestern part of the lake (Überlingersee) (see map given by Ollinger & Bäuerle, III in this volume p. 58). Chlorophyll-*a* [11, 12, and unpublished] was determined spectrophotometrically from hot ethanol extracts and corrected for phaeopigments by acidification according to Nusch. Algal biovolume was obtained by enumerating individual species and morphotypes with the Utermöhl technique and allocating fixed cell volumes to each species [7, 8, and unpublished].

Mean spring values for the physical and biological variables were obtained by averaging measurements from 15 March until the onset of the clear-water phase for each year respectively. In our exploratory analyses we used a variety of techniques to relate short-term periods of unusual weather to fluctuations in algal standing stocks. When many of these quantitative techniques proved difficult to apply we resorted to a simple scoring system where meteorological conditions were classified according to their presumed potential to favour algal growth (exceptionally good (++), relatively good (+), relatively bad (–), and exceptionally bad (– –)). Periods of unusual irradiance or air temperature were then identified by calculating the logarithm of the relative deviation from the long-term mean (i.e. the ratio between the actually observed value and the long-term average) for each day, and summing this value for all consecutive days which had the same sign of deviation. If the sum exceeded a certain threshold value in either direction, the irradiance or temperature during this period was classified as exceptionally or relatively good (++, +) or bad (–, – –). Exceptionally favourable growth conditions (++) with respect to wind speed (smoothed by a moving average of 5 days) were assumed to prevail when the average wind speed was less than 1 m s^{-1}, relatively good conditions (+) prevailed when the average wind speed ranged between

1.0 and 1.2 m s^{-1}, and relatively bad (–) and exceptionally bad (– –) conditions prevailed when the average speeds respectively exceeded 3 and 5 m s^{-1}.

3. Results

Average phytoplankton biovolume and chlorophyll concentrations in spring are unrelated to the average spring irradiance (Fig. 2), air temperature (Fig. 3) or wind speed (Fig. 4). These results suggest that a finer temporal resolution than the spring average (i.e. ca. 8–12 weeks) is required to evaluate the potential impact of weather conditions on algal spring development.

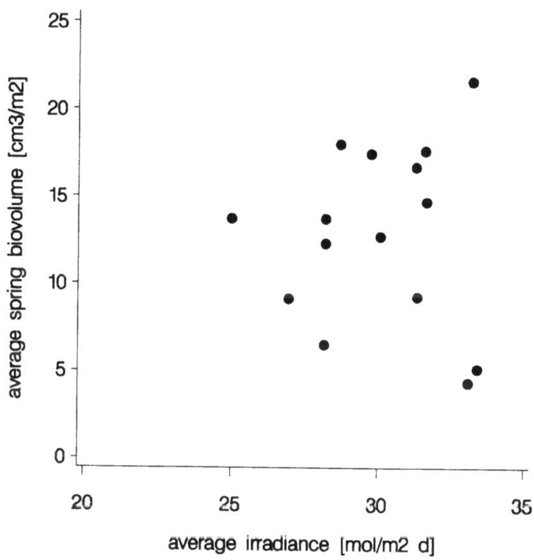

FIGURE 2. Average algal biovolume (cm^3 m^{-2}) in L. Constance and the average irradiance (mol m^{-2} d^{-1}) in spring (15 March until the onset of the clearwater phase).

Relative changes in algal standing stocks were inspected during the periods of time which provided unusually favourable (++, +) or unfavourable (–, – –) growth conditions with respect to irradiance. For each period, the hypothesis that relative low irradiances result in low or decreasing algal abundance and vice versa was tested. This hypothesis was roughly supported by the data but weather conditions were changing too fast to provide unambiguous results based on weekly or biweekly phytoplankton samples and chlorophyll measurements. After evaluating various other statistics with limited success, a simple comparison of time-series by eye, for the daily irradiance smoothed by a moving average of 3 to 5 days and algal standing stocks, was regarded as the most powerful and reliable technique.

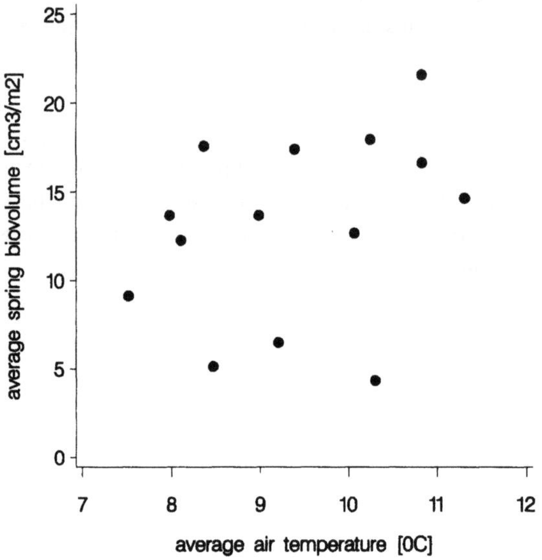

FIGURE 3. Average algal biovolume (cm^3 m^{-2}) in L. Constance and the average air temperature (°C) in spring (15 March until the onset of the clearwater phase).

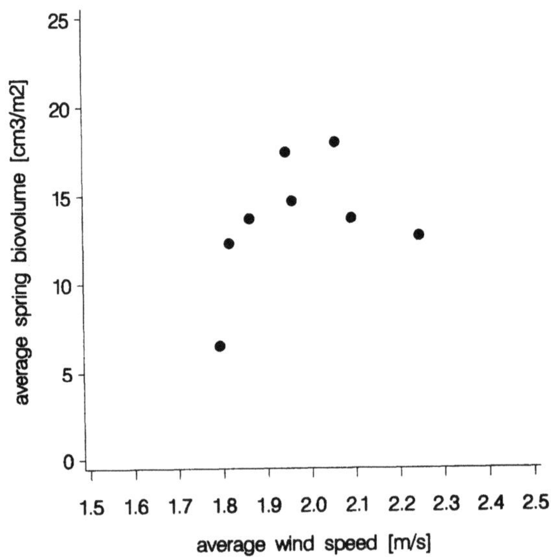

FIGURE 4. Average algal biovolume (cm^3 m^{-2}) in L. Constance and the average wind speed (m s^{-1}) in spring (15 March until the onset of the clearwater phase).

Increases of algal biovolume in the uppermost 20 m of the water column were then relatively closely matched with periods of high irradiance, and periods with low irradiance were typically followed by a decrease in algal standing stocks, e.g. in spring 1982 (Fig. 5). The strong reduction of algal biovolume in June represents the clearwater phase caused by strong grazing that is independent of weather conditions. The peaks of irradiance in March are only slightly above the long-term average and did not result in any increase of algal biomass. However, not all years show such a close relationship as was observed in 1982. For example, in early spring 1985, light conditions similar to 1982 prevailed, but algal development was less responsive and much retarded (Fig. 6). In 1990 spring irradiance was much lower than in 1982 but phytoplankton growth started considerably earlier; i.e. bad light conditions did not prevent a very early bloom (Fig. 7). We conclude that a relationship between the temporal development of spring irradiance and algal development is detectable in most years when using a sufficiently high temporal resolution (i.e. at least weekly measurements). However, it explains only part of the seasonal and interannual variability of algal standing stocks and it is difficult to quantify. These results are independent of the method used to estimate algal standing stocks (biovolume or chlorophyll concentration).

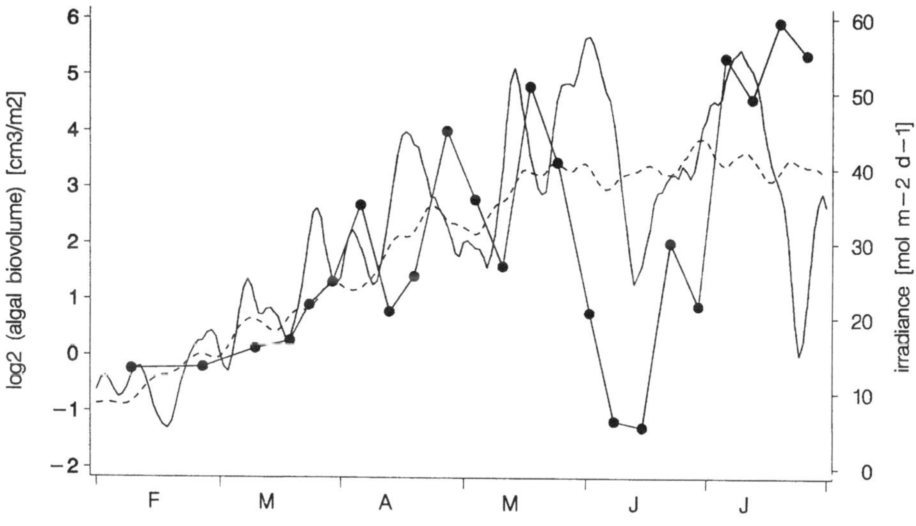

FIGURE 5. Temporal development of algal biovolume (solid circles, $cm^3 m^{-2}$ on a log scale), the moving average (5 days) of daily irradiance ($mol\ m^{-2}\ d^{-1}$; solid line) and the long-term mean of the irradiance (dashed line) in L. Constance in 1982. Strong depletions around June are caused by grazing (clearwater phase).

Investigating the impact of air temperature on phytoplankton development in the same way reveals a moderate covariation between these two parameters. Some of the previously unexplained variability can also be attributed to the synergistic effects of

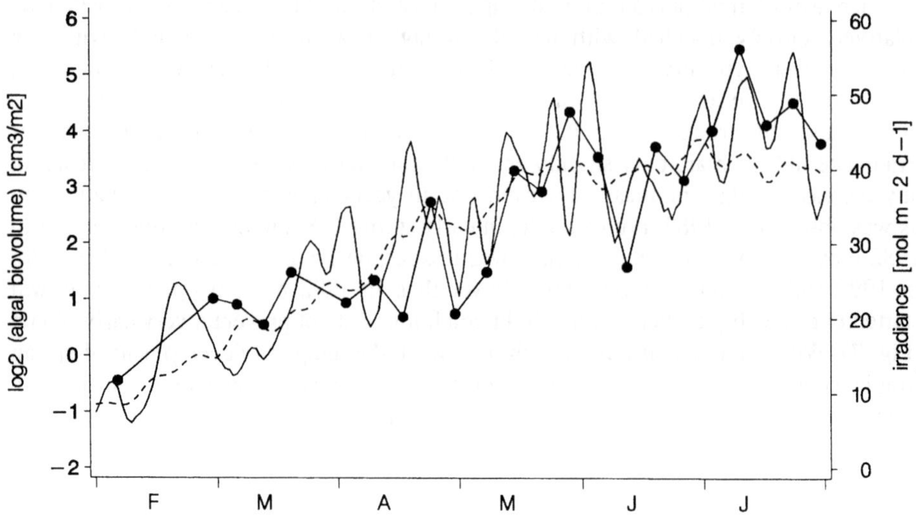

FIGURE 6. Temporal development of algal biovolume (solid circles, $cm^3\,m^{-2}$ on a log scale), the moving average (5 days) of daily irradiance ($mol\,m^{-2}\,d^{-1}$; solid line) and the long-term mean of the irradiance (dashed line) in L. Constance in 1985. Strong depletions around June are caused by grazing (clearwater phase).

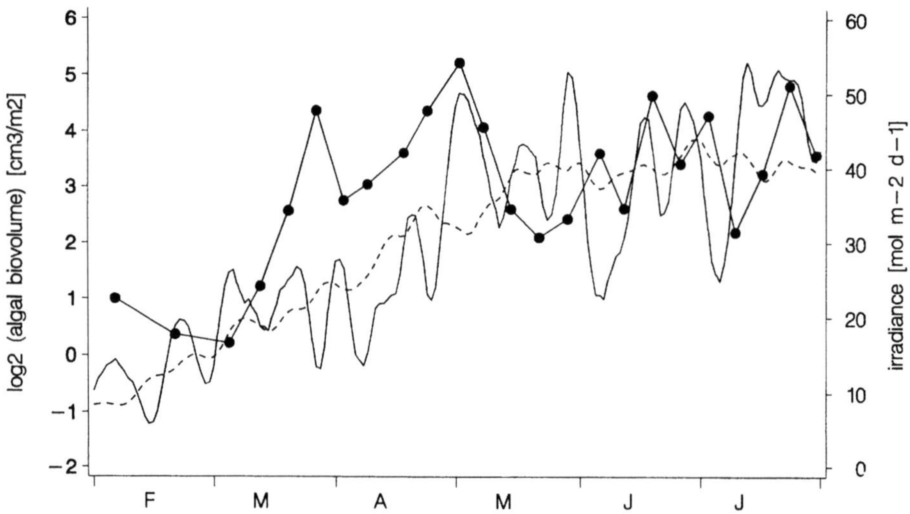

FIGURE 7. Temporal development of algal biovolume (solid circles, $cm^3\,m^{-2}$ on a log scale), the moving average (5 days) of daily irradiance ($mol\,m^{-2}\,d^{-1}$; solid line) and the long-term mean of the irradiance (dashed line) in L. Constance in 1990. Strong depletions around June are caused by grazing (clearwater phase).

irradiance and air temperature. For example, 1985 was characterised by relatively low and 1990 by very warm air temperatures in early spring. Another aspect which may have retarded the development in 1985 is the unusually frequent fluctuations in light conditions; i.e., the rates of change may need to be considered as well as the absolute values.

A third potentially important factor for algal growth in deep Lake Constance is wind, because it influences water column stability and mixing depth. A systematic investigation of the synergistic effects of irradiance, air temperature and wind on phytoplankton dynamics is hardly feasible by comparing the temporal development of the different variables, since the covariation between irradiance, temperature and wind is relatively low. Time-periods with relatively large deviations of the irradiance or air temperature from the long-term average were identified and, in addition, each day was classified according to the average wind speed (Figs 8 and 9). Periods with simultaneously fair, warm and calm conditions, or dark, cold and windy weather, rarely occur; most of the time some conditions are in favour of algal growth whereas others are not. For example, only 12% of the days with relatively high irradiance also have relatively high air temperatures (Table 1). The co-occurrence of high irradiance or high temperature with little wind, or vice versa, is even more unusual.

TABLE 1. Covariation between irradiance, air temperature and wind speed. The table provides the percentages of days on which exceptionally high (++), relatively high (+), relatively low (–), and exceptionally very low (– –) irradiance coincides with corresponding values of the air temperature and very little (++), little (+), moderate (–), and strong (– –) wind (for details of classification criteria see section on methods).

Irradiance	High (+ +)	Relatively high (+)	Relatively low (–)	Low (– –)
Temperature	33%	12%	33%	18%
Wind	20%	9%	10%	0.4%

In late February and early March, algal blooms are observed only if exceptionally favourable weather conditions prevail. They are frequently subject to rapid termination by changes of weather conditions and may then be followed by low standing stocks for several weeks (e.g. in 1992). Under normal circumstances, phytoplankton development starts before the end of March but may be interrupted again by adverse conditions. A delay of the vernal development until mid April occurs only under exceptionally unfavourable weather conditions, and from May onwards phytoplankton dynamics become much less responsive to short-term fluctuations of weather conditions (compare with Gaedke et al. III, pp. 71–84). For example, March and April 1988 are characterised by an unusually distinct change from relatively long-lasting bad weather to favourable

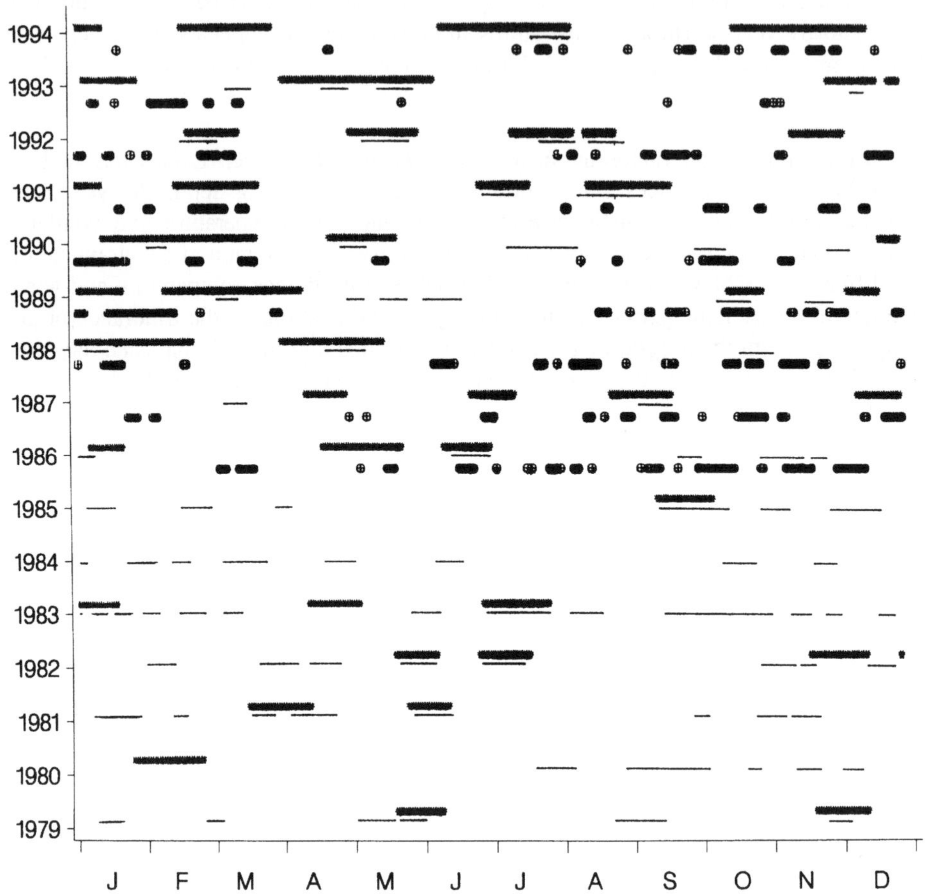

FIGURE 8. Periods of unusual weather conditions from 1979 to 1994 (wind 1987 to 1994). The figure shows days/periods with relatively high (+) irradiance (thin lines), relatively high (+) air temperatures (stars), and relatively low (+) wind (circles). (For details see the section on methods).

weather (Figs 8 and 9), which provided optimal circumstances for studying the impact of meteorological conditions. Bad weather prevailed in March with very low irradiance and a frequent occurrence of strong winds. The air temperature was above the long-term average at the beginning of the year and close to it in March. The weather then changed quickly to favourable conditions in early April, with relatively high air temperatures. The distinct weather pattern resulted in an outburst of algal development after its retardation in March. Algal biomass decreased in March and then increased in April by a factor of 20 within 2 weeks (chlorophyll concentrations), which illustrates the high

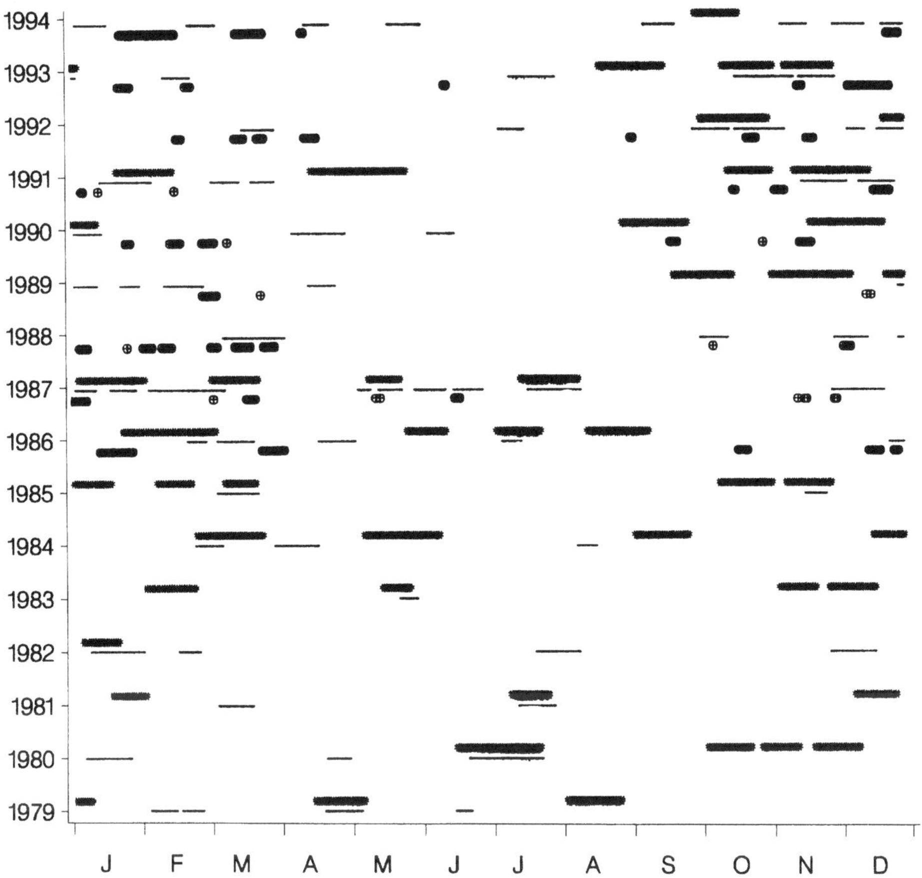

FIGURE 9. Periods of unusual weather conditions from 1979 to 1994 (wind 1987 to 1994). The figure shows days/periods with relatively low (–) irradiance (thin lines), relatively low (–) air temperatures (stars), and relatively strong (–) wind (circles). (For details see the section on methods).

growth potential during this period. A comparable increase was observed only in 1987, a year which is likewise characterised by a strongly retarded spring development. A full understanding of the influence of weather conditions on algal growth requires quantification of the direct and indirect effects of the important meteorological factors affecting algal growth (Fig. 1); this is beyond the scope of the present time-series analysis and requires elaborate modelling studies.

Another way of assessing the impact of weather conditions on algal growth, which does not involve the ambiguities of time-consuming modelling exercises, is to compare the meteorological conditions between years with early and late phytoplankton development. Very early onsets of algal growth were observed during the years 1989, 1990, 1991 and 1993, which again can be contrasted with the more typical years (1979 to 1988, and 1992). During the first-mentioned years, average algal biomass was significantly higher in March and lower in May than during the other years, suggesting there is a statistically significant difference in the beginning of algal development between the two groups of years (Fig. 10). This result holds for algal biovolume and chlorophyll-*a* concentrations (chlorophyll data are available additionally for 1994 (early onset of growth) and 1995 (late onset of growth), but are lacking for 1984 and 1985). Maximum peak values of monthly averages are higher in "late" years which is, however, not necessarily attributable to meteorological factors. Weekly averages of irradiance and air temperature reveal that during years with an early onset of phytoplankton development, light conditions were at or below the mean, but air temperatures around March were significantly higher (Fig. 11). The mean difference between the two groups of years amounts to 3.6°C. The wind regime differed unsystematically between years (Figs 8 and 9). Consequently, our data suggest that an increase of air temperature around March by a few degrees significantly influences the onset of phytoplankton development.

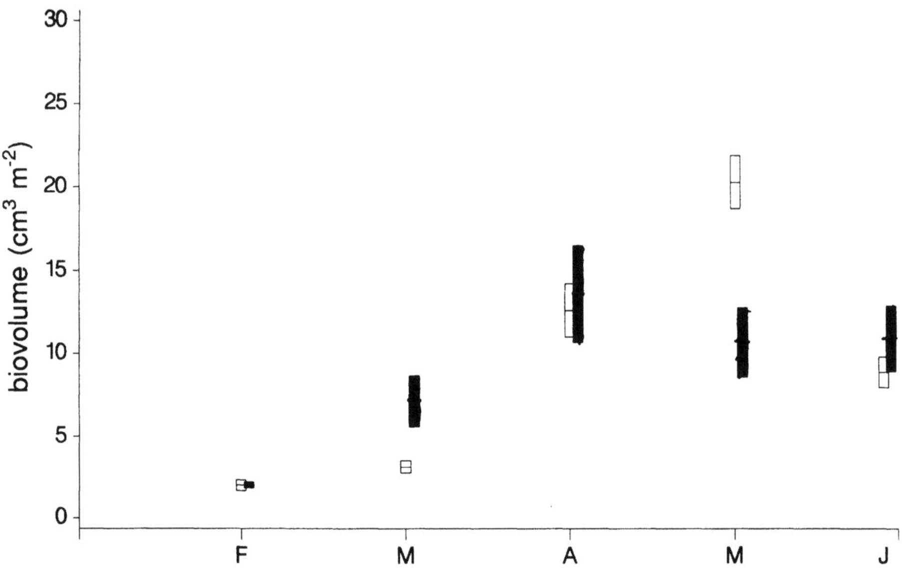

FIGURE 10. Monthly averages of phytoplankton biovolume ($cm^3\ m^{-2}$) in L. Constance for February to June during 1989, 1990, 1991 and 1993 (solid bars) and during the period 1979 to 1988, and 1992 (open bars). The length of the bars indicates the standard deviation of the monthly averages.

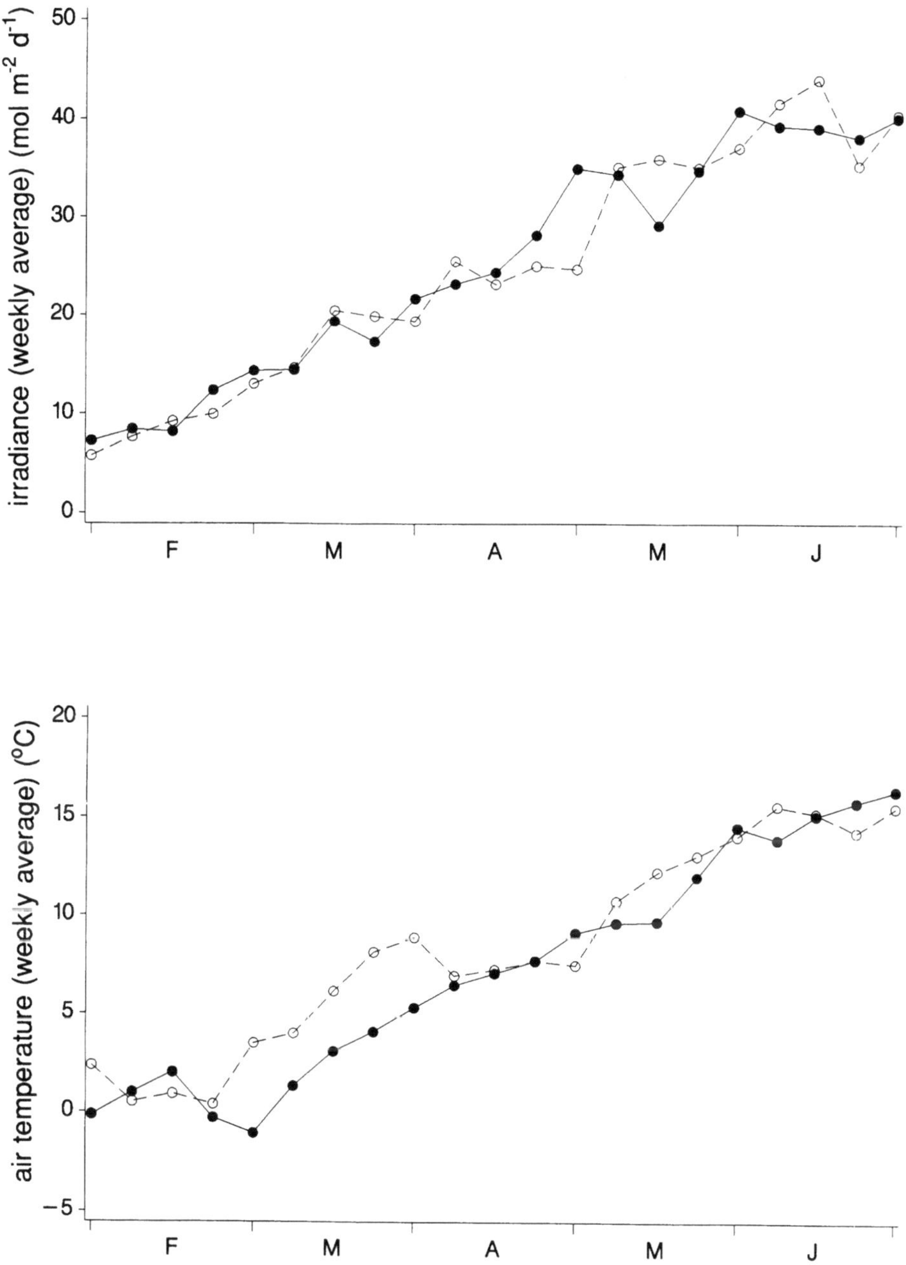

FIGURE 11. Weekly averages of daily irradiance (above) and air temperature (below) during 1989 to 1991 and 1993–1994 (open circles), and 1979 to 1988 and 1992 (solid circles) for January– June.

4. Discussion

The scoring system used here provides a relatively straightforward method of summarising a large amount of data and relating the change in standing stocks to measurable changes in the local weather. The resulting data are easy to summarise and present in a form that can be compared readily with time-series data from similar ecosystems. One major advantage of this empirical approach is the ease with which it can be modified to test new hypotheses. "Scoping" analyses of this kind are particularly useful in the early stages of an investigation, since they do not involve the ambiguities and uncertainties associated with complex model studies that inevitably are based on numerous equations and parameters.

Investigations of phytoplankton development of the kind used here nevertheless require a reliable quantification of algal standing stocks. Although done routinely for many lakes this procedure involves a considerable number of uncertainties. Algal biomass is mostly inferred, either from measurements of chlorophyll or by counting cell numbers of individual species and measuring their average cell volume. Chlorophyll measurements are much less time-consuming than microscopic sample processing which enables a much better vertical resolution. However, the chlorophyll content per unit of biomass (e.g., expressed in units of carbon) is by no means constant but depends on environmental conditions (light, nutrients) and species composition. Thus, variation of algal biomass inferred from chlorophyll measurements may be caused (or damped) partly by these factors. On the other hand, biomass estimates from laborious microscopic determination of algal biovolume are generally based on assumptions like constant species-specific cell volumes, and fixed carbon to volume ratios throughout the season, which are also not met in practice. Additionally, counting volumes are often too small to determine abundances of very large species accurately. In this study, we resolved these problems by using a combination of methods to analyse the general pattern of change.

Time-averaged phytoplankton standing stocks in spring exhibited no relationship with mean spring irradiance, air temperature, or wind speed. In contrast, monthly or bimonthly mean values of water temperature and crustacean biomass covaried in Lake Constance (Straile & Geller IV, pp. 85–92), and this might be attributed to the longer generation times of these organisms. In Constance the average spring algal biomass is influenced by numerous factors, e.g., the timing of the onset of growth, the number and duration of bloom(s) and interruption(s), the timing of the onset of the clearwater phase (Figs 5–7) which, in turn, depend partially on other factors like grazing pressure. Based on average meteorological parameters, a higher predictive power may be achievable for the onset of algal growth rather than for the average spring algal biomass. A simple regression model based on mean annual air temperature, lake surface area (surrogate for wind exposure), and/or the ratio of lake surface area to mean depth, was used to make rough predictions of the date of spring stratification in subtropical and temperate northern hemisphere lakes [13]; (for the relationship between spring stratification and phytoplankton growth in Lake Constance see Gaedke et al. III, pp. 71–84). However, this cross-lake comparison covered very different lakes, with stratification dates differing by more than two months, whereas the interannual variability of the date of stratification within Lake Constance is considerably smaller (about one month) (cf. Ollinger & Bäuerle II, pp. 57–70). Furthermore, such regression analyses are complicated by the dependence on the parameters and measures chosen for the classification of individual years or the identification of particular weather conditions. Considering irradiance, for example, different results were obtained when ranking according to the total sum of absolute irradiance or the relative deviation from the long-term average. The classification becomes even more ambiguous when two or more meteorological parameters are considered simultaneously. Hence, at the present stage of understanding, the kind of meteorological measures to which the algal standing stock should be related remains an open question.

This problem was circumvented by reversing the procedure and classifying the 17 years of investigation into two groups, with an early or later onset of algal growth, and then comparing the average meteorological conditions between these two groups of years. This technique clearly indicated that relatively high air-temperatures gave rise to the exceptionally early onset of growth in some years. It is based on long-term observations, rather than on designed experiments, which prevents a systematic study of the relative importance of different meteorological factors. The aggregation of years into the two groups is to some extent subjective because in some years a minor bloom occurred prior to the spring bloom proper. Actual results are robust against minor changes of the aggregation scheme.

Covariations of phytoplankton standing stocks with daily irradiance or air temperature suggest a substantial impact of meteorological conditions on algal growth. Similarly, analyses of another long time-series on chlorophyll concentrations and diatom abundance in Windermere (English Lake District) indicate irradiance-dependent phytoplankton dynamics during the vernal bloom when using long-term averages [14]. Our weekly or biweekly measurements of plankton biovolume and chlorophyll concentrations provide a high temporal resolution compared to other studies. Nevertheless, this sampling frequency is rather low to recognise unambiguously the reaction of the phytoplankton community to the frequently changing weather conditions. A further complication is that there is only a short time-period in spring when meteorological conditions rule algal growth, before such factors as grazing and nutrient depletion become dominant. The relatively poor correlation between individual meteorological parameters and algal growth indicates that neither irradiance nor air temperature on their own provide a reliable indicator of algal growth conditions in spring in Lake Constance, and that indirect effects of meteorological conditions, like changes of mixing layer depths and water column stability, may play a dominant role in large and deep Lake Constance (Gaedke et al. III, pp. 71–84).

The simplicity of the present evaluations of time-series is achieved at the expense of a low capability to understand sophisticated causal relationships among the key variables. Hence, this study will be supplemented by model studies on the impact of weather conditions on water column stability (see the following contributions (II to IV) in this volume, pp. 57–92), and by an analysis of the subsequent response of higher trophic levels.

5. Conclusions

Average phytoplankton standing stock in spring did not correlate with average spring irradiance, air temperature or wind speed.

A temporal covariation between irradiance and algal development was detectable in most years in spring, when using a high temporal resolution (i.e. weekly measurements). However, it explained only a part of the seasonal and interannual variability of algal standing stocks and it was difficult to quantify.

In addition, algal standing stocks exhibited some covariation with air temperature. By considering the synergistic effects of irradiance and air temperature, the fraction of unexplained variability was reduced.

During years with an early onset of phytoplankton development, air temperatures were a few degrees higher around March than in years with a normal or late spring algal bloom. This indicates that changes in spring temperatures may have far-reaching effects on the seasonal plankton development.

Simple evaluation of time-series data can only begin to explain the complex synergistic direct and indirect impacts of irradiance, temperature and wind on algal growth conditions, and their dependence on the actual water column stability. Further understanding demands more elaborate modelling studies.

Acknowledgements

Data acquisition and the present study were performed within the Special Collaborative Program (SFB) 248 "Cycling of Matter in Lake Constance", supported by Deutsche Forschungsgemeinschaft. Special thanks are due to B. Beese, C. Braunwarth, J. Fürst, U. Sommer, A. Schweizer, and M. Tilzer who conducted measurements of irradiance, algal biovolume and chlorophyll. Meteorological data were partially obtained from the German Weather Service and computations were supported by Astrid Hälbich. We thank the organisers of the workshop and especially Glen George, Gwyn Jones and Pavel Punčochář for stimulation of this work, and Deborah Hart for linguistic corrections.

References

[1]. Carpenter, S. R., Frost, T. M., Kitchell, J. F. & Kratz, T. K. (1993). Species dynamics and environmental change: A perspective from ecosystem experiments. In: *Biotic interactions and global change* (eds. P. M. Kareiva, J. G. Kingsolver & R. B. Huey), pp. 267-279. Sinauer Associates, Sunderland, Massachusetts.

[2]. Kareiva, P. M., Kingsolver, J. G. & Huey, R. D. (Editors) (1993). *Biotic interactions and global change*. Sinauer Associates Inc., Sunderland, Massachusetts.

[3]. Eckmann, R., Gaedke, U. & Wetzlar, H. J. (1988). Effects of climatic and density-dependent factors on year-class strength of *Coregonus lavaretus* L. in Lake Constance. *Can. J. Fish. Aquat. Sci.* **45**, 1088-1093.

[4]. Pusch, M. & Eckmann, R. (1991). Growth patterns of whitefish (*Coregonus lavaretus*) in Lake Constance derived from otolith microstructures. *Verh. Internat. Verein. Limnol.* **24**, 2461-2464.

[5]. Gaedke, U., Barthelmeß, T. & Straile, D. (in press). Temporal variability of standing stocks of individual species, communities, and the entire plankton in two lakes of different trophic state: empirical evidence for hierarchy theory and emergent properties? *Senckenbergiana Maritima*, **27**.

[6]. Sommer, U., Gliwicz, Z. M., Lampert, W. & Duncan, A. (1986). The PEG-model of seasonal succession of planktonic events in fresh waters. *Arch. Hydrobiol.* **106**, 433-471.

[7]. Sommer, U. (1987). Factors controlling the seasonal variation in phytoplankton species composition - A case study for a deep, nutrient rich lake. *Prog. Phycological Res.* **5**, 124-178.

[8]. Gaedke, U. & Schweizer A. (1993). The first decade of oligotrophication in Lake Constance. I. The response of phytoplankton biomass and cell size. *Oecologia* **93**, 268-275.

[9]. Straile, D. (1995). Die saisonale Entwicklung des Kohlenstoffkreislaufes im pelagischen Nahrungsnetz des Bodensees - Eine Analyse von massenbilanzierten Flußdiagrammen mit Hilfe der Netzwerktheorie. Dissertation, Universität Konstanz, 157 S, Hartung-Gorre Verlag Konstanz.

[10]. Gaedke, U. & Schimmele, M. (1991). Internal seiches in Lake Constance: influence on plankton abundance at a fixed sampling site. *J. Plankton Res.* **13**, 743-754.

[11]. Tilzer M. M. & Beese, B. (1988). The seasonal productivity cycle of phytoplankton and controlling factors in Lake Constance. *Schweiz. Z. Hydrol.* **50**, 1-39.

[12]. Tilzer, M. M., Gaedke U., Schweizer A., Beese B. & Wieser T. (1991). Interannual variability of phytoplankton productivity and related parameters in Lake Constance: No response to decreased phosphorus loading? *J. Plankton Res.* **13**, 755-777.

[13]. Demers, E. & Kalff, J. (1993). A simple model for predicting the date of spring stratification in temperate and subtropical lakes. *Limnol. Oceanogr.* **38**, 1077-1081.

[14]. Neale, P. J., Talling, J. F., Heaney, S. I., Reynolds, C. S. & Lund, J. W. G. (1991). Long time series from the English Lake District: Irradiance-dependent phytoplankton dynamics during the spring maximum. *Limnol. Oceanogr.* **36**, 751-760.

THE INFLUENCE OF WEATHER CONDITIONS ON THE SEASONAL PLANKTON DEVELOPMENT IN A LARGE AND DEEP LAKE (L. CONSTANCE)

II. WATER COLUMN STABILITY DERIVED FROM ONE-DIMENSIONAL HYDRODYNAMICAL MODELS

DIETER OLLINGER AND ERICH BÄUERLE
Institut für Umweltphysik, Universität Heidelberg
D-69120 Heidelberg, Germany

Abstract

The physical background for studying the development of spring algal blooms in intermittently turbulent waters is developed. A mathematical model based upon transient Ekman dynamics is used to formulate the conservation of mean momentum and heat energy in their one-dimensional form. The turbulent exchange coefficients are calculated with a two-equation $k - \varepsilon$ model of turbulence. A 10-day episode of observed meteorological conditions at Lake Constance is calculated and compared to observed temperatures in the water column. During calm and moderate winds good agreement is obtained. The effect of the hypothetical insertion of an increased wind speed lasting for 1 day is studied and discussed. The concept of a vertical exchange rate p_{ki} between defined layers is formulated and proposed as a simple but reliable predictor for stratification and, thus, for the onset of spring algal growth in deep lakes.

1. Introduction

The physical properties of the water column in lakes exhibit strong diurnal and seasonal variations. The rate of primary production in the upper near-surface layers and, especially, the onset of the spring phytoplankton bloom, are influenced by these variations. The numerical model presented here simulates the response of the upper water column of a lake to changing meteorological conditions (stabilizing, destabilizing, wind driven) on a time-scale of hours and less, rather than on a seasonal basis. This

D.G. George et al.(eds.), Management of Lakes and Reservoirs during Global Climate Change, 57–70.

time-step is necessary to resolve changes in the phytoplankton's environment if one wishes to explain short-term variations in algal population dynamics. We believe that the observed patterns of vertical mixing are caused by physical processes on very short time-scales, as the dynamics of lakes "are strongly driven by the immediate forcing on the lake" [1].

The model is based on a second-order closure of the Reynolds equations, i.e. it includes calculations of the turbulent energy budget. Second-order models have the potential for providing insight into the interaction of turbulence and stratification. This interaction is an always-changing property. Stirring by wind and waves causes turbulence near the surface. During cold nights or in winter, the water surface cools and there is little or no compensation by absorbed solar radiation. This initiates vigorous convective overturning motions, producing turbulence in the water column.

Solar heating during the day not only provides the light energy for photosynthesis but also heats the near-surface layers by absorption and thus changes the stability of the water column. The stronger the absorption with depth, the larger will be the potential energy deficit of the upper water column and the shallower the depth to which wind mixing can penetrate. The latter is a consequence of the fact that turbulence is suppressed by buoyancy forces arising from the stratification.

Thus the rate of development of an algal bloom as a whole depends on the balance between the rate at which turbulence is created at the surface and distributed in the wind-mixed layer, and the buoyancy created by the surface heat exchange.

In spring it may happen that already existing stratification is nearly totally destroyed by strong winds or storms. Water from the upper metres of the water column is then mixed into depths far below the photic zone. The concentration of phytoplankton cells is diminished to nearly vanishing values; the spring bloom breaks down merely due to physical reasons.

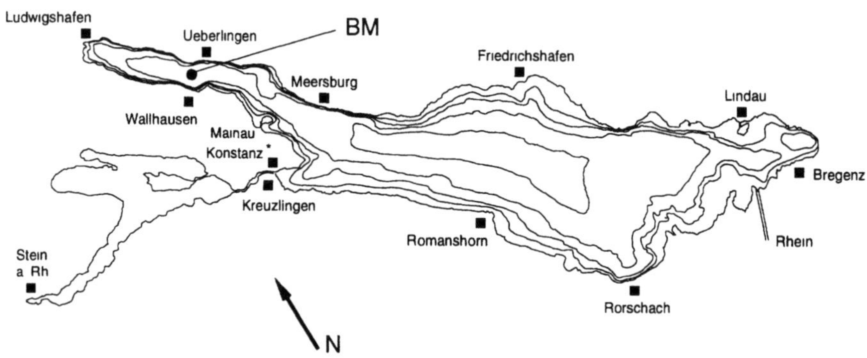

FIGURE 1. Map of the measuring sites at Lake Constance. A permanent thermistor chain with 33 sensors is placed at station BM (close to the central sampling station for biological parameters). Meteorological data originate from the weather station at Konstanz.

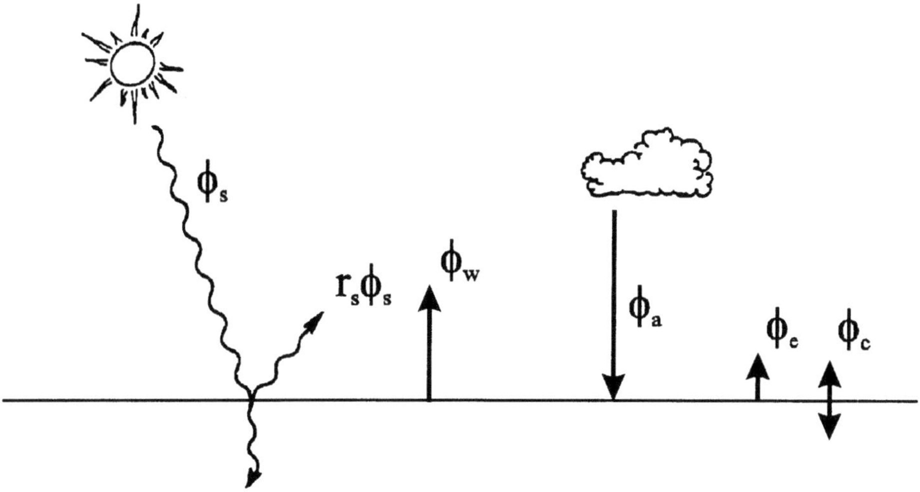

FIGURE 2. Energy fluxes at the air–water interface. Fluxes are taken to be positive when directed into the water. Symbols are explained in the text.

The purpose of this and a following paper by Gaedke et al. (III in this volume, pp. 71–84) is to analyze the correlation between stability of stratification, meteorological forcing and algal standing stock. In the present paper we want to provide the physical and mathematical background of the study. Measuring sites in Lake Constance are shown in Fig. 1, and energy fluxes at the air–water interface of the lake are depicted in Fig. 2.

2. Mathematical formulation

The mathematical model employed herein is based on a model presented by Svensson [2]. The reader is referred to his paper for a detailed discussion of the basic assumptions, boundary conditions, and the turbulence model. We provide a short presentation of the set of equations and restrict our attention to horizontally homogeneous flows, which means that terms containing gradients in the horizontal plane are neglected. It is further assumed that no mean vertical velocities are present. The one-dimensional equations for the conservation of heat and momentum thus read:

$$\delta T/\delta t = (\delta/\delta z)[(\nu_t/\sigma_T)(\delta T/\delta z)] + S_H \quad (1)$$

$$\delta U/\delta t = (\delta/\delta z)[\nu_t(\delta U/\delta z)] + fV \quad (2)$$

$$\delta V/\delta t = (\delta/\delta z)[(\nu_t(\delta V/\delta z)] - fU \quad (3)$$

where z is the vertical space coordinate, positive upward, t is the time coordinate, f is the Coriolis parameter, U and V are the mean horizontal velocities and T is the mean temperature. The kinematic eddy viscosity is denoted by ν_t and σ_T is a Prandtl/Schmidt number for temperature. The source term in the equation for temperature is denoted S_H.

We retain the Coriolis terms in the momentum equations, although it is known (see [3]) that inertial oscillations, which are the consequence of any changes in the wind field in the ocean and in unstratified lakes, are replaced by Poincaré wave-like internal oscillations when stratification has started. As one-dimensional models are not able to incorporate internal oscillations in a straightforward way, we consider the inertial oscillations of the model as representative of all kinds of oscillatory two-dimensional processes. This proceeding produces acceptable results but needs some further thorough investigation.

The turbulent exchange coefficients are calculated from a kinetic energy-dissipation model of turbulence.

2.1. THE TURBULENCE MODEL

Differential transport equations for the kinetic energy of turbulent motion, k, and its dissipation, ε, are formulated in order to account for transport and history effects of turbulence:

$$\delta k/\delta t = (\delta/\delta z)[(\nu_t/\sigma_k)(\delta k/\delta z)] + P_s + P_b - \varepsilon) \tag{4}$$

$$\delta\varepsilon/\delta t = (\delta/\delta z)[(\nu_t/\sigma_\varepsilon)(\delta\varepsilon/\delta z)] + (\varepsilon/k)(c_{1\varepsilon}P_s + C_{3\varepsilon}P_b - C_{2\varepsilon}\varepsilon) \tag{5}$$

$$\text{with } Ps = \nu_t\,[(\delta U/\delta z)^2 + (\delta V/\delta z)^2] \quad \text{and } P_b = [(g/\rho)\nu_t(\delta\rho/\delta z)] \tag{6}$$

P_s is the production due to shear and P_b is production/destruction due to buoyancy. In the standard $k - \varepsilon$ model, the eddy viscosity ν_t is related to k and ε via the relationship:

$$\nu_t = C_\mu k^2/\varepsilon \tag{7}$$

where C_μ is an empirical parameter assumed to be constant ($C_\mu = 0.09$). According to [3], the other empirical constants are taken as standard values ($C_{1\varepsilon} = 1.44$, $C_{2\varepsilon} = 1.92$, $C_{3\varepsilon} = 0.8$, $\sigma_k = 1.0$ and $\sigma_\varepsilon = 1.3$). The value of ν_t as a function of depth and time provides an appropriate measure for the state of turbulence in the water column.

2.2. BOUNDARY CONDITIONS

Surface boundary conditions relate mean flow variables and temperature to wind stress and heat flux, respectively. The surface wind stress, $(\tau_x(t),\tau_y(t))$, specifies the momentum flux at the surface:

$$v_t\ (\delta U/\delta z)\,|_{z=0} = \tau_x(t)/p_0 \tag{8}$$

$$v_t\ (\delta V/\delta z)\,|_{z=0} = \tau_y(t)/p_0 \tag{9}$$

The temperature gradient at the surface is related to the net heat flux $Q(t)$ through the air–water interface:

$$(v_t\,/\sigma_T\,)(\delta T/\delta z) = Q(t)/p_0 c_p \tag{10}$$

p_0 is some reference density and c_p is the specific heat of water.

2.3. EQUATION OF STATE

In the present investigation, variations in density only arise from varying temperature:

$$p(T) = p_0(z) - \alpha(T - T_0)^2 \tag{11}$$

α is a thermal expansion coefficient and T_0 is the temperature of maximum density (ca. 4°C). Salinity is assumed to be a (small) constant incorporated in the reference density $p_0(z)$.

3. Specification of the Fluxes

3.1. SHORT-WAVE AND LONG-WAVE RADIATION

The energy content of a lake is governed primarily by the energy exchange across the water–air interface. The effects of the throughflow of water by inlets and outlets are not important for our considerations of the western part of Lake Constance (see Fig. 1) and, consequently, are neglected. It is of prime interest to evaluate the surface energy budget. At any time, the net energy available to the water is given by:

$$\Phi_n = (1 - \alpha_s)\Phi_s + \Phi_a + \Phi_w + \Phi_l \tag{12}$$

These terms are defined as positive when directed into the water (see Fig. 2). Φ_s is global (direct and diffuse) solar radiation which is partially (α_s) reflected at the lake surface. According to [4], typical values of α_s lie between 0.04 (June) and 0.20 (December). The remaining short-wave radiation is partitioned:

$$(1 - \alpha_s)\Phi_s = \Phi_{ss} + \Phi_{sl} \tag{13}$$

where Φ_{ss} is that fraction of $(1 - \alpha_s)\Phi_s$ which passes through the surface layer and is absorbed exponentially in the water column, whereas Φ_{sl} is absorbed within the top metre of the water column and is incorporated into the heat flux term $Q(t)$ of the boundary conditions (see Eqn 10); an appropriate value is $\Phi_{ss}/[(1 - \alpha_s)\Phi_s] = 0.6$, according to Dake & Harleman [5].

The infrared terms Φ_a and Φ_w are formulated according to the Stefan-Boltzmann law:

$$\Phi_{a,w} = E_{a,w}\sigma(T)^4 \qquad (14)$$

where $\Phi_{a,w}$ is the total amount of energy radiated per unit time and unit area by the atmosphere (Φ_a) or the water surface (Φ_w) at absolute temperature T; σ is the Stefan-Boltzmann constant ($\sigma = 5.67 \times 10^{-8}\ W\ m^{-2}\ K^{-4}$), and $E_{a,w}$ are the emissivities of the atmosphere and water, respectively.

E_w is nearly 1 (ca. 0.97), but E_a varies with, among other things, air temperature and cloud cover. We use the empirical relationship [6]:

$$\Phi_a = [(E_a\sigma(T_a)^4)](1 + 0.17B^2) \qquad (15)$$

with $E_a = 9.36 \times 10^{-6}(T_a)^2$ and with B as a measure for relative cloud cover.

The last term in Eqn 12 comprises the energy loss by evaporation Φ_e , the gain by precipitation Φ_p and transfer of sensible heat Φ_c. But, as the energy exchanges associated with precipitation are very small, we can write the net nonradiative loss as the sum of evaporation and sensitive heat:

$$\Phi_l = \Phi_c + \Phi_e \qquad (16)$$

The evaporative energy loss can by approximated by:

$$\Phi_e = -f(u)(e_{sw} - e_a) \qquad (17)$$

The term $(e_{sw} - e_a)$ is the difference between the saturated vapor pressure at the water temperature and the actual vapor pressure at the air temperature. We adopt the functional form $f(u)$ as given for Lake Sempach [6]: $f(u) = 3.25 + 2u$, where u is the wind speed measured at a height of 10 m. The evaluation of the term $(e_{sw} - e_a)$ demands a knowledge of air temperature, humidity and water temperature. The first two are derived from routine meteorological observations; the last is a result of the model.

The sensible heat flux Φ_c is related to the evaporative energy loss Φ_e by the Bowen ratio B_o (which varies between 0.61 and 0.65; see [7]), resulting in the proportionality:

$$\Phi_c = (B_o \times 10^{-3})p[(T_w - T_a)/(e_{sw} - e_a)]\Phi_e \qquad (18)$$

where p is the atmospheric pressure. Hence if Φ_e is accurately specified, so is Φ_c. But, Eqn 17 and the above specification of $f(u)$ should be improved with respect to specific conditions of Lake Constance. There remains a wide field of research in order to correctly represent the role of the nonradiant fluxes in the surface energy budget.

To summarize the above, the net heat flux $Q(t)$ through the air–water interface is written as:

$$Q\ (t) = \Phi_{sl} + \Phi_a + \Phi_w + \Phi_c + \Phi_e \qquad (19)$$

3.2. ABSORPTION OF SOLAR RADIATION

The light absorption properties of the water determine the depth of the near-surface layer where solar radiation is absorbed (nearly exponentially). The level at which 1% of the solar flux remains, depends strongly on the turbidity of the water and may exceed 20 m in very clear lakes. In the model, the absorbance of downward irradiance is assumed to be an exponential function of depth given by:

$$S_H(z) = (\beta/p_0 c_p)(\Phi_{ss} e^{-\beta z}) \qquad (20)$$

where Φ_{ss} is that part of the solar irradiance which penetrates the upper 1 m, and β is an attenuation coefficient which is assumed to be constant with depth and time. The latter assumption is a poor approximation in the near-surface metres of the lake. Ongoing investigations with an active coupling of physical properties (attenuation coefficient) and biological activities (phytoplankton biomass) show that an increase of surface temperature can be caused by an accumulation of algal cells in a shallow top layer of the lake, where more solar radiation is absorbed due to an elevated absorption coefficient.

3.3 WIND STRESS AT THE SURFACE

Waves and currents are induced by the wind acting at the water surface. The stress τ_0 at the lake surface created by the wind force is commonly parameterized by the square of the wind speed:

$$\tau_0 = p_a C_{10} (W_{10})^2 \qquad (21)$$

where W_{10} is wind speed measured at or transformed to a standard height of 10 m above the water surface, p_a is the density of air, and C_{10} is the so-called drag coefficient which varies between 0.5×10^{-3} (for $W_{10} < 7$ m s^{-1}) and 2.5×10^{-3} (for $W_{10} > 20$ m s^{-1}). It accounts for for the increasing coupling between air and water due to increasing roughness of the air–water interface, i.e., increasing wave heights. Although the flux of mechanical energy induced by the wind is relatively small compared with thermal energy, it is of crucial importance for mixing in lakes. Forcing through the surface, by wind, short-wave radiation and heat fluxes, is taken from data which are available from nearby stations.

It has proved to be extremely difficult to simulate the very beginning of stratification in early spring, especially when several storms have occured during this period. Under such conditions, the thermocline in the model settles too high in the

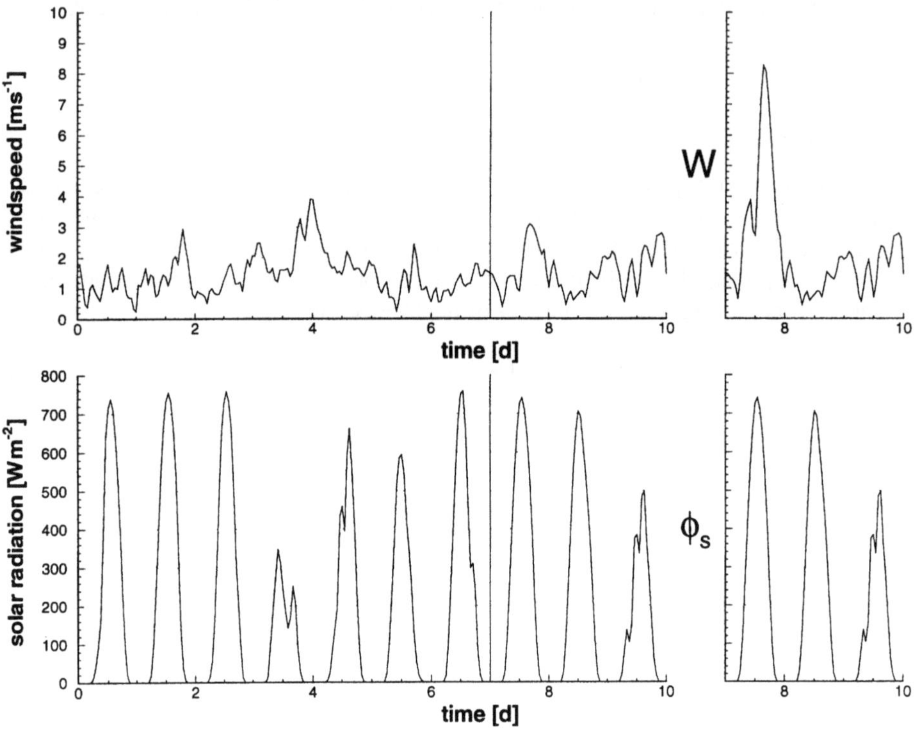

FIGURE 3. Meteorological data for Lake Constance for the period 22 April to 2 May 1994 (left panel). Modified scenario (right panel): amplification of the observed wind speeds by a factor of ca. x3 during day 7 (28 April).

water column, and this mismatch holds throughout the stratified season until late autumn. Such effects as the passage of a storm are far from being adequately modelled with a one-dimensional scheme. The rearrangement of the density field after it is displaced from its equilibrium state, and vastly disturbed by mixing processes, can hardly be described by mixing only in a vertical direction. Resettling of a horizontally homogeneous density field occurs, among other things, via internal surges and large amplitude internal oscillations; both processes are known to be highly non-linear.

4. Comparison of observations and simulations

Plots of wind speed against time are given in Fig. 3, and components of the energy fluxes at the lake surface, computed from meteorological data, are presented against time in Fig. 4, for a 10-day episode in April 1994 (Case 1). In order to study the effects of a storm on the development of stratification, we constructed a second 10-day period (Case 2) identical in meteorological forcing during the first 7 days, followed by a day with about three times the original wind speeds, and then continuing as in Case 1.

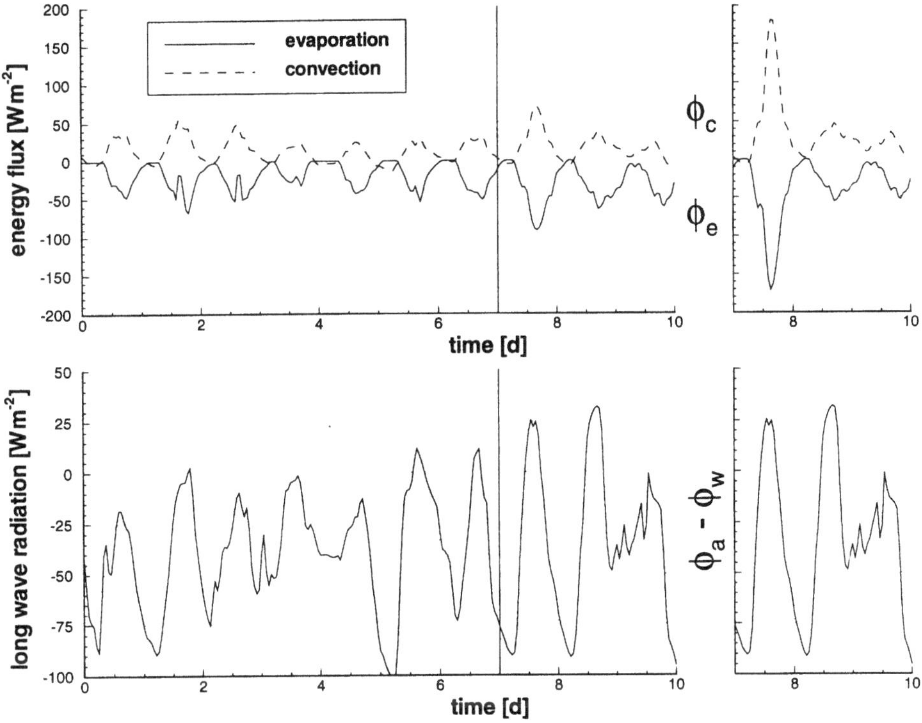

FIGURE 4. Energy fluxes due to evaporation and convection (above) and net long-wave radiation (below) in Lake Constance, for the same period as shown in Fig. 3 (22 April to 2 May 1994).

In both scenarios the solar radiation is the same, and at day 7 even Φ_{ss} and Φ_{sl} are identical, respectively, as the model ignores increased reflectance of short-wave radiation at the water surface due to the changing wind field. Net long-wave radiation did not change significantly on the stormy day. A considerable increase in Φ_e and Φ_c is connected with the hypothetical storm on day 7 (see Fig. 4, upper panel) as evaporation and convection depend on wind speed.

A comparison of the observed temperature structure (Fig. 5, upper panel) at the measuring site in Lake Constance (see Fig. 1) with the output of our numerical model (Fig. 5, lower panel), reveals that observations and model agree quite well with respect to the surface temperature of the lake. Clearly, the model is not able to simulate the observed fluctuations of the subsurface temperatures caused by internal waves and small-scale lateral inhomogeneities of the wind field. The plunging of the isotherms on and after day 8 is due to a temporary downwelling event which, however, is to be considered as a three-dimensional effect caused by the finite extent of the lake. Consequently, the one-dimensional model is not able to simulate this behaviour of the isotherms. But, as can be deduced from the well-developed temperature gradient even

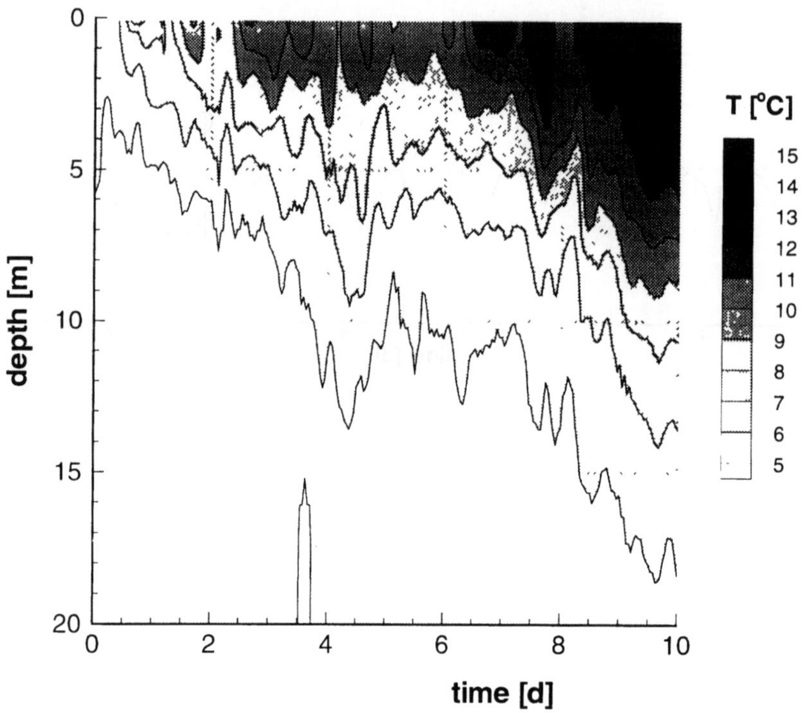
0
5
10
15
20
depth [m]
0
2
4
6
8
10
time [d]
T [°C]
15
14
13
12
11
10
9
8
7
6
5

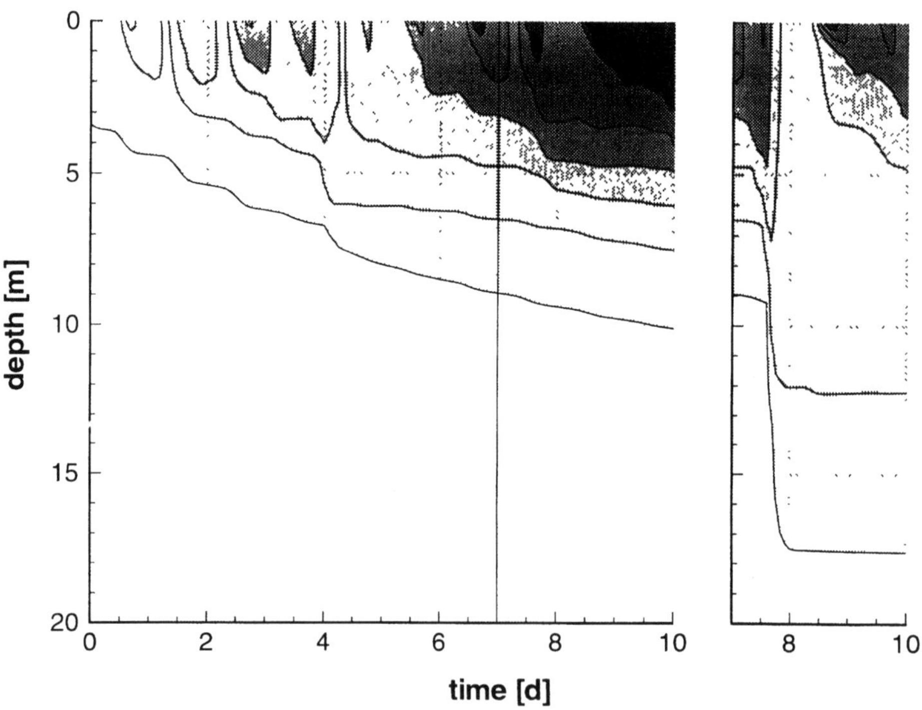
0
5
10
15
20
depth [m]
0
2
4
6
8
10
time [d]
8
10

FIGURE 5. (*On the opposite page*). Contour lines for temperature observed (above) and simulated by a numerical model (below) in the upper 20 m of the water column at station BM in Lake Constance. The time-period is the same as in Figs 3 and 4 (22 April to 2 May 1994).

---

in the uppermost metres (Fig. 5, upper panel), this event has nothing to do with vertical mixing processes. Contrary to this, the storm on day 7 destroys the thermocline and homogenizes the upper 10 m of the water column.

5. The concept of vertical exchange rates

To demonstrate the state of turbulence in the water column, we map variability of the coefficient of turbulent eddy viscosity k_z, with time and depth, as it is computed by the $k - \varepsilon$ model. It is evident from Fig. 6 that the stirring by winds with speeds lower than 3 m s^{-1} is confined to the uppermost 2–3 m of the water column. Only when combined with vertical convection, during the night, may wind-induced turbulence penetrate down to 4–5 m depth. The mixing event at day 8 of the second scenario corresponds with a considerable increase of turbulent diffusivity (see Fig. 6, right panel).

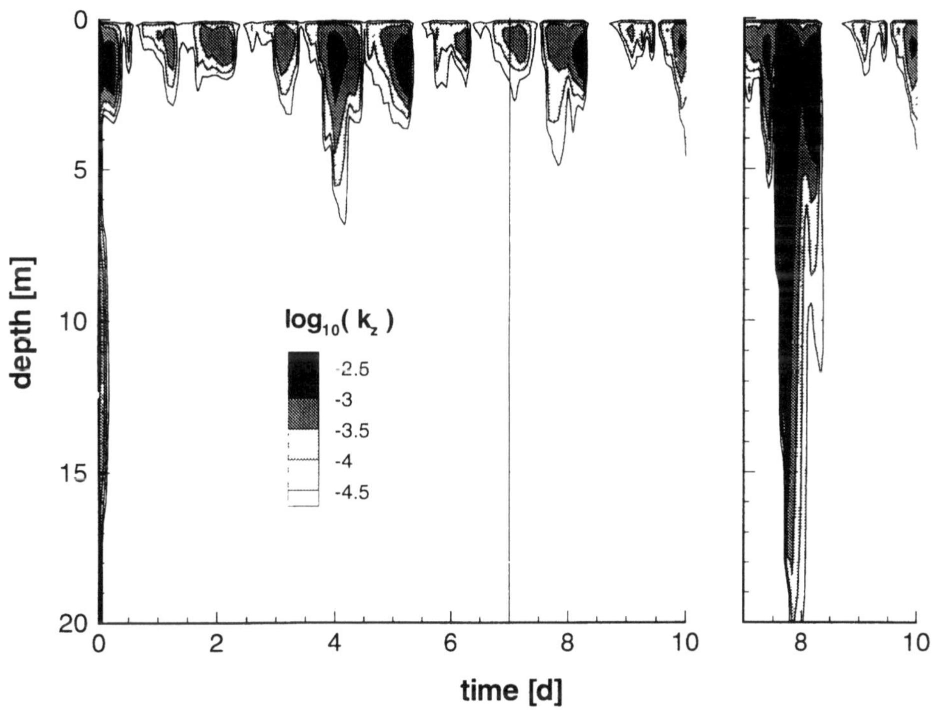

FIGURE 6. Variability of the vertical eddy viscosity (Austausch coefficient) k_z with time and depth during the 10-day episode (left panel) and the last 3 days with stronger wind on day 7 (right panel), as calculated by the numerical model (see the text). Isolines of $\log_{10}$ (k_z) are in metres per second.

The temporal and spatial variability of the state of turbulence in the water column is most comprehensively shown by plotting the coefficient of turbulent eddy viscosity k_z as a function of time and depth, as illustrated in Fig. 6. However, for practical reasons it is desirable to compress information in order to obtain a simple parameter which is easier to handle. We invented the following procedure (see Fig. 7). At the start of a time-interval (e.g. each day) a certain concentration of an imaginary tracer (e.g. passive planktonic cells) is prescribed in a well-defined layer of depth h_z. The gradient of the tracer concentration in the water column develops according to the existing (calculated) turbulent diffusivities. At the end of the time-interval we strike a balance by integrating the depth-dependent concentrations over the individual layers, thus obtaining the amount of tracer mass which has moved from layer k to layer i by turbulent diffusion. The percentage of the original tracer mass that originated from the layer with thickness h_k and then moved to the layer under consideration (with thickness h_i) is called the exchange rate p_{ki}.

This technique, applied to the episode and its modification, yields a series of daily values (see Fig. 8) which can be interpreted as highly compressed information about the state of turbulence in the water column. Because of the method used for measuring biological parameters, the layers chosen in Fig. 8 extend from depths 0 to 8 m (layer 1), 8 to 20 m (layer 2) and 20 to 100 m (layer 3) or, for the sake of clarity, from 0 to 3 m (layer 1a) and 3 to 8 m (layer 1b), respectively. During the original episode with very low winds, neither p_{12}, p_{23} nor p_{13} exhibit values significantly different from background (molecular) diffusion. In view of Fig. 6 (left panel) this result is not very surprising. Subdivision of layer 1 into layer 1a (0–3 m) and layer 1b (3–8 m), leads to an exchange rate p_{1a1b} which offers rather detailed insight into the state of turbulence of the uppermost part of the water column. It is interesting to note that in the case of strong winds at day 7 (right panel of Fig. 8) p_{12} is larger than p_{1a1b}, which expresses the fact that turbulent eddy viscosity is greatest at some distance from the water surface [8]. Nevertheless, even the wind speeds at day 7 are not strong enough to overcome stratification: while p_{12} increases to quite high values, p_{13} remains negligible.

6. Conclusions

A one-dimensional numerical model for the turbulence induced by wind stirring and vertical convection is used to calculate vertical eddy diffusivity. As the model is one-dimensional it is not able to incorporate such two-dimensional effects as tilting of the thermocline and advection, and it ignores energy losses due to internal waves radiating from the thermocline.

Nevertheless, the model simulates the vertical structure of the stratification satisfactorily when compared with observations. Hence, there can be some confidence that the state of turbulence in the water column is calculated realistically. This is a necessary prerequisite for analyzing the relationship between water column stability and the extent and timing of algal blooms in early spring, when the vertical mixing and light history of the algal cells are considered to be the dominant factors for algal growth and standing stock.

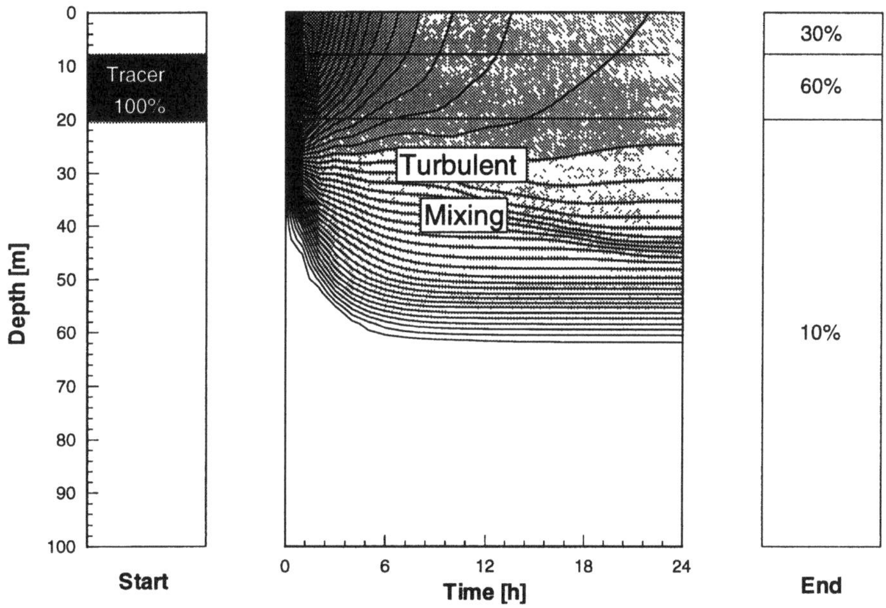

FIGURE 7. A sketch defining the vertical exchange rate p_{ki}. In this example, layer 1 extends from 0 to 8 m, layer 2 from 8 to 20 m, and layer 3 from 20 to 100 m; $p_{21} = 0.3$ and $p_{23} = 0.1$. The time interval is 24 h. Further explanation is given in the text.

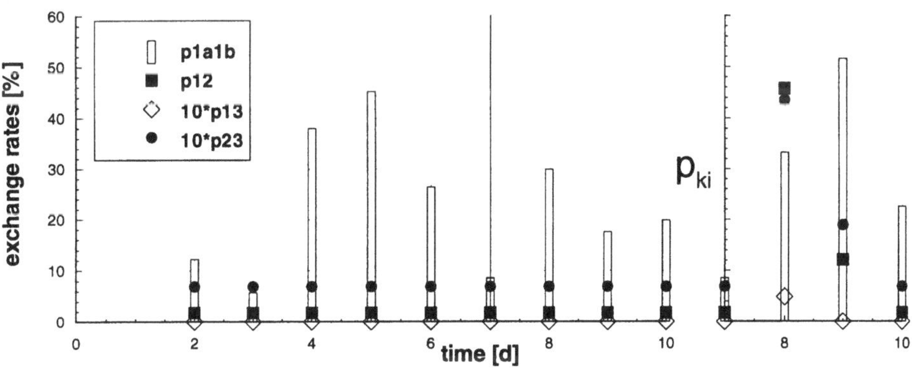

FIGURE 8. Various exchange rates (p_{ki}) calculated for the two episodes shown in Fig. 6. For better resolution the upper 10 m of water is subdivided into layer 1a (0–3 m) and layer 1b (3–8 m). Note the different scales.

Acknowledgements

The present study was performed within the Special Collaborative Program (SFB) "Cycling of Matter in Lake Constance", supported by the Deutsche Forschungsgemeinschaft (DFG). The basic numerical model was provided by U. Svensson from the Swedish Meteorological and Hydrological Institute (SMHI) Norköpping. Meteorological data were partially obtained from the German Weather Service. The authors would like to thank Josef Halder for his help during the data collection and Paul Kirner for his valuable suggestions.

References

[1]. Spigel, R. H. & Imberger, J. (1987). Mixing processes relevant to phytoplankton dynamics in lakes. *New Zealand Journal of Marine and Freshwater Research* **21**, 361-377.

[2]. Svensson, U. (1978). A mathematical model of the seasonal thermocline. Report 1002, pp. 1-187, Dep. Water Resour., Eng. Univ. Lund, Sweden.

[3]. Mortimer, C. H. (1980). Inertial motion and related internal waves in Lake Michigan and Lake Ontario as responses to impulsive wind stresses. *Ctr. Great Lakes Stds., Univ Wisconsin-Milwaukee. Spec. Rept. No.* **37**.

[4]. Imboden, D. M. & Wüest, A. (1995). Mixing mechanisms in lakes. In: *Physics and chemistry of lakes.* Second Edition. Springer.

[5]. Dake, J. M. K. & Harleman, D. R. F. (1969). Thermal stratification in lakes: analalytical and laboratory studies. *Water Resour. Res.* **5**, 484-495.

[6]. Marti, D. E. & Imboden D. M. (1986). Thermische Energieflüsse an der Wasseroberfläche: Beispiel Sempachersee. *Schweiz. Z. Hydrol.* **48**, 196-229.

[7]. Brutsaert, W. H. (1982). *Evaporation into the atmosphere: theory, history, and applications.* D. Reidel, Hingham, Mass. 299 pp.

[8]. Madsen, O. L. (1977). A realistic model of the wind-induced Ekman boundary layer. *J. Phys. Oceanogr.* **7**, 248-255.

THE INFLUENCE OF WEATHER CONDITIONS ON THE SEASONAL PLANKTON DEVELOPMENT IN A LARGE AND DEEP LAKE (L. CONSTANCE)

III. THE IMPACT OF WATER COLUMN STABILITY ON SPRING ALGAL DEVELOPMENT

URSULA GAEDKE[1], DIETER OLLINGER[2], PAUL KIRNER[2] AND ERICH BÄUERLE[2]

[1]*Limnologisches Institut, Universität Konstanz D-78434 Konstanz, Germany*

[2]*Institut für Umweltphysik, Universität Heidelberg D-69120 Heidelberg, Germany*

Abstract

Under non-stratified conditions in deep lakes such as Lake Constance, algal growth is regarded as being primarily light limited. This is mostly a result of the large mixing depth in relation to the euphotic zone. Consequently, the beginning of spring algal development is expected to depend largely on water column stability. The temporal development of water column stability in spring was estimated for six years of investigation, using elaborate hydrodynamical models, and expressed as two vertical exchange rates: (1) between the uppermost water layer (0–8 m) and the intermediate layer (8–20 m), and (2) between the uppermost and deep layers (20–100 m) (Ollinger & Bäuerle, II in this volume, pp. 57–70). The rates are compared with phytoplankton dynamics. Pronounced spring algal blooms developed in some years despite strong mixing between the uppermost and intermediate water layers. In contrast, the exchange rate between the uppermost and deep water layers proved to be a reliable predictor for the onset of spring algal growth. Substantial increases in algal standing stock never occurred before the algae were largely safe from being mixed below 20 m. In contrast, low exchange rates were invariably associated with enhanced growth rates even when they occurred as early as February, i.e. the reduction of the mixing depth to less than 20 m is a necessary and sufficient prerequisite for vernal blooms in Lake Constance. The predictability of vertical exchange rates from meteorological conditions, by the

D.G. George et al.(eds.), Management of Lakes and Reservoirs during Global Climate Change, 71–84.

hydrodynamical models, and the close relationship between vertical exchange rates and algal growth, enable rough predictions to be made about the response of phytoplankton dynamics to changes of weather conditions. Our results suggest that, for example, an increase in air temperature by 2°C at the beginning of the year, or the absence of strong winds for a few days, has significant effects on the timing of the vernal algal bloom.

1. Introduction

Weather conditions directly and indirectly influence the population dynamics of plankton organisms in numerous ways (e.g. [1–3]). Our capacity to understand, quantify and predict the impact of meteorological factors on the seasonal plankton succession and food web regulation is, however, very limited when compared with the temporally dominant role of these processes in most inland waters (but see, e.g., [4]). It has been established that phytoplankton development in late winter and early spring is dominated by physical factors in temperate, deep, non-ultraoligotrophic lakes such as Lake Constance. During this period, grazing pressure on phytoplankton is at its minimum and nutrient concentrations reach maximum (non-limiting) values. This suggests that algal net growth rates are most strongly influenced by light limitation. Incoming average irradiance in early spring is still low (8–20 mol m^{-2} d^{-1}) when compared with typical summer values (around 40 mol m^{-2} d^{-1}). More importantly, the average light history experienced by individual algal cells is also very low under non-stratified conditions, where the majority of planktonic cells spend most of the time below the euphotic zone (mean depth of Lake Constance: 100 m, euphotic depth ca. 20 m). Hence, it is commonly assumed (e.g. [5]) that the onset of the spring phytoplankton bloom depends strongly on water column stability in large and deep lakes such as Lake Constance. This point of view was indirectly supported by a time-series analysis of the impact of irradiance, air temperature and wind on spring algal development in Lake Constance, which revealed only a moderate covariation of algal standing stocks with one of these meteorological parameters (Gaedke et al., I in this volume, pp. 39–55). Neither irradiance nor air temperature alone provide a useful indicator of algal growth conditions. This indicates that the complex synergistic direct and indirect impacts of meteorological conditions on algal growth are primarily mediated by water column stability which, in turn, depends on the actual stratification of the water column. Consequently, comprehensive hydrodynamical models were established to estimate the seasonal course of water column stability and vertical exchange rates, and their sensitivity to changes of individual meteorological parameters (Ollinger & Bäuerle, II in this volume).

The objective of the present investigation is to evaluate the effect of water column stabilisation on the spring development of phytoplankton in Lake Constance, in order to contribute to a functional understanding of the processes by which weather conditions influence plankton dynamics. This aim is achieved by analysing the temporal covariation of vertical exchange rates and algal growth in spring during six years of investigation (1987–89, 1991, 1992 and 1994).

2. Study site, materials and methods

Lake Constance (in German: Bodensee) is a large (476 km^2) and deep (z_{max}= 252 m, mean depth 100 m), prealpine, monomictic lake bordered by Austria, Germany and Switzerland. The central sampling station is located in the fjord-like northwestern part of the lake (Überlingersee, z = 147 m) (see Fig. 1 in Ollinger & Bäuerle, II in this volume). Further details are given by Gaedke et al. (I in this volume) and by Tilzer & Beese [6].

A one-dimensional hydrodynamical model was used to simulate the turbulent momentum, heat and mass transports, which are induced by the direct influence of the wind at the lake's surface, and by shortwave and longwave radiation (to mention only the most important effects). The downward diffusion and dissipation of turbulent kinetic energy depends strongly on stratification and is parameterised by the turbulent eddy viscosity or the so-called Austausch coefficient, k_z. The model calculates this coefficient by solving two differential equations for the kinetic energy (k) of the turbulent motion, and for the dissipation (ε) of turbulent energy, respectively. Thus, the temporal and vertical distributions of the turbulent eddy viscosity are known as a result of meteorological conditions. Subsequently, the water column was divided into three strata (0–8, 8–20, and 20–100 m). For each day a tracer is introduced into the uppermost stratum (0–8 m) and it is calculated which fraction of the tracer is transferred from this layer into the stratum between 8 and 20 m within one day. The resulting exchange rate from the first to the second layer is denoted by p_{12}. Correspondingly, the fraction of tracer transferred from the uppermost layer into the stratum below 20 m is called p_{13}. Computations were performed for the years 1987, 1988, 1989, 1991, 1992 and 1994, due to lack of meteorological data (obtained from the local weather station) for other years. Further details on the structure and reliability of the hydrodynamical model, and the techniques used to translate turbulent Austausch coefficients to exchange rates, are described by Ollinger & Bäuerle (II in this volume).

Phytoplankton standing stock was measured weekly or biweekly at different depths from 1979 to 1993 (biovolume: Gaedke & Schweizer [7] and unpublished) at the central sampling site, and chlorophyll-*a* was determined from 1980 to 1982 and 1986 to 1995 ([6, 8] and unpublished). Algal biovolume was determined microscopically and chlorophyll-*a* spectrophotometrically (for details and references see Gaedke et al., I in this volume).

3. Results

3.1. SEASONAL CHANGES OF EXCHANGE RATES BETWEEN DIFFERENT STRATA

The daily exchange rate between the uppermost (0–8 m) and intermediate (8–20 m) water layers, p_{12}, is generally high (ca. 30%) in January, February, March and often also in April (Figs 1–4). In early 1987 it was rather low, due to an inverse stratification that occurred during this exceptionally cold winter. In most years, p_{12} decreases rapidly

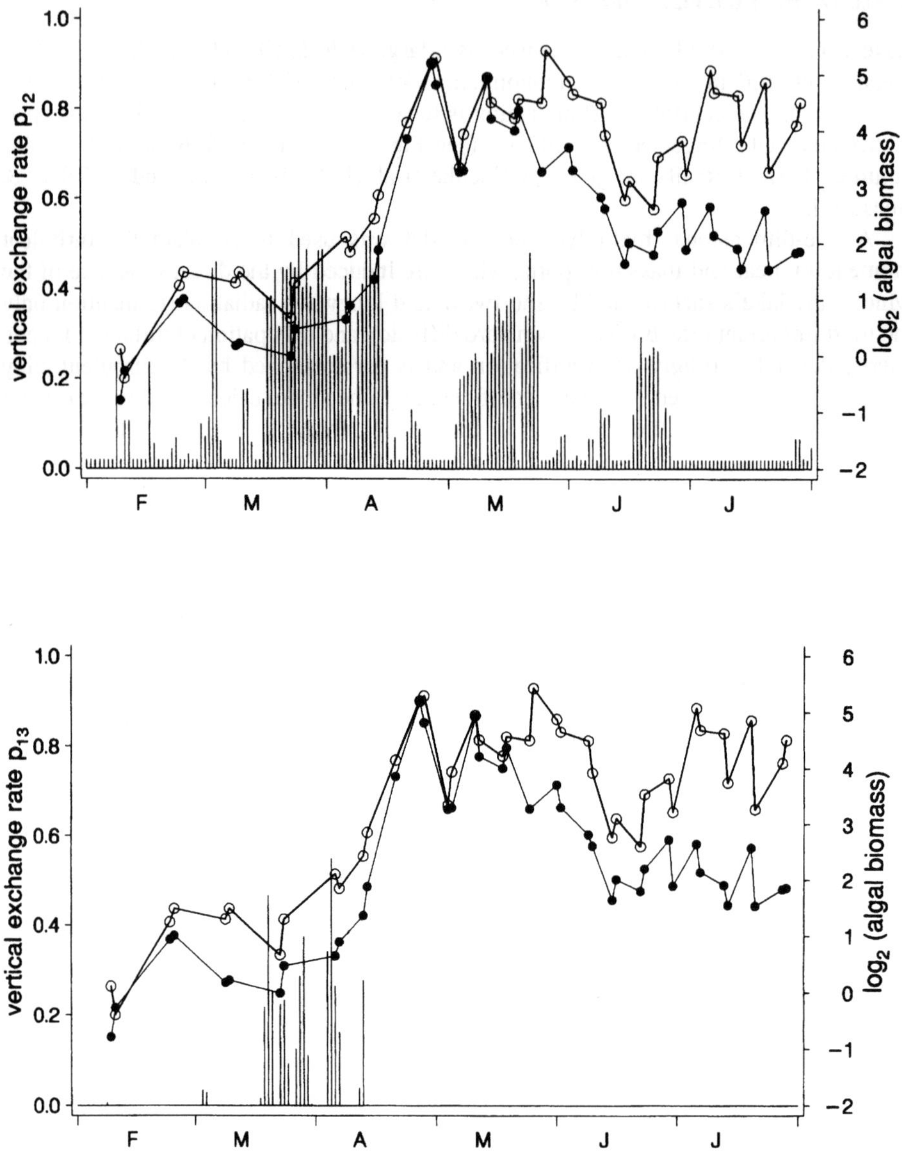

FIGURE 1. Temporal changes in daily vertical exchange rates (vertical bars; p_{12} and p_{13}) between three major strata in the water column of Lake Constance during February to July, chlorophyll concentrations (µg l^{-1} ; solid points) and algal biovolume (g FW m^{-2}; open circles) in spring 1987. *Above:* the exchange rates between the uppermost (0–8 m) and intermediate (8–20) water layers. *Below:* the exchange rates between the uppermost (0–8 m) and deep (20–100 m) water layers.

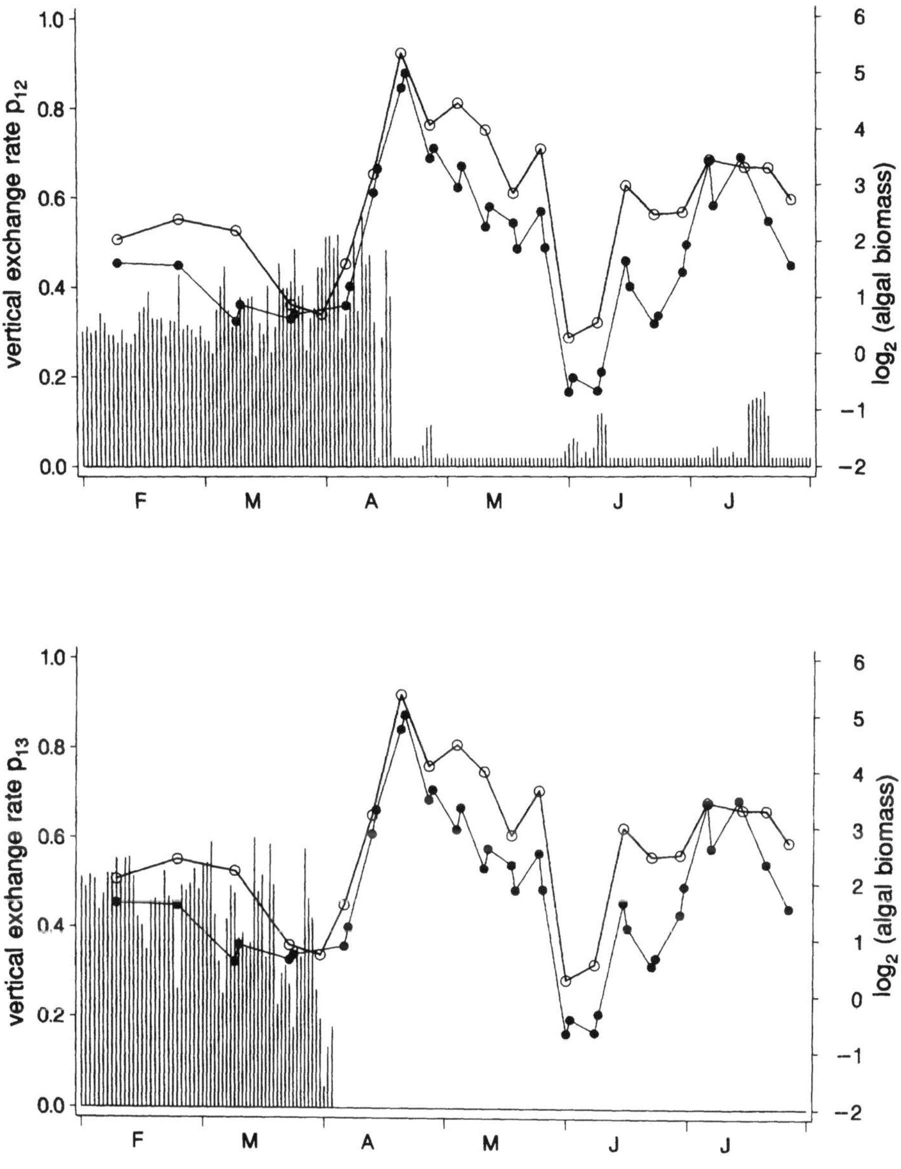

FIGURE 2. Temporal changes in daily vertical exchange rates (vertical bars; p_{12} and p_{13}) between three major strata in the water column of Lake Constance during February to July, chlorophyll concentrations (µg l^{-1} ; solid points) and algal biovolume (g FW m^{-2}; open circles) in spring 1988. *Above:* the exchange rates between the uppermost (0–8 m) and intermediate (8–20) water layers. *Below:* the exchange rates between the uppermost (0–8 m) and deep (20–100 m) water layers.

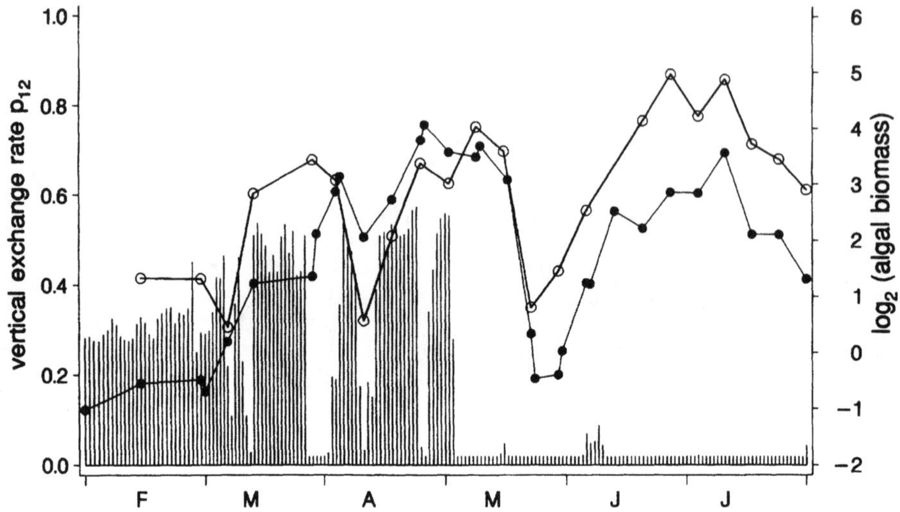

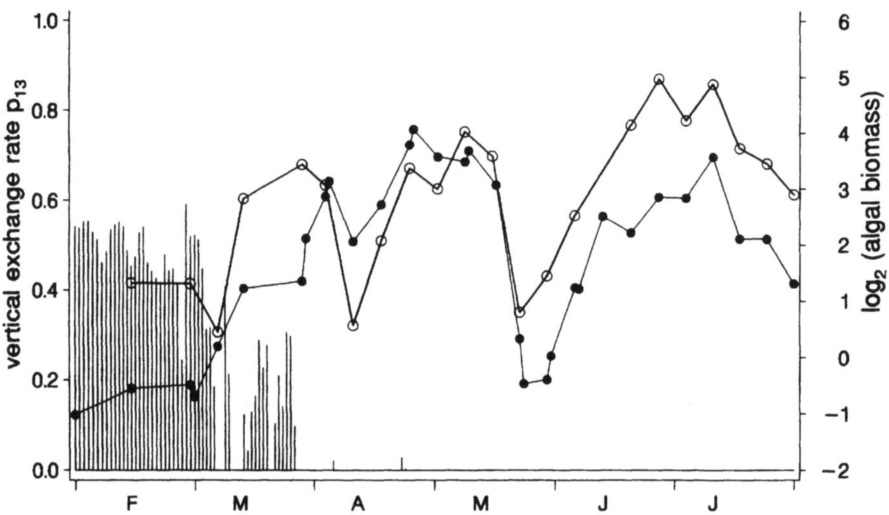

FIGURE 3. Temporal changes in daily vertical exchange rates (vertical bars; p_{12} and p_{13}) between three major strata in the water column of Lake Constance during February to July, chlorophyll concentrations (μg l^{-1} ; solid points) and algal biovolume (g FW m^{-2}; open circles) in spring 1989. *Above:* the exchange rates between the uppermost (0–8 m) and intermediate (8–20) water layers. *Below:* the exchange rates between the uppermost (0–8 m) and deep (20–100 m) water layers.

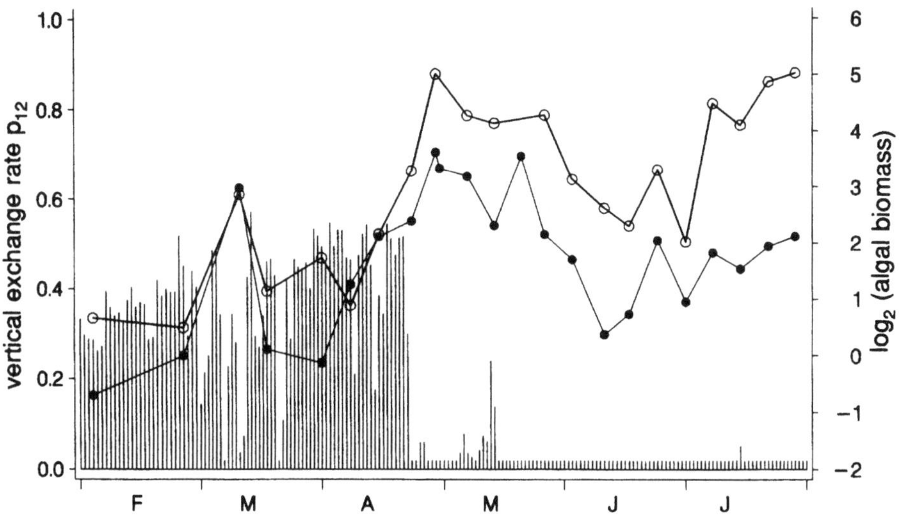

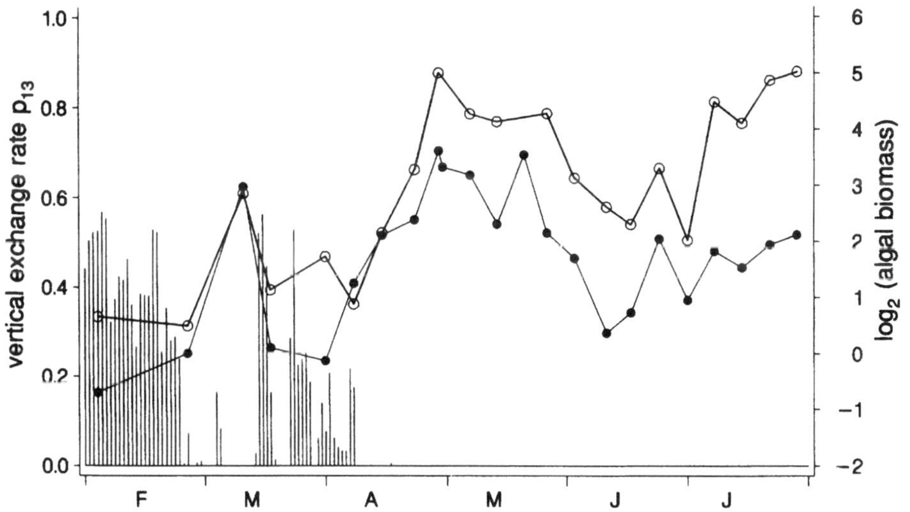

FIGURE 4. Temporal changes in daily vertical exchange rates (vertical bars; p_{12} and p_{13}) between three major strata in the water column of Lake Constance during February to July, chlorophyll concentrations (µg l^{-1} ; solid points) and algal biovolume (g FW m^{-2}; open circles) in spring 1992. *Above:* the exchange rates between the uppermost (0–8 m) and intermediate (8–20) water layers. *Below:* the exchange rates between the uppermost (0–8 m) and deep (20–100 m) water layers.

in March or April and remains uninterruptedly low or even close to zero until about the end of August. The unusually bad weather during almost all of 1987 (cf. Fig. 4 in Gaedke et al., I in this volume) is reflected by p_{12} values of 10 to 20% and more in May–June (Upper Fig. 1); also in August and September. Spring 1988 was characterised by a very distinct change of weather conditions, from bad weather in March to good weather in April, which is reflected in a very marked decrease of exchange rates in early April (Upper Fig. 2). Each year in autumn, p_{12} increases strongly and reaches maximum values of 60% per day at the end of October or in the first half of November. This value then steadily decreases to ca. 40% in December. Thus, from September to March a large fraction of plankton organisms is transferred from the uppermost 0–8 m into deeper and less illuminated water layers (8–20 m) each day in Lake Constance. The water exchange that occurs simultaneously from the lower to the upper stratum does not fully compensate for these losses. It is quantitatively of less importance for the entire population since the lower water layers are larger than the uppermost stratum.

The exchange rate (p_{13}) between the uppermost 0–8 m and the deep layer below 20 m, is always close to zero from May to October, and increases sharply in early or mid November when the bottom of the mixed layer passes a depth of 20 m. Maximum values of ca. 50 to 70% per day are reached in January and February (except for 1987 owing to the inverse stratification) (Figs 1–4). Values in March and April vary considerably within and between years. Thus, epilimnetic plankton organisms experience hardly any losses by vertical mixing in summer but, from November until about March, approximately half of the algal cells occurring in the uppermost 0–8 m are mixed below 20 m each day. Only 3 to 5% of these cells below 20 m return to the uppermost 8 m each day.

3.2. COVARIATION OF VERTICAL EXCHANGE RATES AND ALGAL BIOMASS

A comparison of the temporal development of spring algal biomass, indicated here by concentrations of the biovolume, chlorophyll-*a* and p_{12}, reveals that the spring development starts immediately after a reduction of p_{12} in mid April in 1987 (Upper Fig. 1). Further mixing within the uppermost 20 m in May and June does not strongly affect algal standing stocks (in June, a weakly expressed clearwater phase and a mixing event coincide most probably by chance).

In the spring of 1988, which was characterised by a sudden change to favourable weather conditions in early April, algal growth had already began when p_{12} was still high (Upper Fig. 2). However, maximum biomass was reached when p_{12} was close to zero. In the spring of 1989, considerable algal growth occurred during extensive mixing within the uppermost 20 m of the water column (Upper Fig. 3). Changes in p_{12} match closely the changes in irradiance. Observations in 1991, 1992 and 1994 yield similar results (e.g. Upper Fig. 4). To conclude, the exchange rate between 0–8 and 8–20 m does not have a high predictive value for the beginning of the spring algal development. Strong blooms may develop despite strong mixing within the uppermost 20 m.

Considering p_{13}, the vertical exchange rate between the uppermost 0–8 m and the deep layer between 20 and 100 m, gives a different result. In 1987, inverse stratification caused low values of p_{13} until mid March. From then to mid April, p_{13} reached values of up to 40% and algal biomass remained low. Substantial spring algal growth started immediately after p_{13} decreased to low values, indicating there was a reduction of mixing below 20 m. In the spring of 1988, p_{13} was continuously high until the end of March, when mixing stopped rapidly. The onset of a strong net increase of algae coincided exactly with the termination of mixing below 20 m (Lower Fig. 2). In 1989, algal biomass again responded immediately to the first occurrence of low values of p_{13}. Deep mixing events in mid March coincided with a more-or-less unchanged biomass, so far as can be evaluated from biweekly measurements. In accordance with previous years, a coincidence of low p_{13} values and the beginning of algal growth, and vice versa, were observed for spring 1991, 1992 and 1994 (Lower Figs 3 and 4). For example, vernal algal growth in 1992 was characterised by a very early peak in early March which later collapsed to reach winter levels; these were followed by a second pronounced growth period from April onwards (Fig. 4). This pattern can be directly related to the weather pattern, with exceptionally high irradiances, exceptionally low wind and relatively warm air temperatures at the end of February and the beginning of March (cf. Fig. 4 in Gaedke et al., I in this volume). The second half of March was characterised by higher wind speeds and relatively low irradiance, and periods of extensive mixing below 20 m. Thus, exceptional weather conditions may give rise to an onset of algal growth as early as the beginning of March but such growth rates are seldom sustained, since such premature periods of thermal stratification are very easily reversed. Vertical mixing below 20 m was rapidly interrupted in late February and provided a striking explanation for the changes in algal standing stocks, whereas exchange rates within the uppermost 20 m remain relatively high. Measurements and computations for spring 1991 support these findings.

To conclude, a high correlation exists between the exchange rate from the uppermost 0–8 m to the deep water layers (20–100 m) and the algal development in early spring in Lake Constance. A spring bloom cannot develop until most algae are safe from being mixed below 20 m for a number of consecutive days. In contrast, pronounced mixing within the uppermost 20 m does not prevent substantial increases in or maintenance of algal biomass. Peak values of algal biomass are mostly observed in the absence of high values of p_{12} which is, however, most probably caused by a covariation of p_{12} and p_{13}.

3.3. PREDICTION OF WATER COLUMN STABILITY AND ALGAL RESPONSE TO ALTERNATIVE METEOROLOGICAL CONDITIONS

Identification of the dominant mechanism by which meteorological conditions influence spring algal growth permits a systematic study of the potential effect of weather conditions on algal dynamics. The hydrodynamical model allows computation of vertical exchange rates based on moderate deviations of the meteriological parameters from the observed ones (Ollinger & Bäuerle, II in this volume).

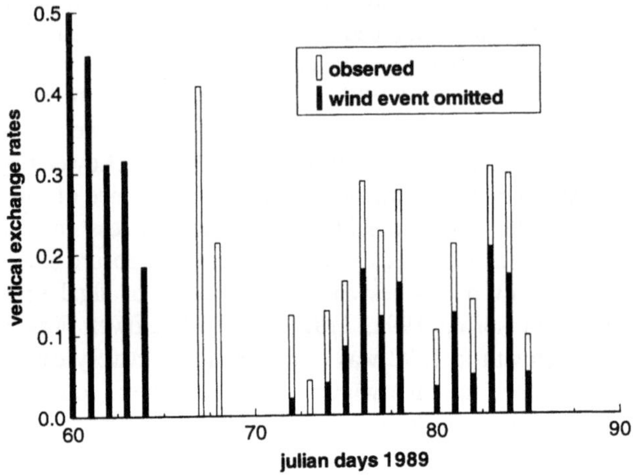

FIGURE 5. The temporal variability of vertical exchange rate p_{13} in Lake Constance, comparing the effects of observed wind speeds (open bars) and a calculated constant wind speed (solid bars; see text) for 6–9 March 1989.

For example, the model was run with a modification of the wind speed during three days (6–9 March 1989) which include a strong wind event (up to 9 m s^{-1}). The high temporal variability of the wind speed during this time was replaced by a constant wind speed of 2.5 m s^{-1} for the three days; this corresponds to the average value observed during this period as well as being typical of wind speeds in March. The model suggests that the exchange between the surface and the deep water layer would have been interrupted during the days with modified wind speed (Fig. 5). The stabilising effect of the smoothed wind speed on the water column would have had a lasting effect; the model indicates a pronounced reduction of p_{13} during the following 2 to 3 weeks and, hence, an earlier onset of algal growth under the alternative wind regime.

Another model run was done with an increase of air temperature by 2°C above the one observed from 1 January 1989 onwards. The impact on p_{13} is small in January and increases only slowly in February (Fig. 6). During this non-stratified period, the additional long-wave radiation is dispersed over the entire water column. From March onwards surface temperatures increase slightly more, which yields less mixing and, hence, a further concentration of the incoming radiation close to the surface. This results in a pronounced decrease of p_{13} (Fig. 3) which is considerably enhanced by the increase in air temperature (Fig. 6). The latter suggests that an increase of air temperature by 2°C would cause an earlier start to the spring algal development in March 1989. Thus, within the given range of experience, we can roughly predict the onset of phytoplankton development and the impact of individual meteorological parameters, using the computations of vertical exchange rates between the uppermost and the aphotic water layers.

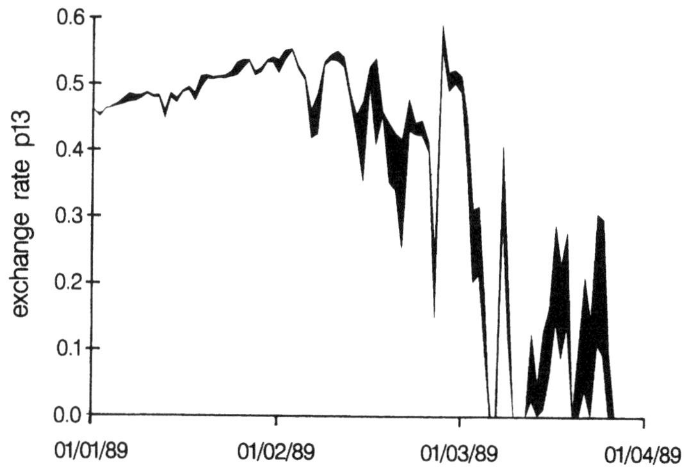

FIGURE 6. The effect that a calculated 2°C increase in air temperature would have on the vertical exchange rate p_{13} in Lake Constance during the period 1 January to 1 April 1989. The upper line represents calculated values for p_{13} at the observed air temperatures, and the lower line represents calculated values for p_{13} at temperatures 2°C higher. The solid black area indicates reductions in the vertical exchange rates due to the increase in temperature.

4. Discussion

For a long time it has been presumed that the beginning of phytoplankton spring growth depends on water column stability in deep temperate lakes and, hence, on the meteorological and hydrodynamical conditions which dominate it [9, 10]. However, verifications of this hypothesis (including quantification of the impact of individual meteorological parameters) have seldom been attempted. Our results provide quantitative evidence in support of the concept that in deep lakes significant phytoplankton blooms do not develop prior to the onset of thermal stratification and the subsequent reduction in the depth of the mixed layer. In Lake Constance, a spring bloom cannot develop until most algae are safe from being mixed below 20 m. This result is in accordance with general rules on the ratios of euphotic to mixing depth which permit algal development (e.g. [11, 12]). In Lake Constance, Secchi-disc depth and euphotic depth are respectively ca. 15 and 18–22 m in winter, decreasing to 3–4 and 8 m during a pronounced spring bloom ([6, 13] and also Häse, personal communication). Thus, the mixing depth of algae has to be considerably smaller than the depth of the lake prior to the onset of vernal blooms.

The hydrodynamical model provides insight into the complex synergistic effects of irradiance, air temperature, wind and other meteorological factors, on water column stability (compare Ollinger & Bäuerle, II in this volume). This allows us to identify which kind of deviations in meteorological conditions, from the long-term average, have most effect on phytoplankton development, and to identify periods when these effects are most intense. Furthermore, the predictability of stratification depths, from meteorological parameters, enabled us to predict the timing of the vernal bloom under various weather conditions which deviate moderately from those during the period of investigation. For example, the model suggested that an increase of winter air temperatures by 2°C can have an impact on p_{13} in March and thus, presumably, on spring algal development. This result is in accordance with findings derived directly from time-series analyses (Gaedke et al., I in this volume). As such changes in air temperature roughly match the actual predictions of global warming, our findings may improve our predictive capabilities on future changes of our lake's ecosystems (for responses of higher trophic levels see Straile & Geller, IV in this volume). A long-term study of the small hypertrophic shallow Heiligensee also revealed a significant impact, by several consecutive unusually mild winters, on the plankton development throughout almost of the year [14]. However, details of the responses and the underlying mechanisms differ greatly between this shallow lake and our deep lake.

A final step towards an understanding of the influence of weather conditions on phytoplankton dynamics in its entirety demands an integration of the direct effects of (for example) irradiance and temperature, and the indirect effects via (for example) water column stability, which requires a coupling of hydrodynamical and biological models. Consequently, the hydrodynamical model is currently enlarged by adding phytoplankton dynamics. The vertical trajectories of numerous algal cells are tracked and form the basis for computing primary production and changes in standing stocks. The effects of self-shading and feed-back of the algae on the distribution of the incoming radiation in the water column are accounted for (Ollinger & Bäuerle, in preparation).

5. Conclusions

In Lake Constance, the onset of stratification varies strongly between years. The first water column stabilisation, where only a small proportion of algae in the surface layer is mixed below 20 m, occurred between late February and mid April.

Strong mixing within the uppermost 20 m did not prevent the development of spring algal blooms. The exchange rate between 0–8 and 8–20 m had little predictive capacity for the beginning of the spring algal development.

In contrast, pronounced mixing of surface water into deep water layers (20–100 m) always prevented spring blooms. Vice versa, spring blooms developed in each year of investigation immediately after the termination of significant mixing below 20 m (i.e. p_{13} close to zero).

A close relationship between vertical exchange rate p_{13} and the onset of the algal bloom, combined with a knowledge of the sensitivity of p_{13} to changes of meteriological conditions, enabled us to make preliminary predictions of the reaction of the phytoplankton spring community to alternative weather conditions. For example, computations of vertical exchange rates and time-series analysis (Gaedke et al., I in this volume) suggest that individual one-day storm events, or an increase of air temperature by a few degrees about March-time, significantly influence the onset of phytoplankton development.

The final step towards an understanding of the influence of meteorological factors on phytoplankton dynamics demands a dynamical simulation model which includes the major hydrodynamical and biological processes and their interplay.

Acknowledgements

Data acquisition and the present study were performed within the Special Collaborative Program (SFB) 248 "Cycling of Matter in Lake Constance" supported by Deutsche Forschungsgemeinschaft. Special thanks are due to B. Beese, C. Braunwarth, J. Fürst, U. Sommer, A. Schweizer, and M. Tilzer who conducted measurements of irradiance, algal biovolume and chlorophyll. Meteorological data were obtained from the German Weather Service. Computations were supported by Astrid Hälbich, and Angelika Seifried and Deborah Hart improved the English.

References

[1]. Schindler, D. W., Beaty, K. G., Fee, E. J., Cruikshank, D. R., DeBruyn, E. R., Findlay, D. L., Linsey, G. A., Shearer, J. A., Stainton, M. P. & Turner, M. A. (1990). Effects of climatic warming on lakes of the central boreal forest. *Science* **250**, 967-970.

[2]. George, D. G. & Taylor, A. H. (1995). UK lake plankton and the Gulf Stream. *Nature* **378**, 139.

[3]. Seip, K. L. & Reynolds, C. S. (1995). Phytoplankton functional attributes along trophic gradient and season. *Limnol. Oceanogr.* **40**, 589-597.

[4]. Seip, K. L. (1991). The ecosystem of a mesotrophic lake - I. Simulating plankton biomass and the timing of phytoplankton blooms. *Aquatic Sciences* **53**, 239-262.

[5]. Sommer, U., Gliwicz, Z. M., Lampert, W. & Duncan, A. (1986). The PEG-model of seasonal succession of planktonic events in fresh waters. *Arch. Hydrobiol.* **106**, 433-471.

[6]. Tilzer M. M. & Beese, B. (1988). The seasonal productivity cycle of phytoplankton and controlling factors in Lake Constance. *Schweiz. Z. Hydrol.* **50**, 1-39.

[7]. Gaedke, U. & Schweizer A. (1993). The first decade of oligotrophication in Lake Constance. I. The response of phytoplankton biomass and cell size. *Oecologia* **93**, 268-275.

[8]. Tilzer, M. M., Gaedke U., Schweizer A., Beese B. & Wieser T. (1991). Interannual variability of phytoplankton productivity and related parameters in Lake Constance: No response to decreased phosphorus loading? *J. Plankton Res.* **13**, 755-777.

[9]. Sverdrup, H. U. (1953). On conditions for the vernal blooming of phytoplankton. *J. Cons. Int. Explor. Mer.* **18**, 287-295.

[10]. Erga, S. R. & Heimdal, B. R. (1984). Ecological studies on the phytoplankton of Korsfjorden, western Norway. The dynamics of a spring bloom seen in relation to hydrographical conditions and light regime. *J. Plankton Res.* **6**, 67-90.

[11]. Reynolds, C. S. (1973). The seasonal periodicity of planktonic diatoms in a shallow eutrophic lake. *Freshwater Biology* **3**, 89-110.

[12]. Horn, H. & Paul, L. (1984). Interactions between light situation, depth of mixing and phytoplankton growth during the spring period of full circulation. *Int. Revue ges. Hydrobiol.* **69**, 507-519.

[13]. Tilzer, M. M. (1984). The quantum yield as a fundamental parameter controlling vertical photosynthetic profiles of phytoplankton in Lake Constance. *Arch. Hydrobiol. Suppl.* **69**, 169-198.

[14]. Adrian, R., Deneke, R., Mischke, U., Stellmacher, R. & Lederer P. (1995). A long term study of the Heiligensee (1975-1992). Evidence for effects of climatic change on the dynamics of eutrophied lake ecosystems. *Arch. Hydrobiol.* **133**, 315-337.

THE INFLUENCE OF WEATHER CONDITIONS ON THE SEASONAL PLANKTON DEVELOPMENT IN A LARGE AND DEEP LAKE (L. CONSTANCE)

IV. THE RESPONSE OF CRUSTACEAN ZOOPLANKTON TO VARIATIONS IN WATER TEMPERATURE AND ALGAL GROWTH IN SPRING AND EARLY SUMMER

DIETMAR STRAILE[1] AND WALTER GELLER[2]

[1] *Limnological Institute, University of Constance*
D-78434 Constance, Germany

[2] *UFZ-Centre for Environmental Research,*
Institute for Inland Water Research, Magdeburg
D-39104 Magdeburg, Germany

Abstract

Statistical analysis of long-term (14 years) data on phytoplankton and zooplankton biomass in large and deep Lake Constance showed that the timing of the onset of vernal stratification has pronounced and long-lasting effects on the development of the crustacean zooplankton during spring. The spring biomass of herbivorous cladocerans and cyclopoid copepods differs between years, associated with an "early" or "late" onset of stratification and algal development. The increased biomass of herbivorous cladocerans, usually recorded in May, is most probably caused by the direct effect of temperature on cladoceran growth, rather than its indirect effect on food supply or predation intensity. Multivariate regression analysis suggests that a reduction in nutrient loading by a factor of three did not affect the temporal course of zooplankton spring development.

1. Introduction

Research on zooplankton ecology has a long history in Lake Constance. Following the pioneering work of Hofer [1], various aspects of seasonal population dynamics [2–5], diurnal vertical migration [6, 7], grazing impact on algae [8, 9], and the responses to eutrophication [10], were examined. However, studies on the influence

D.G. George et al.(eds.), Management of Lakes and Reservoirs during Global Climate Change, 85–92.

of weather conditions on the population dynamics and species composition of the zooplankton are lacking for Lake Constance and, with a few exceptions [11, 12], also for other lake ecosystems. Climatic change, i.e. increasing average annual temperatures on a global scale [13], will affect the thermal structure of lakes [14] and subsequent various ecological processes [15]. To predict the effects of global warming on lake ecosystems, a better understanding of the impact of present-day climate variability on population dynamics is required. In Lake Constance, the onset of stratification (Ollinger & Bäuerle, II in this volume, pp. 57–70) and consequently the onset of the algal spring bloom (Gaedke et al., III in this volume, pp. 71–84), varies between years according to differences in irradiance, air temperature and wind regime. Usually, the increase of phytoplankton biomass during spring is followed by the growth of herbivorous zooplankters [16]. Crustaceans respond to favourable weather conditions in spring either directly or indirectly, by their temperature-mediated responses to changes in the food supply or predation intensity [17]. In this paper, we examine the relative importance of these direct and indirect effects on the spring growth of crustacean zooplankters in Lake Constance. A close coupling between temperature and crustacean growth would imply that crustacean biomass will be affected by the timing of the onset of spring stratification, i.e. warm or cold water temperatures during spring should lead respectively to a higher or lower biomass of crustacean zooplankton.

2. Study site and methods

Lake Constance is a large and deep lake situated at the northern fringe of the Alps. Due to improvements of sewage treatment, the eutrophication process was reversed in the late 1970s and the winter concentration of phosphorus has subsequently decreased from a maximum of 87 $\mu g\ l^{-1}$ in 1979 to 28 $\mu g\ l^{-1}$ in 1994 (International Commission for the Protection of Lake Constance (IGKB)). Total phosphorus concentrations during winter mixis were provided by the IGKB. For details on measurements of water temperature and algal biomass see Ollinger & Bäuerle (II, this volume) and Gaedke et al. (III, this volume). Crustacean zooplankton was sampled with a 30-cm Clark-Bumpus sampler (mesh size 140 µm) once a week at the sampling station in the central part of the Überlinger See, a fjordlike arm of Lake Constance. The biomass of crustaceans was calculated using length–weight relationships established for Lake Constance (for details see [5] and [18]). Based on the onset of algal growth, the biomass of zooplankton in years with an early onset of algal growth (1989, 1990, 1991 and 1993) was compared with the biomass of zooplankton in "late" years, i.e. from 1979 to 1988, and 1992 (Gaedke et al., III in this volume).

3. Results

Until April, monthly averages of herbivorous cladoceran biomass (*Daphnia hyalina, Daphnia galeata* and *Bosmina*) are very similar in "early" and "late" years but, in May, two to three times as much biomass is built up during early years (Fig. 1a).

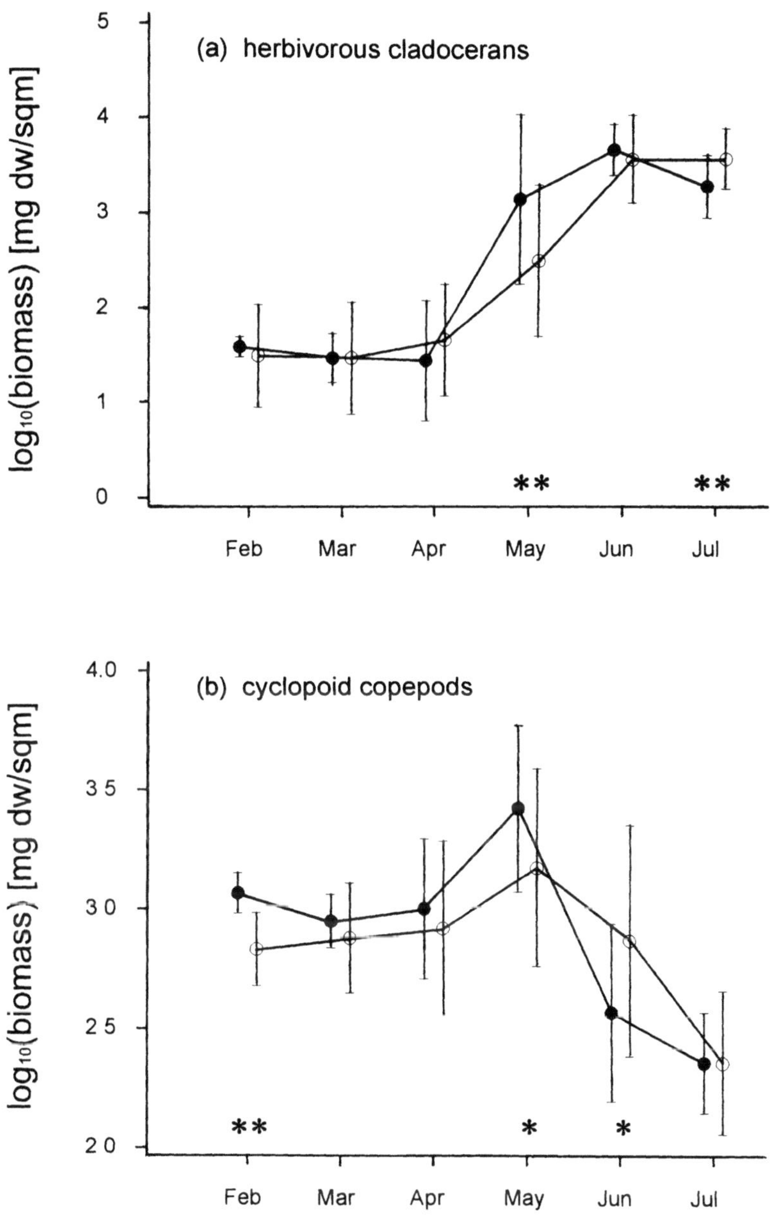

FIGURE 1. Mean crustacean biomass ($\log_{10}$ [mg dry wt m^{-2}]) and standard deviations (vertical bars) in Lake Constance during years with an early onset of phytoplankton growth (solid circles: 1989 to 1992, and 1994) and a late onset of phytoplankton growth (open circles: 1979 to 1988, and 1993). (a) Herbivorous cladocerans; (b) cyclopoid copepods. * $p<0.05$; ** $p<0.01$.

In June, the biomasses are again similar in early and late years whereas in July, late years have higher biomasses. Monthly averages of omnivorous cyclopoid copepod biomass (predominantly *Cyclops vicinus* during spring) are higher in February and May, but lower in June in the years with a later onset of phytoplankton development (Fig. 1b). Identifying the effects of climatic variability on zooplankton biomass is hampered during our 16-year period of investigation (1979 to 1994) by the re-oligotrophication process of the lake, which covered approximately the same period. Years with an early onset of the algal bloom also happen to be the most oligotrophic, with the exception of 1993. In recent years, phosphorus depletion increased during late spring [19] and may have had an effect on zooplankton biomass from May onwards. This complicates a distinction between the effects of temperature and re-oligotrophication, when examined by simple time-series analysis.

TABLE 1. Some statistical results from individual and multiple regressions of the mean biomass of herbivorous cladocerans (dependent variable *Y*) in Lake Constance, versus four environmental variables (*X*): the mean water temperature at 8 m depth in May (*Temp*), mean algal biovolume in March (*Algae 3*), April (*Algae 4*), May (*Algae 5*) and phosphorus concentrations during winter mixis (*Phosph*). d.f. indicates the degrees of freedom; P is the probability for regression slopes differing from zero (positive or negative, NS is not significant); r^2 and R^2 are values for the regression coefficients.

X variables	d.f	P for regression slopes		Sign of slopes	r^2 or R^2
Temp	1	0.005		+	0.56
Algae 3	1	0.02		+	0.38
Algae 4	1	NS		0	0.0005
Algae 5	1	NS		0	0.06
Phosph	1	NS		0	0.15
Temp + Algae 3	2	Temp = 0.05;	Algae 3 = NS	+; 0	0.56
Temp + Algae 4	2	Temp = 0.0005;	Algae 4 = 0.03	+; –	0.74
Temp + Algae 5	2	Temp = 0.01;	Algae 5 = NS	+; 0	0.59
Temp + Phosph	2	Temp = 0.01;	Phosph = NS	+; 0	0.59

To analyse the confusing effects of early stratification and enhanced nutrient depletion in recent years, regression statistics were performed. They reveal that mean herbivorous cladoceran biomass in May is closely related to mean water temperature in May (Table 1). Additionally, a significant relationship exists between the average herbivorous cladoceran biomass recorded in May and the average algal biovolumes recorded in March, but not those recorded in April and May. There is, however, no relationship between the herbivorous cladoceran biomass recorded in May and the winter concentration of phosphorus – which provides a general measure of basin productivity. Multiple regression of herbivorous cladoceran biomass on water temperatures in May, phytoplankton biovolume in March or May, and winter phosphorus concentrations, did not improve the regression coefficient substantially, compared with that from a regression on water temperature alone. In contrast, a multivariate regression against May water temperature and average algal biovolume in April, explains 74% of the variability in herbivorous cladoceran biomass, and mean herbivorous cladoceran biomass is inversely related to average algal biomass in the multivariate regression ($R^2 = 0.74$, and negative slope for Algae 4; Table 1).

4. Discussion

The differences in average crustacean biomass observed in "early" and "late" years may have been produced by the direct or indirect effects of changes in the weather, i.e. from changes in the timing of the onset of stratification, or from indirect effects of the re-oligotrophication process. The timing of stratification onset may influence crustacean zooplankters via higher water temperatures (Ollinger & Bäuerle, II in this volume) or an early increase of food supply (Gaedke et al., III in this volume). The re-oligotrophication process may affect zooplankters by altering abundance, species composition and production of their food organisms [19]. For herbivorous cladoceran biomass in May, regression analysis suggests that the effects of early stratification are more important than the effects of the re-oligotrophication process. These direct and indirect effects cannot be distinguished readily by simple statistical analyses, but at least two arguments can be advanced for the dominant role of temperature. First, the mean cladoceran biomass in May (*Y*) is significantly related to algal biovolume in March (*X*) in a simple regression of *Y* on *X*, but not in a multivariate regression that includes water temperature as a second independent variable. This suggests that the effect of water temperature is more important for crustacean growth than the effect of algal biovolume. Second, the average algal biovolume in April can increase the explained variability in a multivariate regression. However, there is an inverse relationship between herbivorous cladoceran biomass in May and algal biovolume in April. This suggests there is no direct causal link between the two variables.

Algae are able to react faster to favourable conditions than crustaceans can due to the differences in size between predator and prey, and the allometric nature of the body size/growth rate relationship [20]. This may result in abundant food for cladocerans shortly after the onset of phytoplankton growth. Thereafter, cladoceran growth may be limited by temperature [8]. The temperature sensitivity of crustacean egg development and growth rates is well established in laboratory cultures [21, 22]. Temperature was also invoked to explain the timing of the population increase of daphniids during spring in Schöhsee [23]. Additionally, Lampert [8] suggests that there is a predation control by adult cyclopoid copepods, resulting in a delay in the daphniid population increase. However, cyclopoid copepods also have higher biomasses in May during the years with an early onset of stratification. If cyclopoid copepods control daphniids, the biomass of the daphniids should be negatively affected during these years. As this is not the case, the predation control hypothesis becomes questionable.

Both herbivorous cladocerans and cyclopoid copepods show significantly higher biomasses in May and lower biomasses in subsequent months in the "early" (as distinct from "late") years. However, the mechanisms responsible for these year-to-year variations were probably very different for the two groups. High biomasses of herbivorous cladocerans imply high grazing rates which can severely suppress phytoplankton biomass [23–25]. Consequently, a significantly higher herbivorous cladoceran biomass during May causes a significantly lower biomass of phytoplankton in May in early years, compared with late years (see Gaedke et al., III in this volume). This may result in an insufficient food supply and a subsequent population decline of herbivorous cladocerans. Thus, an early start for the population development of herbivorous cladocerans is followed by an early decline in their population biomass. In contrast, the decline of cyclopoid copepod biomass from May to June is not a result of decreasing food supply, but is due to the diapause migration of *Cyclops vicinus* into the sediments [26]. The timing of diapause migration is triggered by the photoperiod, and the sensitivity to light differs among copepodite stages. It is highest within the 4th copepodite stage [26]. Increased temperature in years with an early onset of stratification will result in a faster development and an earlier timing of the photo-sensitive 4th copepodite stage. The timing of vernal stratification in March influences the biomasses of herbivorous cladocerans and cyclopoid copepods during early summer. This demonstrates the long-lasting effects of spring climatic conditions.

To conclude, the timing of the onset of stratification has important and long-lasting consequences for the population dynamics of crustaceans during spring. Earlier onset of stratification affects the metabolic rates of crustacean zooplankters, via increased water temperatures. This analysis suggests that the direct effect of water temperature is more important than the indirect effects mediated by food supply and predation. The analysis of long-term data series with respect to the influence of weather conditions provides information necessary to predict the ecological consequences of climate change and obtain valuable insights into food web regulation.

Acknowledgements

Data acquisition and the present study were performed within the Special Collaborative Program (SFB) 248 "Cycling of matter in Lake Constance", supported by Deutsche Forschungsgemeinschaft. We thank G. Schulze for counting zooplankton samples and U. Gaedke and D. Hart for improving the style and content of this manuscript.

References

[1]. Hofer, B. (1896). Die Verbreitung der Tierwelt im Bodensee nebst vergleichenden Studien in einigen anderen Süßwasserbecken. *Schr. V. G. Bodensee* **28**, 281-366.

[2]. Elster, H.-J. (1954). Über die Populationsdynamik von *Eudiaptomus gracilis* und *Heterocope borealis* im Bodensee Obersee. *Arch. Hydrobiol. Suppl.* **20**, 546-614.

[3]. Elster, H.-J. & I. Schwoerbel (1970) Beiträge zur Biologie und Populationsdynamik der Daphnien im Bodensee. *Arch. Hydrobiol. Suppl.* **38**, 18-72

[4]. Geller, W. (1985). Production, food utilization and losses of two coexisting, ecologically different *Daphnia* species. *Arch. Hydrobiol. Beih. Ergebn. Limnol.* **21**, 67-79.

[5]. Geller, W. (1989). The energy budget of two sympatric *Daphnia* species in Lake Constance: productivity and energy residence times. *Oecologia* **78**, 242-250.

[6]. Geller, W. (1986). Diurnal vertical migration of zooplankton in a temperate great lake (L. Constance): a starvation avoidance mechanism? *Arch. Hydrobiol. Suppl.* **74**, 1-60.

[7]. Stich, H.-B. & Lampert, W. (1981). Predator evasion as an explanation of diurnal vertical migration by zooplankton. *Nature* **293**, 396-398.

[8]. Lampert, W. (1978). Climatic conditions and planktonic interactions as factors controlling the regular succession of spring algal bloom and extremely clear water in Lake Constance. *Verh. Internat. Verein. Limnol.* **20**, 969-974.

[9]. Pinto-Coelho, R. M. (1991). The importance of *Daphnia* for zooplankton grazing in Lake Constance. *Arch. Hydrobiol.* **121**, 319-342.

[10]. Einsle, U. (1988). The long-term dynamics of crustacean communities in Lake Constance (Obersee, 1962-1986). *Schw. Z. Hydrol.* **50**, 136-165.

[11]. George, D. G. & Harris, G. P. (1985). The effect of climate on long-term changes in the crustacean zooplankton biomass of Lake Windermere, UK. *Nature* **316**, 536-539.

[12]. George, D. G., Hewitt, D. P., Lund, J. W. G. & Smyly, W. J. P. (1990). The relative effects of enrichment and climate change on the long-term dynamics of *Daphnia* in Esthwaite Water, Cumbria. *Freshwat. Biol.* **23**, 55-70.

[13]. Schneider, S. H. (1993). Scenarios of global warming. In: *Biotic interactions and global change* (eds. P. M. Kareiva, J. G. Kingsolver & R. B. Huey). Sinauer Associates Inc. Sunderland, Massachusetts, pp. 9-23.

[14]. McCormick, M. J. (1990). Potential changes in thermal structure and cycle of lake Michigan due to global warming. *Trans. Am. Fish. Soc.* **119**, 183-194.

[15]. Schindler, D. W., Beaty, K. G., Fee, E. J., Cruikshank, D. R., DeBruyn, E. R., Findlay, D. L., Linsey, G. A., Shearer, J. A., Stainton, M. P. & Turner, M. A. (1990). Effects of climatic warming on lakes of the central boreal forest. *Science* **250**, 967-970.

[16]. Sommer, U., Gliwicz, Z, M., Lampert, W. & Duncan, A. (1986). The PEG-model of seasonal succession of planktonic events in fresh waters. *Arch. Hydrobiol.* **106**, 433-471.

[17]. Threlkeld, S. T. (1987). *Daphnia* population fluctuations: patterns and mechanisms. In: *Daphnia* (eds. R. H. Peters & R. de Bernhardi). *Mem. Ist. Ital. Idrobiol.* **45**, 367-388.

[18]. Wölfl (1991). The pelagic copepod species in Lake Constance: abundance, biomass, and secondary productivity. *Verh. Internat. Verein. Limnol.* **24**, 854-857.

[19]. Gaedke, U. & Schweizer, A. (1993). The first decade of oligotrophication in Lake Constance. I. The response of phytoplankton biomass and cell size. *Oecologia* **93**, 268-275.

[20]. Peters, R. (1983). *The ecological implications of body size.* Cambridge University Press.

[21]. Botrell, H. H. (1975). Generation time, length of life, instar duration and frequency of moulting, and their relationship to temperature in eight species of cladocera from the River Thames, Reading. *Oecologia* **19**, 129-140.

[22]. Vijverberg, J. (1980). Effect of temperature in laboratory studies on development and growth of Cladocera and Copepoda from Tjeukemeer, The Netherlands. *Freshwat. Biol.* **10**, 317-340.

[23]. Lampert, W., Fleckner, W., Rai, H. & Taylor, B. E. (1986). Phytoplankton control by grazing zooplankton: a study on the spring clear-water phase. *Limnol. Oceanogr.* **31**, 478-490.

[24]. Geller, W. (1980). Stabile Zeitmuster in der Planktonsukzession des Bodensees (Überlinger See). *Verh. Ges. Ökol.* **8**, 373-382.

[25]. Markager, S., Hansen, B. & Søndergaard, M. (1994). Pelagic carbon metabolism in a eutrophic lake during a clear-water phase. *J. Plankt. Res.* **16**, 1247-1267.

[26]. Einsle, U. (1967). Die äußeren Bedingungen der Diapause planktisch lebender *Cyclops*-Arten. *Arch. Hydrobiol.* **63**, 387-403.

FLUCTUATIONS OF PHYTOPLANKTON PRODUCTION AND CHLOROPHYLL CONCENTRATIONS IN A SMALL HUMIC LAKE DURING SIX YEARS (1990–1995)

J. KESKITALO AND K. SALONEN
University of Helsinki
Lammi Biological Station
FIN-16900 Lammi, Finland

1. Introduction

Despite the variable nature of the lake environment, relatively few long-term studies comparable to those described in [1] have been reported hitherto. This is surprising, because long-term studies offer important advantages in the interpretation of biological changes in systems where irregular fluctuations and cyclical variations are common. Although this variation renders the detection of trends more difficult, such interannual changes can also help to identify the critical driving variables. In many instances, these long-term datasets allow us to separate biogeochemical effects that operate within a catchment from climatic effects that operate on a regional scale. Long-term studies of this kind are, however, logistically very demanding, since most biological processes in lakes are highly dynamic. For example, in studying phytoplankton production or chlorophyll concentration at least weekly sampling is needed. The necessity for frequent sampling makes hydrobiological monitoring rather expensive and is therefore one of the reasons for the rarity of long-term data series.

In this paper we analyse the relative importance of catchment and climatic effect on the seasonal and interannual variations of primary production and phytoplankton biomass in a small headwater lake of the boreal forest zone. The lake chosen for the study (Valkea-Kotinen) is strongly coloured by organic matter leaching from the soil [2–5] and has a shallow epilimnion that responds rapidly to changes in the weather. The interpretation of our results further benefits from the fact that the whole catchment area of the lake is under continuous, multidisciplinary monitoring [6].

D.G. George et al.(eds.), Management of Lakes and Reservoirs during Global Climate Change, 93–109.

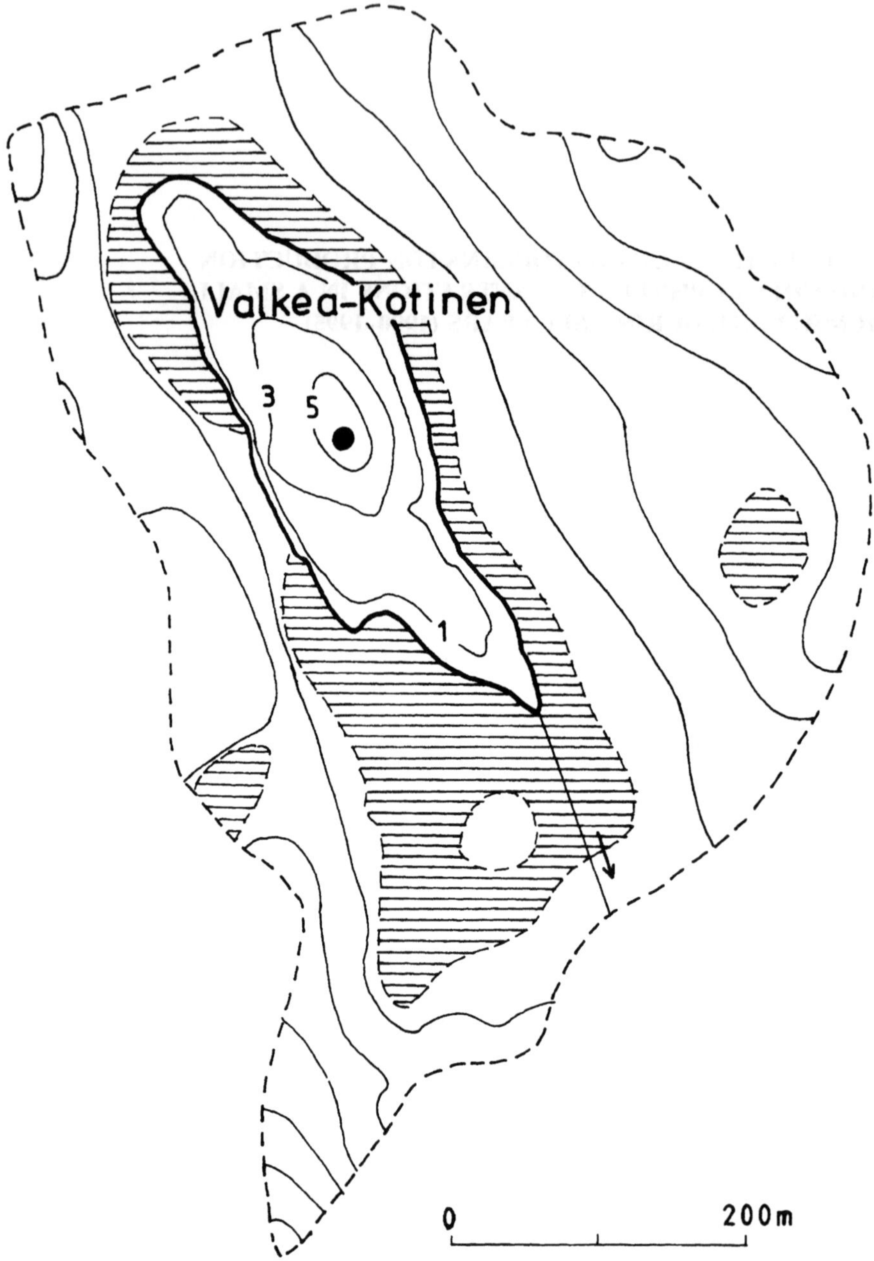

FIGURE 1. Lake Valkea-Kotinen. Height interval of the contour lines is 5 m in the catchment area. Shaded area = peatland; central dot in the lake = sampling point.

2. Study area

Lake Valkea-Kotinen is situated in the Kotinen nature reserve in southern Finland (61°15'N, 25°04'E). The catchment area mainly consists of old virgin forest with coniferous trees and small areas (21%) of peatland (Fig. 1). The annual mean air temperature is 3.1°C and the mean precipitation is 618 mm [7]. Valkea-Kotinen is a relatively shallow lake (maximum depth 6.5 m) with one outlet stream at the southeastern end but with no definite inlets. The catchment area is small (30 ha), and compared to this the lake surface area is proportionally high (4 ha). Valkea-Kotinen is a headwater lake without any local pollution but it is periodically influenced by the deposition of long-range atmospheric pollutants.

3. Material and methods

The plankton of Valkea-Kotinen has been sampled weekly during the ice-free season and monthly during the winter since 1990. The variables covered by this monitoring programme include phytoplankton species composition, phytoplankton biomass, metabolic measurements on the plankton and a range of physicochemical measurements on lake water [8–10]. The sampling and laboratory methods used are those documented in the Manual for Integrated Monitoring [8]. Water samples were taken at the deepest point of the lake (Fig. 1) in the morning, between 09.30 and 12.00 hours. Surface water samples were collected in the storage vessels and deep-water samples with a Teflon-coated Limnos sampler (height 30 cm). In the Limnos sampler the top and bottom lids are vertical and allow an unimpeded water passage through the sampler whilst it is being lowered through the water column [8]. Chlorophyll samples were taken with a 1-m tube sampler. The sampling depths were 0, 1, 2, 3 and 5 m for physicochemical samples, 0, 0.25, 0.5, 1, 2, 3 and 5 m for dissolved inorganic carbon (DIC) and 0–1 m for dissolved organic carbon (DOC). Analytical methods are given in Table 1. The nutrient concentrations were determined by Helsinki Water and Environment District (presently Uusimaa Regional Environment Centre) in 1990–1994, and by Häme Regional Environment Centre in 1995. Other physicochemical determinations were carried out by the laboratory of the Lammi Biological Station.

Chlorophyll samples were taken from 0–1, 1–2, 2–3, 3–4 and 4–5 m water columns and primary production was determined at 0, 0.25, 0.5, 1 and 2 m. Primary production and dark fixation of inorganic carbon were determined with the acidification and bubbling modification of the ^{14}C method [11, 12]. The samples were incubated in situ for 24 h in summer and 48 h in winter, after which formaldehyde solution was added to stop the carbon assimilation. Radioactivities were measured with a liquid scintillation counter, and primary production was calculated as a difference between light and dark fixation of carbon.

For the chlorophyll determination algae were filtered on glass microfibre filters (Whatman GF/C). Chlorophyll was extracted in 94% ethanol, and the absorbances were measured with a spectrophotometer at wavelengths of 665 nm (the absorption maximum of chlorophyll-*a*) and 750 nm (turbidity). The results were calculated by using an absorption coefficient of 83.4. In 1990 the extraction time was 18 h at a temperature of 4°C. In 1991–1995, chlorophyll was extracted in hot ethanol (5 min in a 75°C bath) and measured immediately. In four parallel sample series no statistically significant differences were found between the results of these two methods (multifactor ANOVA).

Respiration of plankton was determined as an increase of DIC in dark bottles and net community production as a decrease of DIC in light bottles during an incubation time of 24 h in situ.

TABLE 1. Analytical methods for physicochemical variables monitored in Lake Valkea-Kotinen.

Variable	Analytical method
Temperature	YSI (Yellow Springs Instruments) probe
pH	Potentiometry, pH meter
Specific conductivity	Conductimetry
Oxygen	YSI probe
Alkalinity	Acid titration, Gran plot
Colour value	Spectrophotometry at 420 nm
Total phosphorus	Spectrophotometry, after persulphate oxidation
Phosphate phosphorus	Spectrophotometry, molybdate method
Total nitrogen	Spectrophotometry, after persulphate oxidation
Ammonium nitrogen	Spectrophotometry, indophenol method
Nitrate nitrogen	Spectrophotometry, as nitrate after cadmium reduction
Dissolved inorganic carbon	Acidification and bubbling method with IR detection
Dissolved organic carbon	High temperature combustion with IR detection

4. Results

4.1. THE IMPACT OF YEAR-TO-YEAR CHANGES IN THE WEATHER ON THE PHYSICOCHEMICAL PROPERTIES OF VALKEA-KOTINEN

During the period of study there were large year-to-year changes in the weather which had a major effect on the biology of the lake. The growing season in 1993 was rather cold and rainy (Table 2) but the late summer of 1995 was warm and sunny. Throughout the period of study the ice melted between late April and early May but there were large year-to-year variations in the intensity of the spring circulation which oxygenated the deep water (Table 3). In 1993 the spring circulation did not even extend to the bottom so the physicochemical characteristics of the deep water changed very little from winter to autumn. In general, the near-bottom water remained aerobic for 2–3 weeks after the spring circulation but in 1995 this aerobic period continued for more than one month (Table 3). In addition to the spring circulation, the water column was oxygenated by the autumn circulation (Table 3), but during the winter stagnation the deeper water layers were again depleted of oxygen (Fig. 2). The uppermost 2.5–3.0 m under ice remained aerobic in late winters 1991–1992, but in 1993–1995 this layer was reduced to only 1.5–2.0 m.

TABLE 2. Mean air temperature and precipitation during the growing seasons (May–September) and summer months (June–August) in 1990–1995 at the observation station of the Lammi Biological Station (25 km south of Kotinen).

Month	1990	1991	1992	1993	1994	1995
Temperature (°C)						
May–September	12.3	12.3	13.7	11.6	12.2	12.8
June–August	14.2	15.2	15.3	13.2	14.8	15.4
Precipitation (mm)						
May–September	258	294	268	326	286	293
June–August	156	194	193	297	156	151

TABLE 3. Date of ice break-up, duration of aerobic periods and maximum oxygen concentrations in the near-bottom water (depth 6.0–6.5 m) in spring and autumn/early winter, and durations of the ice-free and ice-covered periods (the last mentioned until the ice break-up in the next year).

		Spring		Autumn/early winter			
Year	Ice break-up	Aerobic (weeks)	O_2 max. ($g\ m^{-3}$)	Aerobic (weeks)*	O_2 max. ($g\ m^{-3}$)	Ice-free (weeks)	Ice-covered (weeks)
1990	20 Apr	2	6.4	12	12.3	30	24
1991	5 May	2	2.7	8	9.8	29	24
1992	5 May	3	2.4	5	9.0	24	28
1993	4 May	0	0.0	5	1.0	24	27
1994	30 Apr	3	6.2	8	11.4	27	26
1995	8 May	6	6.5	11	9.5	26	26

* Rough estimates.

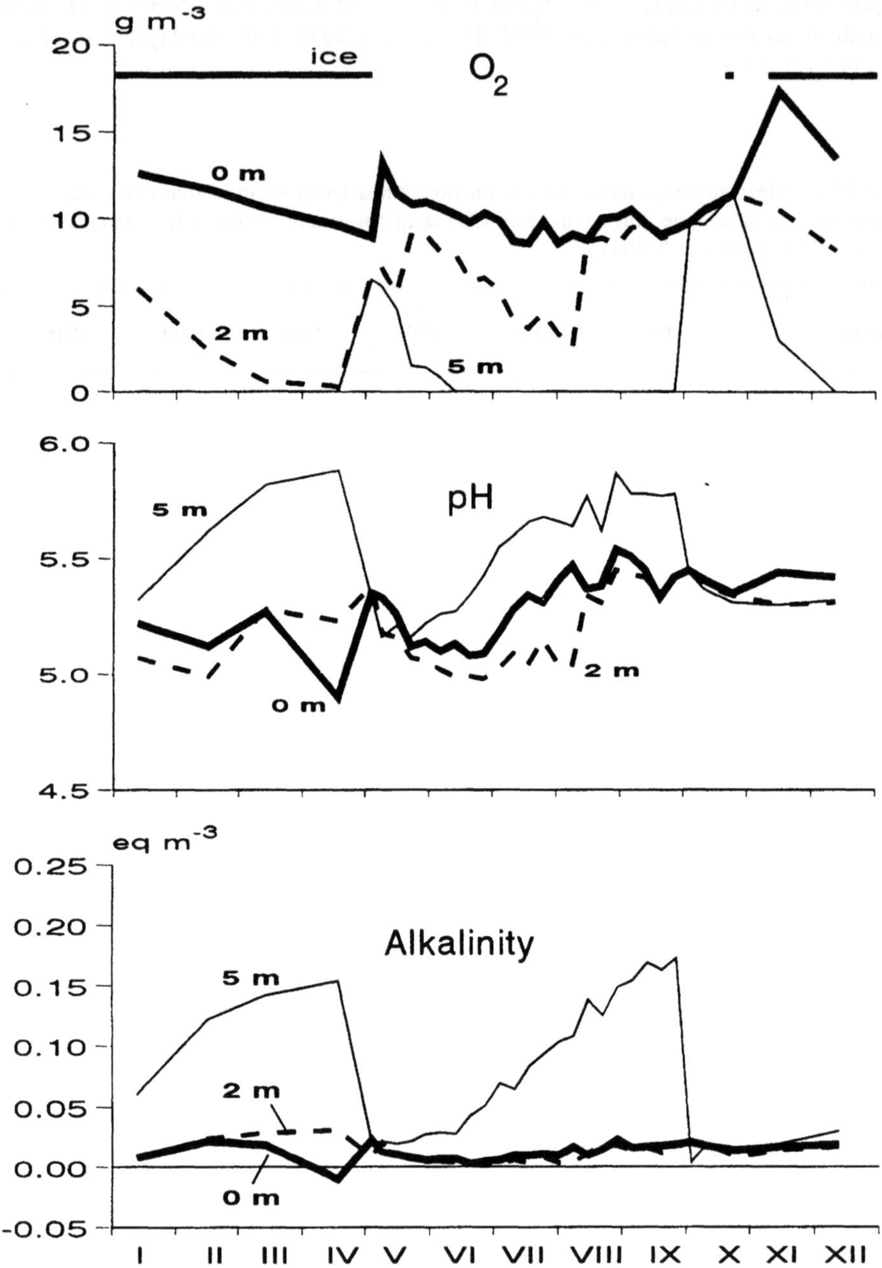

FIGURE 2. Variation of oxygen concentration (g m^{-3}), pH, and alkalinity (equivalents m^{-3}) at 0, 2 and 5 m depths in Lake Valkea-Kotinen in 1994.

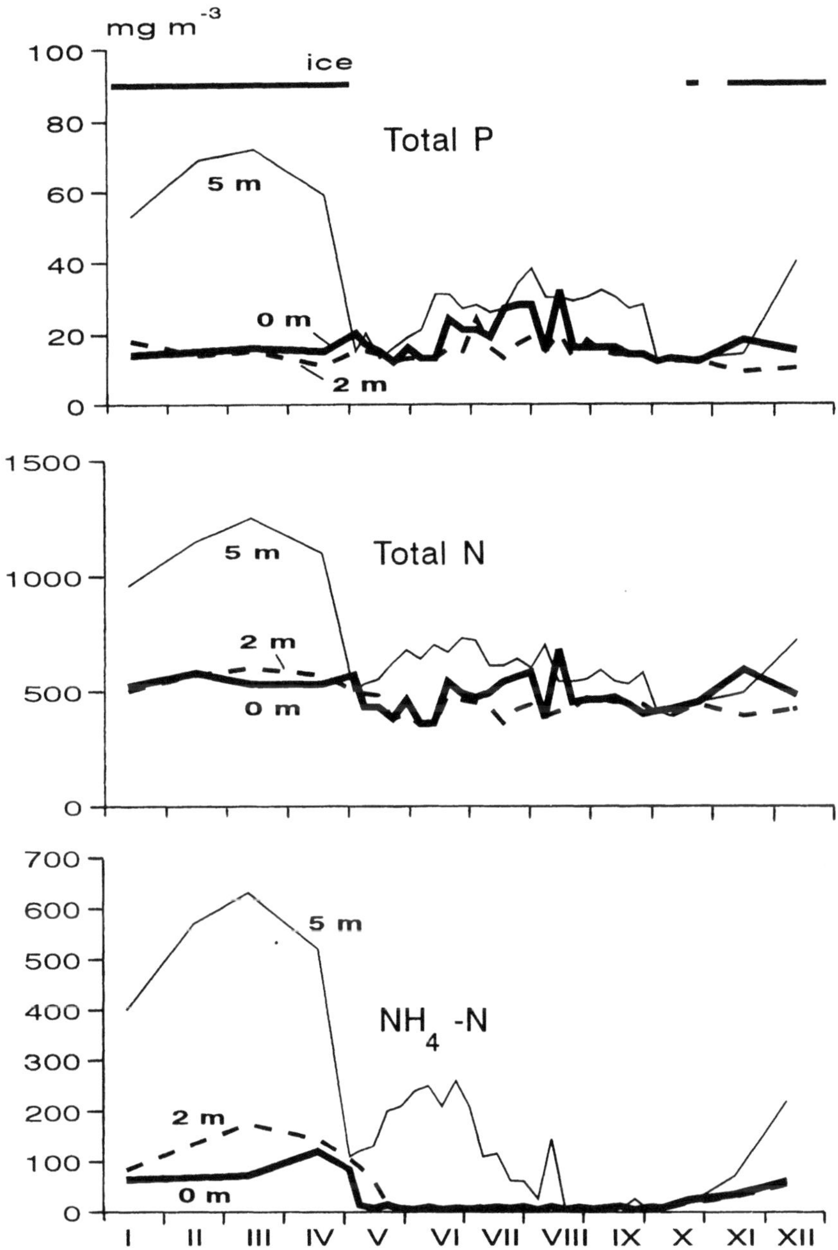

FIGURE 3. Variation of total phosphorus, total nitrogen and ammonium nitrogen (mg m^{-3}) at 0, 2 and 5 m depths in Lake Valkea-Kotinen in 1994.

TABLE 4. Means of physico-chemical properties during the growing season (May-September) for epilimnion (0-1 m) and hypolimnion (3-5 m) in Lake Valkea-Kotinen

	Epilimnion						**Hypolimnion**					
Variable	1990	1991	1992	1993	1994	1995	1990	1991	1992	1993	1994	1995
Water temperature, °C	15.3	15.0	15.8	14.5	15.2	15.2	7.6	7.5	7.1	6.8	7.2	7.3
O_2, g m^{-3}	10.0	9.6	9.1	9.5	9.9	8.8	1.2	2.1	1.6	1.4	1.7	2.5
pH	5.2	5.4	5.3	5.3	5.3	5.1	5.3	5.3	5.2	5.3	5.3	5.2
Conductivity, mS m^{-1}	3.0	2.8	3.1	3.0	2.9	3.1	3.5	3.2	3.3	3.4	3.1	3.3
Colour, g Pt m^{-3}	103	128	141	136	140	149	154	141	172	178	161	168
Total N, mg m^{-3}	462	560	499	436	495	485	643	607	678	651	547	682
NH_4-N, mg m^{-3}	4	9	12	7	12	25	133	89	181	152	74	160
NO_3-N, mg m^{-3}	-	-	-	4	5	7	-	-	-	6	10	10
Total P, mg m^{-3}	19	22	19	18	20	17	22	23	22	28	23	23
PO_4-P, mg m^{-3}	-	-	-	<2	<2	<2	-	-	-	<2	<2	<2
Alkalinity, eq m^{-3}	-0.007	0.009	0.002	0.007	0.011	0.002	0.054	0.050	0.044	0.068	0.056	0.045
DIC, g m^{-3}	0.44	0.63	0.64	0.68	0.54	0.88	-	5.20	4.93	5.95	5.55	5.01
DOC, g m^{-3}	9.6	11.3	12.2	11.5	11.9	11.4	-	-	-	-	-	-
Chlorophyll, mg m^{-3}	27.2	31.4	15.1	15.6	34.7	12.9	42.7	24.7	16.4	40.7	25.9	17.9

The water in Valkea-Kotinen is always strongly coloured by humic substances and its reaction is acid (Table 4). These humic substances have a major effect on the attenuation of light, and the Secchi-disc transparency during the growing season typically fluctuates between 1.4 and 1.6 m. The depth of the epilimnion was ca. 2 m, and there were large differences in physicochemical properties between the epilimnion and the hypolimnion (Table 4, Figs 2 and 3). In summer, temperature of the uppermost metre was on the average 16 to 18°C in the afternoon, with a highest measured value of 24.1°C in July 1994. Due to the shallowness of the epilimnion its temperature closely followed the air temperature, and increases of several degrees were periodically recorded in the afternoon.

For most of the year, the carbonate buffering system (alkalinity) was almost wholly lost in the epilimnion, but towards late winter and in late summer it increased in the hypolimnion. Lowest pH values were recorded in April (Fig. 2), when meltwater from the catchment area spread as a distinct layer immediately under the ice-cover.

Phosphate, nitrate and ammonium were exhausted from the epilimnion usually in early June. For most of the summer, the concentrations are at or near the limits of detection but increase rapidly during the period of autumn circulation. Similarly, DIC concentrations were very low in the epilimnion, typically ranging between 0.1 and 0.3 g m^{-3}.

In general, no clear trends were evident in the physicochemical properties during the study period. Epilimnetic DOC concentrations, water colour, alkalinity and DIC, however, were lower in 1990 (Table 4), presumably because there was less leaching from the catchment during the dry summer.

The example profiles in Fig. 4 show the vertical distribution of primary production, phytoplankton biomass and a series of physicochemical properties in Valkea-Kotinen in late summer. Most of the primary production occurs in the uppermost 1 m and the productive layer as a whole is only < 2 m. In spite of this thin productive layer, annual production was fairly high (25–38 g C m^{-2}) and very high levels were periodically recorded in the top 0.5 m where they frequently exceeded 200 mg C m^{-3} d^{-1}. The vertical distribution of phytoplankton biomass was influenced by the concentration of photosynthetic bacteria found in the hypolimnion as well as the concentration of phytoplankton in the upper layers. In Fig. 4, the absorbance ratio between wavelengths of 665 and 656 nm is used as a measure of phytoplanton biomass and shows that the concentration of phytoplankton chlorophyll in the hypolimnion is relatively low.

Net production (according to the DIC method) was often negative below a depth of 1 m; i.e. respiration of the phytoplankton, zooplankton and bacterioplankton communities was higher than carbon fixation.

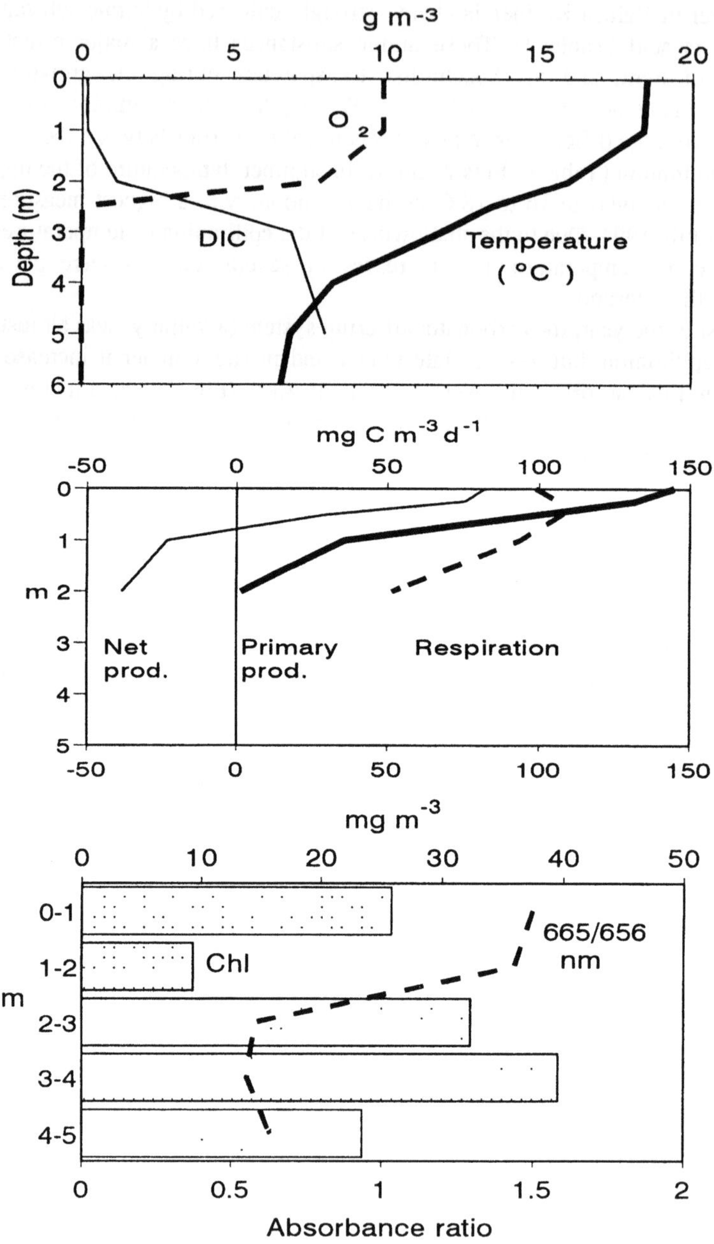

FIGURE 4. An example of the vertical distribution of some physicochemical properties, plankton metabolism and chlorophyll concentrations in late summer in Lake Valkea-Kotinen.

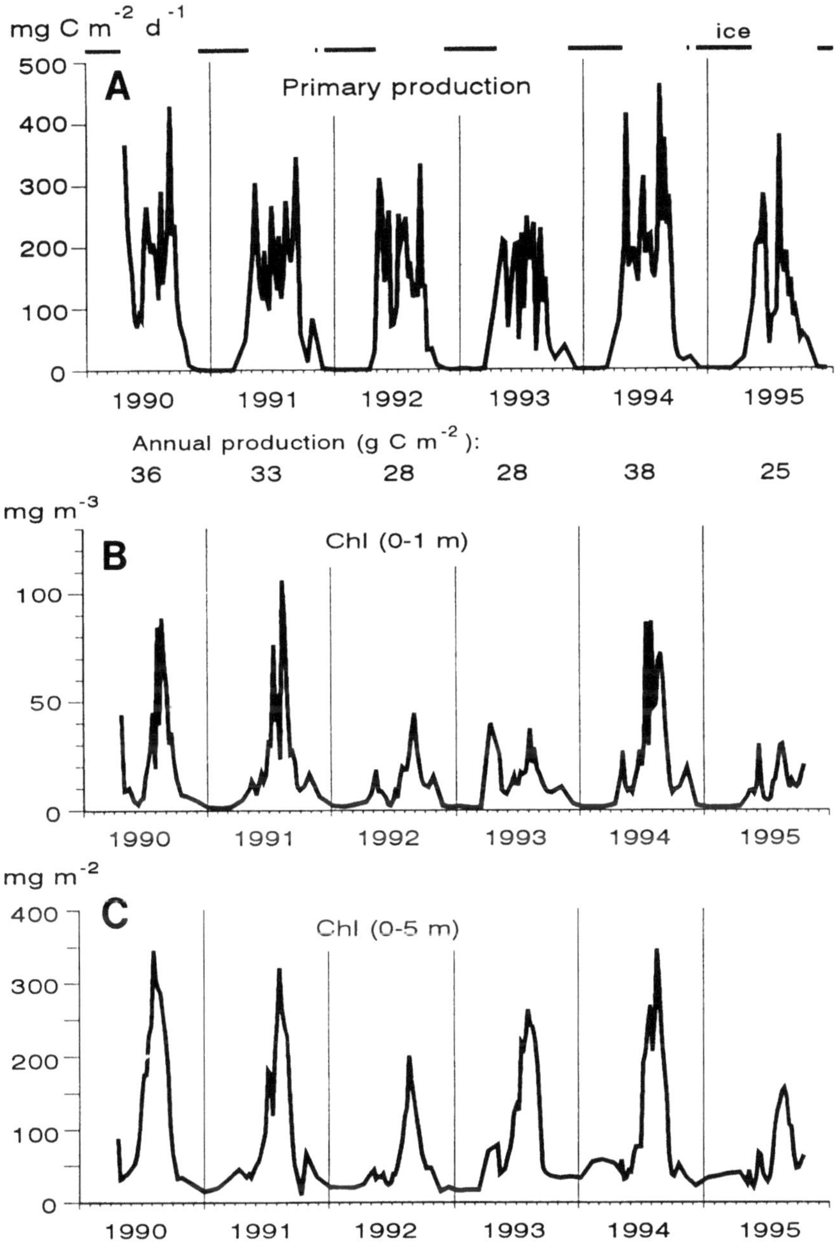

FIGURE 5. Primary production (A, mg C m^{-2} d^{-1}) and chlorophyll concentrations (B, C, mg m^{-3}) in Lake Valkea-Kotinen in the period 1990 to 1995.

4.2. LONG-TERM CHANGES IN THE PRIMARY PRODUCTION OF PHYTOPLANKTON

Fig. 5 shows the changes in primary production and the concentrations of chlorophyll at two depth-ranges in Valkea-Kotinen between the beginning of 1990 and the end of 1995. Fig. 5A shows the seasonal and year-to-year variations in the measured levels of primary production. High levels of primary production (> 300 mg C m^{-2} d^{-1}) were recorded in the spring soon after the break-up of the ice. There was then a pronounced decrease, followed by a rising trend which culminated in a second maximum in September. In 1993, both the spring and autumn maxima were exceptionally weak and did not exceed mid summer production. In 1995, summer production fell to an exceptionally low level (< 40 mg C m^{-2} d^{-1}) at the beginning of July, before increasing to a late summer maximum (380 mg C^{-2} d^{-1}) in August. These periodic reductions in phytoplankton production were closely corelated with changes in the weather. In 1993, the air temperature was very low throughout the growing season. In 1995, a period of cold weather in mid summer led to a gradual deepening of epilimnetic circulation and a subsequent increase in nutrient availability. In the autumn, primary production levels were strongly influenced by year-to-year variations in underwater light. Small increases in production were, however, recorded quite late in the year and some were even recorded when the lake was covered with ice (e.g. as in 1993).

In general, annual primary production was correlated poorly with the precipitation and the means of the physicochemical variables given in Tables 2 and 4. Significant correlation coefficients were obtained only with epilimnetic oxygen (0.916; $p<0.05$; n = 6), epilimnetic DIC (–0.856; $p<0.05$) and hypolimnetic total nitrogen (–0.840; $p<0.05$).

4.3. LONG-TERM CHANGES IN THE CONCENTRATION OF CHLOROPHYLL

Fig. 5B shows the seasonal and year-to-year variations in the concentration of chlorophyll measured in the top metre of the epilimnion. Epilimnetic chlorophyll concentrations in the spring were relatively low but several well-defined spring maxima were recorded during the period of study (Fig. 5: 0–1 m). Concentrations invariably decreased in early summer, but increased again in late summer. After the late summer decline, a slight increase in concentration was frequently recorded in the autumn. Average epilimnetic chlorophyll concentrations during the growing season (Table 4) generally followed the year-to-year variations in primary production (Fig. 5) but there was some disparity in the two variables in 1992. Epilimnetic chlorophyll concentration correlated significantly with epilimnetic oxygen (0.819; $p<0.05$; n = 6), epilimnetic electrical conductivity (–0.847; $p<0.05$), hypolimnetic total nitrogen (–0.904; $p<0.05$) and hypolimnetic ammonium nitrogen (–0.938; $p<0.01$). The correlation coefficient between annual primary production and epilimnetic chlorophyll was significant (0.934; $p<0.01$; n = 6).

Fig. 5C shows the seasonal and year-to-year variations in the concentration of chlorophyll measured in the top 5 metres. In summer, these concentrations are strongly influenced by the concentration of bacteriochlorophyll originating from green sulphur bacteria in the hypolimnion. Average chlorophyll concentrations in the 0–5 m layer therefore reach a maximum during the anaerobic period and decreased rapidly in late August and early September. The interannual variations of bacteriochlorophyll in general followed those of primary production and epilimnetic chlorophyll. However, due to the long-lasting depletion of oxygen the hypolimnetic bacterial biomass was relatively high in 1993 (Tables 3 and 4; cf. Fig. 5).

5. Discussion

Lake Valkea-Kotinen is acid (mean pH 5.1–5.4), has a relatively low buffering capacity and is strongly coloured with humic materials (colour 103–149 g Pt m^{-3}; DOC 10–12 g m^{-3}). The epilimnion is consequently thin (2 m) and the hypolimnion is depleted of oxygen for most of the growing season. The water column is usually mixed in spring and autumn but in the spring of 1993 the circulation did not reach the bottom in the deepest point (6.5 m). An unusual pattern of mixing was also recorded in 1995 when the near-bottom water remained aerobic for 6 weeks in May and June, instead of the usual 2–3 week period.

An investigation of sedimented diatoms suggests that the pH of Valkea-Kotinen decreased by almost one unit between the turn of the century and the early 1970s [13]. The pH of the lake has, however, remained relatively stable throughout the 1990s. This period of stability can be explained partly by the reduction in atmospheric loading [15, 16] and partly by the stabilising effect of the anoxic hypolimnion. Prolonged summer anoxia generates alkalinity by sulphate reduction and retards acidification [14] but this trend could be reversed if there was a climatically-induced change in the pattern of vertical mixing.

Many small forest lakes in Finland are highly coloured, and have shallow epilimnia that are strongly stratified; a significant proportion of these waters have anoxic hypolimnia [17]. In most cases, the consumption of oxygen in the hypolimnion is due to sedimentation of allochthonous organic matter rather than internal production, but planktonic production rates are often surprisingly high.

Despite its shallow productive layer ($\leq$ 2 m), annual primary production in Valkea-Kotinen was two or even three times greater than that reported in several neighbouring lakes [3]. This high production rate consumed most of the nutrients present in the epilimnion in early summer, but high production rates were recorded later in the growing season, even when nutrient concentrations were low. In Valkea-Kotinen, summer production rates are largely sustained by the diel migration of flagellated algae from the epilimnion into the nutrient-rich hypolimnion [10]. Two flagellate genera (*Gonyostomum* and *Cryptomonas*) are common in Valkea-Kotinen during the summer. Both species migrate into the anaerobic hypolimnion at night in order to acquire nutrients before returning to the epilimnion in early morning [18].

The year-to-year variations in primary production and epilimnetic chlorophyll concentrations were not closely correlated with measured physical and chemical driving variables but could often be linked to changes in the weather. Phytoplankton production was, however, significantly positively correlated with epilimnetic oxygen concentration and negatively correlated with the concentration of dissolved inorganic carbon. Both primary production and chlorophyll concentration were negatively correlated with hypolimnetic total nitrogen and the concentration of chlorophyll was also negatively correlated with hypolimnetic ammonium nitrogen. One reason for the negative correlations of primary production and chlorophyll with nitrogen (ammonium, total N) may be a more thorough utilisation during those growing seasons when primary production was high and a large biomass of *Gonyostomum semen* developed. This is in agreement with the suggestions of Jones [19] that in humic waters nitrogen – in addition to phosphorus – is an important regulator of phytoplankton production.

This would also explain the low production in summer 1995, where the long-lasting spring circulation will have delayed the flux of nutrients from the anaerobic sediment into the hypolimnion. It is, however, not clear why production in 1993 was also fairly low, even though the sediment was anaerobic through the spring. One reason may be cold and rainy weather and thus a weakened illumination.

Variations in the means of the physical and chemical variables were not, however, closely correlated with year-to-year variations in primary production or the epilimnetic chlorophyll concentration. Any nutrients released into the water from the sediments are rapidly consumed by the algae and concentrations therefore remain low [3]. Some flagellates are also able to alter their migration behaviour with changing nutrient conditions [20]. Other survival strategies for algae are storage of nutrients in the cells [21–23], and mixotrophy, which means a smaller dependence on inorganic nutrients compared with pure autotrophy [24, 25].

The phytoplankters that dominate this small lake have clearly developed physiological and behavioural strategies that allow them to thrive in a hostile environment with strong acidity, poor illumination, poor nutrients and severe oxygen conditions. Year-to-year variations in primary production, chlorophyll concentration and physicochemical variables were marked, but no clear trends were observed during 1990–1995. The recorded seasonal changes in primary production and chlorophyll concentration were not only strongly influenced by the physical and chemical environment but also influenced this environment (e.g. by controlling DIC and O_2 concentrations). The years differed from each other in those environmental factors that affect productivity in Valkea-Kotinen (e.g. weather, washout, spring and autumn circulations), and consequently annual production varied markedly. The results from 1990–1995 give a good basis for understanding the ecosystem of Valkea-Kotinen, but much longer datasets are needed to make conclusions about possible long-term trends.

Acknowledgements

This study is part of the International Co-operative Programme on Integrated Monitoring of Air Pollution Effects on Ecosystems, and it was financed by the Finnish Ministry of the Environment. We wish to thank the staff of the Lammi Biological Station for providing facilities and technical assistance.

References

[1]. Talling, J. F. (1993). Comparative seasonal changes, and inter-annual variability and stability, in a 26-year record of total phytoplankton biomass in four English lake basins. *Hydrobiologia* **268**, 65-98.

[2]. Arvola, L. (1983). Primary production and phytoplankton in two small, polyhumic forest lakes in southern Finland. *Hydrobiologia* **101**, 105-110.

[3]. Arvola, L. (1984). Vertical distribution of primary production and phytoplankton in two small lakes with different humus concentration in southern Finland. *Holarctic Ecology* **7**, 390-398.

[4]. Arvola, L., Metsälä, T.-R., Similä, A. & Rask, M. (1990). Phyto- and zooplankton in relation to water pH and humic content in small lakes in Southern Finland. *Verh. Internat. Verein. Limnol.* **24**, 688-692.

[5]. Eloranta, P. (1978), Light penetration in different types of lakes in Central Finland. *Holarctic Ecology* **1**, 362-366.

[6]. Bergström, I., Mäkelä, K. & Starr, M. (Eds.) (1995). *Integrated monitoring programme in Finland, first national*¬ *report.* Ministry of the Environment, Environmental Policy Department, Helsinki. Report **1**, 138 pp. + appendices.

[7]. Mäkelä, K. (1995). Valkea-Kotinen; general features of the monitoring area. In: *Integrated monitoring programme in Finland, first national report* (eds. I. Bergström, K. Mäkelä & M. Starr). Ministry of the Environment, Environmental Policy Department, Helsinki. Report **1**, pp. 16-18.

[8]. Keskitalo, J. & Salonen, K. (1994). Manual for Integrated Monitoring. Sub-programme Hydrobiology of Lakes. *Publications of the Water and Environment Administration - Series B* **16**. 41 pp. National Board of Waters and the Environment, Helsinki.

[9]. Keskitalo, J. & Salonen, K. (1995). Water chemistry, Lake Valkea-Kotinen. In: *Integrated monitoring programme in Finland, first national report* (eds. I. Bergström, K. Mäkelä & M. Starr. Ministry of the Environment, Environmental Policy Department, Helsinki. Report **1**, pp. 76-78.

[10]. Keskitalo, J., Salonen, K. & Holopainen, A.-L. (1995). Plankton. In: *Integrated monitoring programme in Finland, first national report* (eds. I. Bergström, K. Mäkelä & M. Starr. Ministry of the Environment, Environmental Policy Department, Helsinki. Report **1**, pp. 102-110.

[11]. Schindler, D. W., Schmidt, R. V. & Reid, R. A. (1972). Acidification and bubbling as an alternative to filtration in determining phytoplankton production by the ^{14}C method. *Journal of the Fisheries Research Board of Canada* **29**, 1627-1631.

[12]. Niemi, M., Kuparinen, J., Uusi-Rauva, A. & Korhonen, K. (1983). Preparation of algal samples for liquid scintillation counting. *Hydrobiologia* **106**, 149-156.

[13]. Liukkonen, M. (1989). Latvajärvien happamoituminen Suomessa sedimentoituneen piilevästön osoittamana [In Finnish]. MSc Thesis, Dept. Bot., University of Helsinki. 147 pp. + appendices.

[14]. Cook, R. B., Kelly, C. A., Schindler, D. W. & Turner, M. A. (1986). Mechanisms of hydrogen ion neutralization in an experimentally acidified lake. *Limnology & Oceanography* **31**, 134-148.

[15]. Söderman, G. & Dahlbo, K. (1990). *Tuloksia Suomen ympäristön yhdennetystä seurannasta kaudelta 1988/89* [In Finnish]. National Board of Waters and the Environment, Helsinki. 25 pp.

[16]. Leinonen, L. (Ed.) (1994). *Ilmanlaatumittauksia — Air quality measurements 1993*. Finnish Meteorological Institute, Helsinki. 245 pp.

[17]. Salonen, K., Järvinen, M., Kuoppamäki, K. & Arvola, L. (1990). Effects of liming on the chemistry and biology of a small acid humic lake. In: *Acidification in Finland* (eds. P. Kauppi, K. Kenttämies & P. Anttila). Springer-Verlag, Berlin, Heidelberg, pp. 1145-1167.

[18]. Salonen, K., Arvola, L. & Rosenberg, M. (1993). Diel vertical migration of phyto- and zooplankton in a small steeply stratified humic lake with low nutrient concentration. *Verh. Internat. Verein. Limnol.* **25**, 539-543.

[19]. Jones, R. I. (1992). Phosphorus transformations in the epilimnia of small humic forest lakes. *Hydrobiologia* **243/244**, 105-111.

[20]. Arvola, L., Ojala, A., Barbosa, F. & Heaney, S. I. (1991). Migration behaviour of three cryptophytes in relation to environmental gradients: an experimental approach. *British Phycological Journal* **26**, 361-373.

[21]. Mackereth, F. J. H. (1953). Phosphorus utilization by *Asterionella formosa* Hass. *Journal of Experimental Botany* **4**, 296-313.

[22]. van Donk, E. & Kilham, S. S. (1990). Temperature effects on silicon- and phosphorus-limited growth and competitive interactions among three diatoms. *Journal of Phycology* **26**, 40-50.

[23]. Chapman, A. D. & Pfiester, L. A. (1995). The effects of temperature, irradiance, and nitrogen on the encystment and growth of the freshwater dinoflagellates *Peridinium cinctum* and *P. willei* in culture (Dinophyceae). *Journal of Phycology* **31**, 355-359.

[24]. Salonen, K. & Jokinen, S. (1988). Flagellate grazing on bacteria in a small dystrophic lake. *Hydrobiologia* **161**, 203-209.

[25]. Bird, D. F. & Kalff, J. (1987). Algal phagotrophy: regulating factors and importance relative to photosynthesis in *Dinobryon* (Chrysophyceae). *Limnology and Oceanography* **32**, 277-284.

SUDDEN AND GRADUAL RESPONSES OF PHYTOPLANKTON TO GLOBAL CLIMATE CHANGE: CASE STUDIES FROM TWO LARGE, SHALLOW LAKES (BALATON, HUNGARY, AND THE NEUSIEDLERSEE, AUSTRIA/HUNGARY)

JUDIT PADISÁK
Balaton Limnological Institute
Hungarian Academy of Science
H-8237 Tihany, Hungary

Abstract

This paper analyses two phytoplankton long-term datasets; both are from large, temperate shallow lakes. The main difference between them is that phytoplankton growth in Lake Balaton remained severely P-limited despite P-driven eutrophication during the last 30 years, whereas extremely high turbidity causes a permanant light limitation in Neusiedlersee and therefore an increase in P-loadings did not result in a similar increase in phytoplankton biomass. Neusiedlersee is a (slightly) saline inland lake. In Lake Balaton, the blue-green alga *Cylindrospermopsis raciborskii* blooms invariably if the July–August temperature deviates positively from a 30-year average by ca. 2°C. A supposed global warming is predicted to cause a higher frequency (but not intensity!) of these blooms. Neusiedlersee is very shallow and therefore regulation techniques cannot prevent water levels sinking in successive dry years. Annual averages of phytoplankton seem to follow quite a regular, wave-like cyclicity. Such cycles can be recognised in the population records of the characteristic species. Similar changes were seen in changes of water level, conductivity, inorganic-P, inorganic N-forms and nutrient ratios. How phytoplankton species can follow a climatic cycle that covers 200 to 500 generations has not yet become clear. Because of reasons discussed in the paper, neither of the two cases can be generalised; each is quite individual.

1. Introduction

Understanding the effects of environmental change on different kinds of ecosystems is hardly possible without good scaling, which involves matching biotic responses to the

D.G. George et al.(eds.), Management of Lakes and Reservoirs during Global Climate Change, 111–125.

appropriate level of environmental variability. "Global climate change" is a relatively new term. Consequently, it is frequently imprecisely defined and is rarely qualified temporally or spatially. It can be accepted that global change has large spatial dimensions: roughly, many or most of the world's ecosystems, both terrestrial and aquatic, are affected. Concerning time-scales, global change is expected to influence the life of many future human generations, at scales of decades to centuries or even millenia. Global climate change is usually anticipated to be a slow, trendlike process.

Phytoplankton species live for some days. Population changes occur in time-spans of days to weeks. Some months are sufficient to observe major successions. In the time-span of a decade or longer, floristic changes in phytoplankton can take place and evolutionary adaptation may occur. If we accept that global change occurs over a period of some years, we should therefore expect the reactions of the phytoplankton to be expressed at the levels of "floristic change" or "evolutionary adaptation" [1]. Logic says that global change is too slow and gradual to have any impact at the level of phytoplankton population dynamics or of seasonal succession.

In this paper, I attempt to predict phytoplankton responses to global climate change, by extrapolation from observations in two ongoing, long-term studies. One of them (Lake Balaton, Hungary) shows us that despite the considerations presented above, a global warming can have effects on phytoplankton population dynamics within individual calendar years. Moreover, this signal is distinguishable from a prevalent effect of eutrophication and subsequent recovery on the phytoplankton. The other (Neusiedlersee, Austria/Hungary) invokes a 27-year study of phytoplankton and the changes which follow a 6- to 10-year cyclicity related to fluctuations in the hydrological conditions.

The data used in this paper, many of which have been published previously, were obtained by widely accepted standard samplings and methods (see [2] to [6]).

2. Lake Balaton: occurrence and population dynamics of *Cylindrospermopsis raciborskii*

Lake Balaton (Hungary) is the largest shallow lake in central Europe. The lake has a surface area of 593 km^2, is 77.9 km long, 7.2 km wide on average (maximum width 15 km) and has a mean depth of 3.14 m (maximum 11 m). The theoretical retention time is 3 to 8 years. The lake was originally mesotrophic but underwent rapid eutrophication during the 1960s and 1970s [7] as a consequence of increased P-loadings. Because the majority of the nutrient load is received by the western basin, a sharp trophic gradient developed in the elongated lake. Even in the less eutrophic part of the lake, annual average phytoplankton biomass increased significantly when compared with early records (Fig. 1). Nevertheless, phytoplankton growth in the lake remains strongly P-deficient throughout the year [8]. The lake sediment is significantly enriched with phosphorus [9].

Summer phytoplankton of Balaton used to be dominated by *Snowella lacustris, Ceratium hirundinella* and centric diatoms (especially *Aulacoseira* spp.). As eutrophication proceeded, heterocytic blue-green algae became increasingly prominent in the late summer phytoplankton. The first to become prominent had been recorded while the lake was still mesotrophic (*Aphanizomenon flos-aquae, Anabaena* spp.). Since the beginning of the 1970s, new species of heterocytic blue-green algae (*Aphanizomenon issatschenkoi, Rhaphidiopsis meditarranea*) have appeared in the flora [10, 11].

A large-scale restoration programme in the early 1980s has arrested and begun to reverse the eutrophication. So far, the programme has resulted in about 50–60% reduction of the biologically available P-load of the lake. In the hypertrophic (western) basin of the lake, the P-load reduction is almost 80%. Annual average biomass of the above heterocytic blue-green algae in the hypertrophic part of the lake decreased in parallel with the load-reduction [6], suggesting that these species utilize P-sources originating mainly from external (catchment) sources.

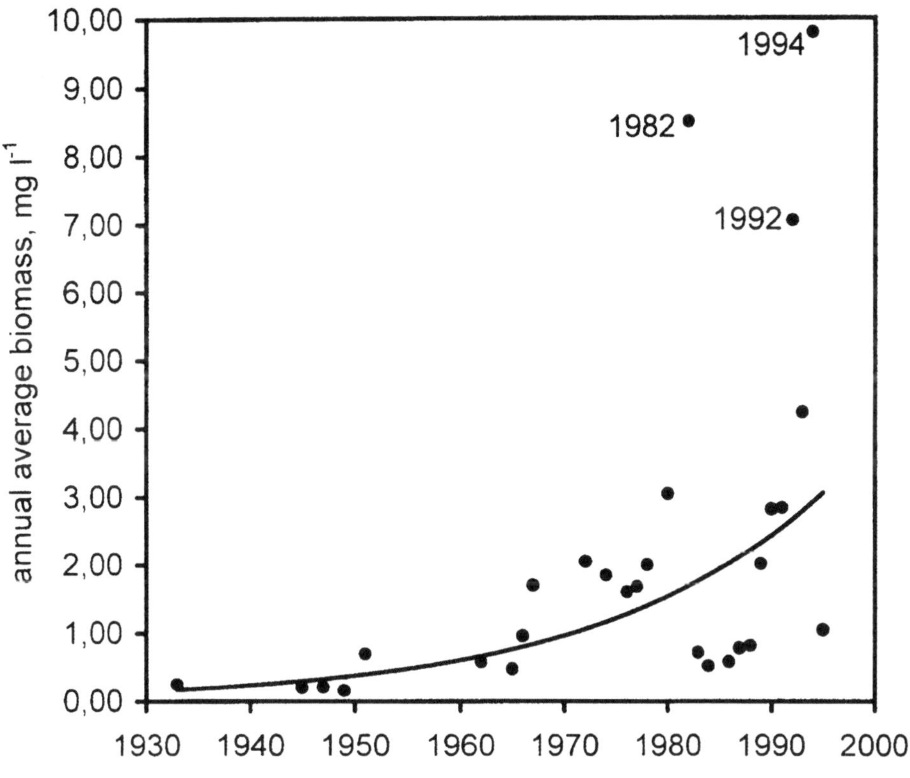

FIGURE 1. Annual average phytoplankton biomass (mg l^{-1}) between 1933 and 1994 in the northeastern part of Lake Balaton. A continuous trendline indicates the eutrophication process.

Cylindrospermopsis raciborskii Wolosz. is a heterocytic, nitrogen-fixing blue-green alga. The species was recorded almost simultaneously at several localities in Hungary during the 1970s [12–14]. *C. raciborskii* appeared in Balaton in 1979 [15]. In late summer and autumn of 1982, a heavy bloom swept through the lake, peaking at 10^8 trichomes l^{-1} in the northeastern part of the lake [16]. Similar blooms occurred in the summers of 1992 and 1994 (Fig 2; [6]). In these years, late summer phytoplankton maxima of 35–50 mg l^{-1} were observed in the northeastern part of the lake. In other years summer maxima did not exceed 15 mg l^{-1} [4], a level consistent with mesotrophy.

Concerning the ecology of *C. raciborskii* in Balaton, two peculiar features need explanation. One is the irregular nature of blooms and the other is its apparent independence of prevailing biologically available P concentrations and external P-loadings. So far as meeting its phosphorus requirement is concerned, *Cylindrospermopsis* blooms may draw sufficient from the sediment store of phosphorus during the period of its overwintering [6]. Indirect evidence supports the hypothesis that, as in the case of *Gloeotrichia echinulata* [17], phosphorus assimilation and planktonic growth of *C. raciborskii* may become spatially and temporally uncoupled.

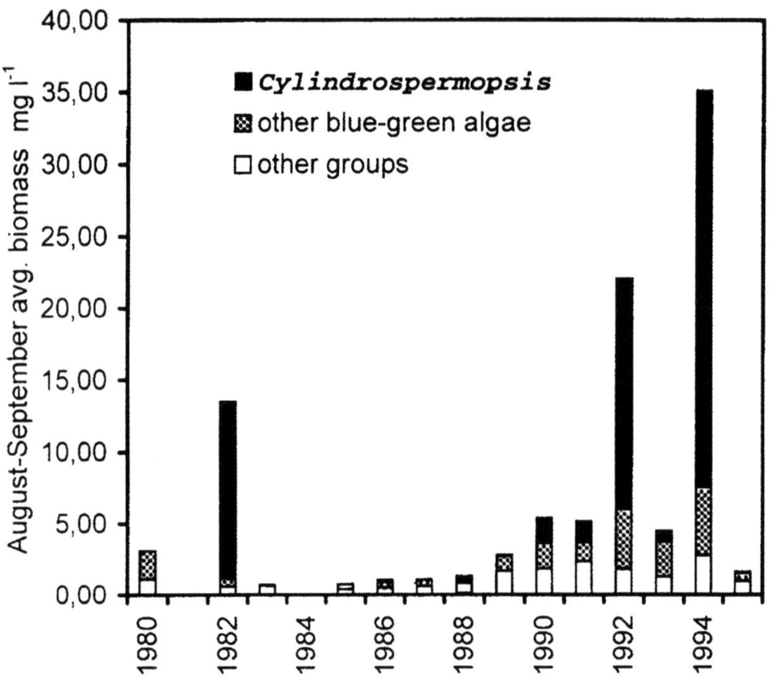

FIGURE 2. August–September average biomass (mg l^{-1}) of *Cylindrospermopsis raciborskii*, other blue-green algae and other phytoplankton groups in the northeastern basin of Lake Balaton between 1980 and 1995.

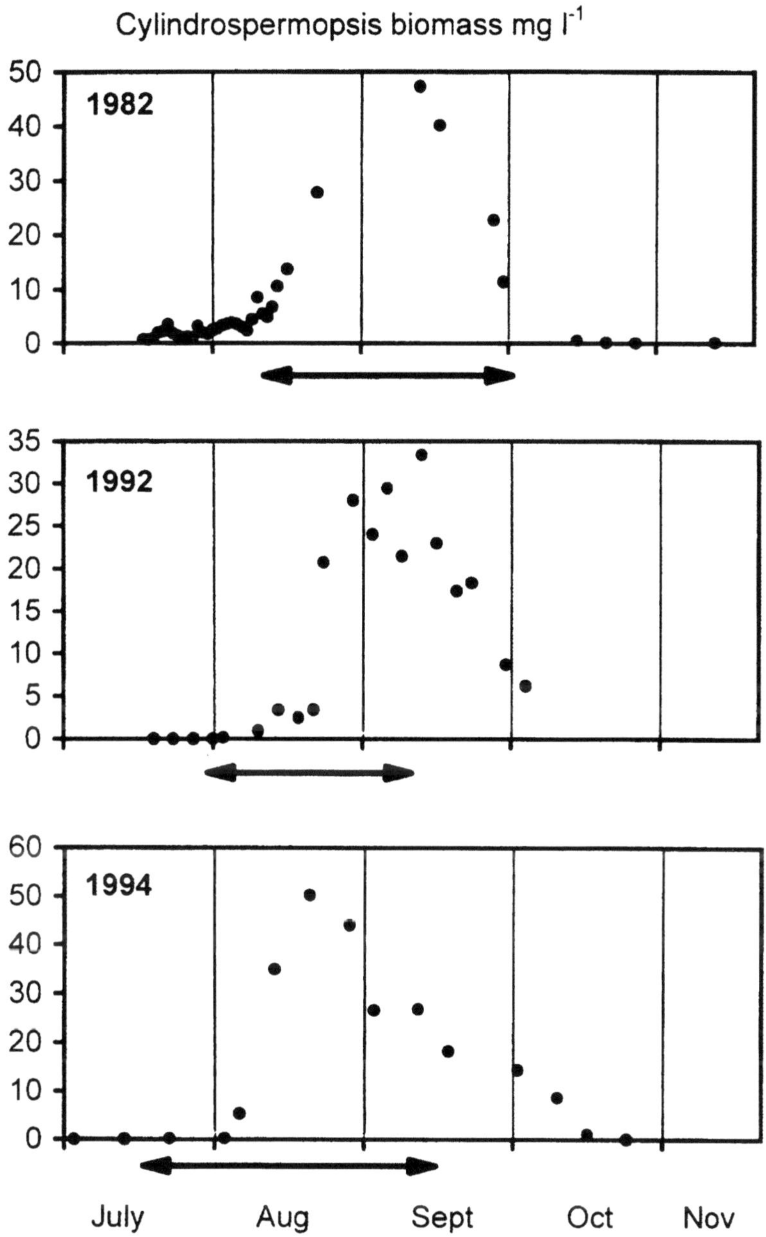

FIGURE. 3. Development and decline of *Cylindrospermopsis raciborskii* blooms (biomass, mg l^{-1}) in the northeastern part of Lake Balaton in the summers of 1982, 1992 and 1994. Arrows indicate the extension of the continuously calm, hot periods

Irregularity of the *C. raciborskii* blooms in Balaton owes more to interannual variations in summer weather conditions. The summers when the species bloomed were unusually hot. Numerous records, both from field observations and laboratory studies, point to the high temperature requirement of *C. raciborskii*. Indeed the species, first described in the Indo-Malayan flora, has for long been regarded as a tropical species.

In the *Cylindrospermopsis* summers, the exponential phase of growth started roughly two weeks after the onset of the hot, calm weather (Fig. 3). The August–September average biomass of *Cylindrospermopsis* was high only in years when the average air temperatures of two subsequent summer months (July–August or August–September) exceeded the 30-year average by at least 2°C (Fig. 4). There is no linear correlation between temperature deviation and *Cylindrospermopsis* biomass; if the former exceeds a certain degree the bloom occurs invariably (see arrows on Fig. 4).

Laboratory studies [18] showed that akinetes of *Cyilindrospermopsis* differ from those of other blue-green algae in Balaton. Akinetes of other species germinate over a broad temperature range (16 to 27°C). The temperature optimum of *Cylindrospermopsis* is narrow; akinetes germinate best at 21–22°C. Such high sediment temperatures occur in the temperate Balaton only in exceptionally hot years. There is thus strong circumstantial evidence for regarding the occurrence of *Cylindrospermopsis* blooms in Balaton as a primary response to above-average temperatures at the location.

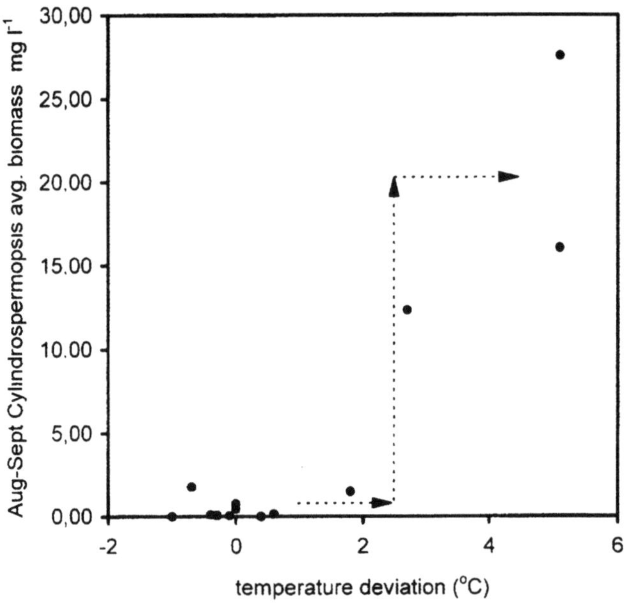

FIGURE 4. August–September averages (1982–1995) of *Cylindrospermosis raciborskii* biomass (mg l^{-1}) in Lake Balaton, plotted against the deviation of August–September (1982–1995) average air temperatures from a 30-year average. Zero on the abscissa corresponds to average summer; negative values indicate cold summers and positive values indicate warm summers.

3. Neusiedlersee: long-term population dynamics of dominant species

The Neusiedlersee (Fertô) is a large shallow lake on the Austrian-Hungarian border. Its surface area is 300 km^2; it is 35 km long, 8.6 km wide on average, and has a mean depth of 1.3 m (maximum 1.8 m). The theoretical retention time is 3 years. Because the lake does not have a natural outlet and most of the lakewater is resupplied from precipitation and groundwater, the water level has fluctuated within a wide range during the 20 thousand years of lake history. The last occasion on which the lake completely dried out was between 1868 and 1872. After the basin refilled, the reed-belt reappeared on the shores and progressively increased towards the centre of the lake basin during the last 100 years. For the last 30 years or so, approximately one-third of the lake area has been covered by reed-stands. The water level of the lake can be regulated by opening and closing the artificial outlet (Hanság csatorna/Einser Kanal) in the Hungarian part of the lake. This regulation can prevent high water levels during rainy years, but fails to avoid low water levels during dry periods. Consequently, each biotic community, from higher plants to plankton, has been subjected to fluctuations in water levels caused by climatic changes.

The mesotrophic lake has a high salt-content, is alkaline and very turbid. Conductivity ranges from 2000 to 3500 μS cm^{-1}, alkalinity is 8.0–10.5 meq l^{-1}, pH is 7.5–10. Secchi-disc transparency in the open water is characteristically ca. 0.2 m (range 0.06 to 0.8 m; higher values occur only under ice). The lake sediment is characterised by small, slowly settling, fine-grained inorganic particles. The average seston content of the lake is very high (80–100 mg l^{-1}; comparable high values are found only in flooding lowland rivers). The water is rich in silica, and the concentrations of dissolved N and P (average for 1987–1992: 297 μg-N l^{-1} and 15 μg-P l^{-1}) are usually higher than the level that is considered limiting in other lakes. Because of the inherently high turbidity of the open water of the lake, growth of phytoplankton is presumably subject to frequent or continuous light limitation, with only brief exceptions.

The short-term and seasonal changes of dominant species in Neusiedlersee are unusual, really corresponding to the dynamics of meroplankton [5]. Although the lake is quite a large one, horizontal differences in phytoplankton distribution are slight and temporary (see Figs 2 and 3 in [5]).

Hundreds of algal species are described from different localities in the extended reed-belt of Neusiedlersee but the open water of the lake is relatively poor in phytoplankton species: no more than several dozen species are noted in the annual cycle; for most lakes, the list might run to one or two hundreds. The exacting environmental constraints of turbidity and salinity are doubtless implicated. The most important and regularly occurring species are listed in Table 1.

There are two notable features in the long-term development of phytoplankton. One is the regular cyclicity of biomass development, and the other is a sharp change in the composition during the early 1970s. The overwhelming contribution of diatoms and coccal green algae, and the insignificant fraction of buoyant species, are the most characteristic quantitative features of the phytoplankton composition (Fig. 5). In this respect, the phytoplankton resembles that of a lowland river [19]. Diatoms usually contribute the larger part of the annual average biomass, but in two of the last 27

TABLE 1. Characteristic species from the phytoplankton of Neusiedlersee.

Campylodiscus clypeus Ehr.	*Koliella* sp.
C. clypeus var. *bicosta* (W. Smith) Hustedt	*Lobocystis planktonica* (Tiff. & Ahlstr.) Fott
Chaetoceros muellerii Lemm.	*Microcystis aeruginosa* Kütz.
Chroococcus limneticus Lemm.	*M. wesenbergii* Komárek
Ch. minutus (Kütz.) Nägeli	cf. *Merismopedia minima* Beck.
Cryptomonas erosa/ovata	*Monoraphidium contortum* (Thur.) Kom.-Legn.
Crucigenia quadrata Morren	*M. pseudobraunii* (Belch. & Schwale) Heynig
Chrysochromulina parva Lackey	Small *Navicula* spp.
Cyclotella meneghiniana Kütz.	*Neglectella peisonis* Schagerl
Dicttyosphaerium pulchellum (Wood)	Small *Nitzschia* spp.
Elakatothrix lacustris Kors.	*Oocystis lacustris* Chodat
Euglena oxyuris Schmarda	*Pediastrum duplex* Meyen
E. tripteris (Duj.) Klebs	*Phacus pyrum* (Ehr.) Stein
Fragilaria acus Kütz.	*Planktosphaeria gelatinosa* G. M. Smith
F. brevistriata Grunow	*Rhodomonas minuta/lacustris*
F. construens (Ehr.) Grunow	*Sphaerocystis schroeterii* Chodat
Small *Fragilaria* spp.	*Surirella peisonis* Pantocsek
Small centric diatoms	

years (1975, 1976), green algae (especially *Pediastrum duplex*) were dominant. In the late 1970s, blue-green algae (first of all *Microcystis* spp.) contributed significantly to the annual biomass. These changes appear quite abruptly on compositional plots (see Fig. 13a in [5]) and they may reflect human impacts on the catchment area. Until the late 1960s, phytoplankton growth was most probably limited both by phosphorus deficiency and insufficient light. P-loadings to the lake increased sharply in the 1970s (this trend was reversed by the 1980s); moreover, large open-water stands of submerged macrophytes were extinguished. At the same time, there have been unusual variations in the annual distribution of rainfall, with summer rains tending to maintain water levels into autumn, thus offsetting the normal pattern of an autumnal diminution. It is not yet possible to separate the physical and chemical influences sufficiently to interpret the precise mechanisms underpinning these short-term compositional responses. These fluctuations are nevertheless superimposed upon a larger-scale cycle of biomass fluctuations in Neusiedlersee. In the period 1968 to 1994, there have been three distinct peaks in the phytoplankton biomass supported in the lake: the first was in 1975, the second in 1983 and the third in 1991 (Fig. 5). The 8-year cycle is distinctly wave-like: in this way, years with above average biomass do not appear stochastically or "out of trend", as they do in Balaton (see above). The cycle is echoed in the variations in the annual averages of each of the constant taxa of phytoplankton (Fig. 6). However, they do not each coincide precisely with the 8-year cycle of total biomass maxima. This observation might explain the findings of previous investigations to the effect that phytoplankton composition in Neusiedlersee varies enormously with spontaneous year-to-year appearances and disappearances [20].

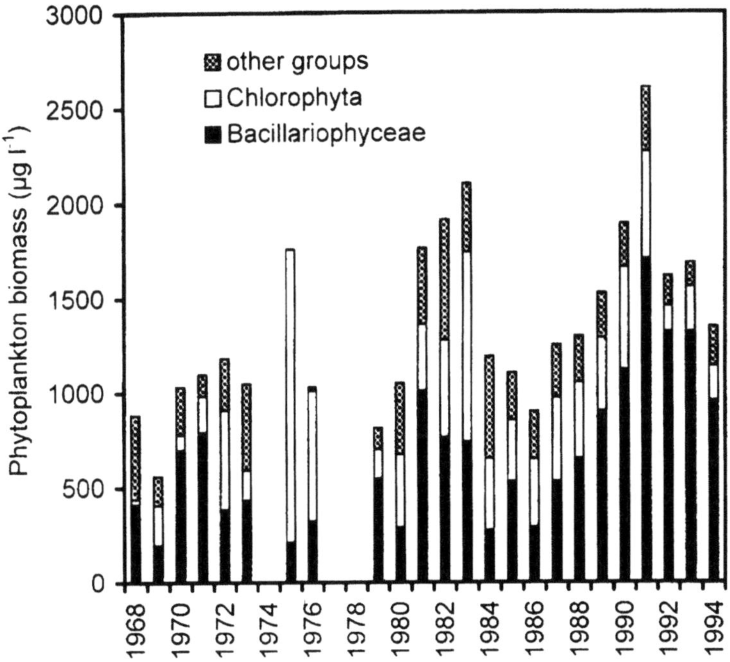

FIGURE. 5. Annual average biomass (µg l^{-1}) of diatoms, green algae and other groups in Neusiedlersee between 1968 and 1994.

Some descriptions of the plankton prior to 1968 are available. For example, *Monoraphidium contortum* was the dominant green alga in 1956–58 with cell numbers up to 3 million cells l^{-1} [21]; this corresponds to a biomass of ca. 300 µg l^{-1} (cf. Fig 2a, b). The species rose progressively to dominance at the beginning of the 1970s, with a peak of 8500 µg l^{-1} in June 1972. During subsequent years its abundance fell significantly, large populations being found only in the colder seasons. In October 1982, however, the species was found to be dominant in many of the brown-water lagoons enclosed within the reed-belt of the lake [22]. *Pediastrum duplex* was found only in these reed-belt habitats during the 1960s [20], the first record of its planktonic occurrence not coming until 1969. This species predominated in 1975 when it represented some 80% of the total annual biomass. A peak abundance of 6440 µg l^{-1} was noted for 28 July 1975. Since that time no comparable high biomasses have been recorded, although several specimens were recovered in every year since. Hustedt [23] commented that the diatom, *Chaetoceros muelleri,* was one of the most important species in both the plankton and the littoral zone of the lake. More recently it has been found exclusively in littoral samples [22, 24, 25]. Between 1987 and 1994, however, quite large planktonic populations of *Chaetoceros* developed almost every year in

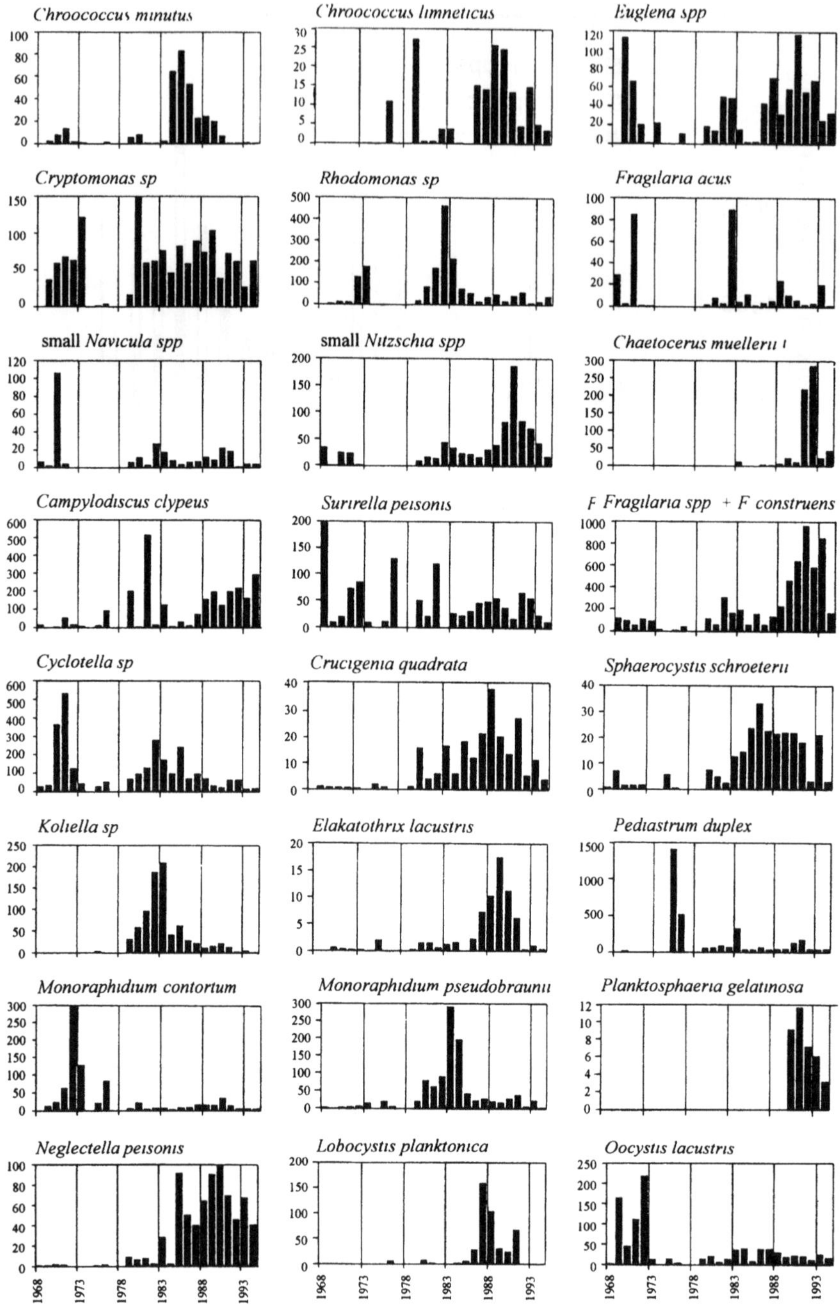

FIGURE 6. Annual average biomass (μg l^{-1}) of the most important phytoplankton species in Neusiedlersee between 1968 and 1994.

one or more of the reed-belt lagoons. Detailed data on *Chaetoceros* population dynamics are given in [5].

This is the best of several examples suggesting that many algal species have refuges in the littoral swamps but periodically penetrate the open water to constitute large or dominant planktonic populations (Fig. 6). Both pilot [22] and systematic [2, 3] studies have shown that phytoplankters in small enclosed lakes differ significantly not only from those in the open water but from each other as well. With more intensive studies, probably, more "*Chaetoceros*-like" examples would be revealed. The testable hypothesis is advanced that the supra-annual cycles of abundances of phytoplankton in Neusiedlersee are related to large-scale hydrological changes which alternately reduce the range of species to the reedswamp refuges and expand them periodically to the open water of the lake.

This hypothesis is not invalidated by attempts to correlate the phytoplankton data with the physical and chemical variability (these have failed to show any compelling relationship, save one with the volume of suspended matter: see discussion in [5]), and by the fact that the wave-like trend in annual average phytoplankton biomass mirrors the fluctuations in water level (Fig. 7). The possibility of another proximate variable is not excluded, for instance, through the influence of salinity changes, although these are known to fluctuate simultaneously (albeit with differing damping) in the swamps and the open water. Changes in turbidity and the relative penetration of light may be crucial to the success of species in the open water, in which case these too could be related primarily to changes in water level.

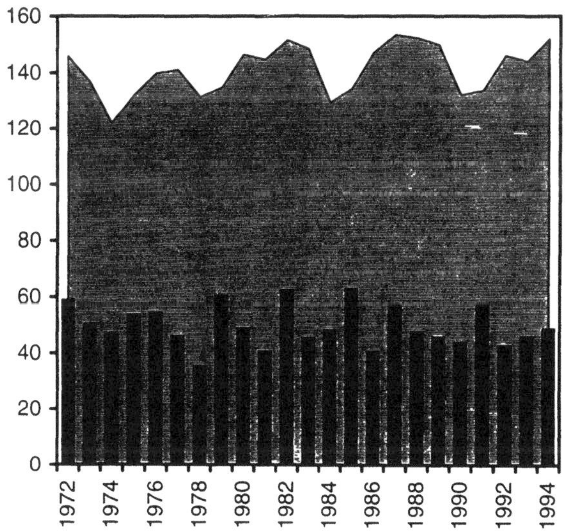

FIGURE 7. Annual average water level (m a.s.l.; shaded area) in Neusiedlersee and monthly mean precipitation (mm; solid verical bars). Data from the Meterorological Station Illmitz, 1972 to 1994.

4. General discussion

The above two case studies describe quite different phytoplankton responses to external climatic forcing, supposed to be subject to revision as a result of global modification. It is reasonable to suppose that global warming might mean an increasing frequency of years with positive temperature deviations of greater magnitudes. In this case, the observations on the ecology of *Cylindrospermopsis* blooms in Lake Balaton permit a trend of higher and more persistent biomass peaks to be forecast. July–August temperatures 2°C higher than present might well sustain regular *Cylindrospermopsis* blooms, independent from changes in fertility so long as the P-pool in the sediment remains active. However, this might be rather an exceptional example to take: not all of the species of phytoplankters in the lake have such a distinctive and crucial autecology or respond everywhere for the same reason and with the same sensitivity. Thus it would be unreasonable to apply the same explanation to the dominance of *C. raciborskii* in the tropical, oligotrophic Lake Kariba (Africa) in the rainy season (see [26]).

The behaviour of phytoplankton in Neusiedlersee is, or at least used to be, possibly more generalised. The lake is astatic with no natural outlet. Although a canal was constructed to regulate top water level, it does not protect the lake from volume fluctuations attendant upon its shallowness and large surface area, especially in dry years. Astatic lakes were more common in the past (for example, Lake Balaton was one of them), until drainage works in the 19th and 20th centuries brought some stabilisation about the lower natural limits. In consequence, the case of Neusiedlersee is yet more exceptional. The oscillation of drier to wetter years in this region has a period of 6 to 10 years. The doubling times of characteristic species of phytoplankton are mainly less than 5 to 7 days during the vegetation period [5]. Thus, the cycle of climatic change accommodates 200 to 500 generations of planktonic algae. It is far from clear how phytoplankton species might adapt to such gradual climatic changes so precisely or how the dominance of any might be driven by competition; it is far more likely that different and remotely separated climatic conditions will favour the performances of algae from among those present at the time.

It has been demonstrated several times, for example in Tilman's [27] experiments on competitive exclusion and, quite recently, in testing and extending Connell's intermediate distrubance hypothesis to the short life-times of planktonic organisms [28–31], that the rapidity of phytoplankton population responses to environmental forcing lent great advantages to their study, in nature and in experimental manipulations. Responses to global climate change may be exceptional in this respect: the 7-to 10-year cycles in the abundance and rarity of species are not abrupt, equalling in real time the establishment of pioneer terrestrial vegetation. The dataset from Neusiedlersee also shows us that slow, small and gradual global changes, drive changes in the dynamics of phytoplankters (wave-like overall trend), but these are nevertheless sensitive to local human activities.

Acknowledgements

I thank Mrs Zsuzsa Z.-Doma, Mr Csaba Marton, Mr István Báthory, Mr Géza Dobos (Balaton Limnological Institute, Tihany), Mr Robert Klein and Mr Franz Rauchwarter (Biologische Station, Illmitz), for their essential support during the field, laboratory and computer work. The water chemical data (Neusiedlersee) used in this paper are from the archives of the Biologische Station Illmitz, Burgenland, Austria. Phytoplankton data (Balaton) between 1983 and 1988 were made available by the Middle Transdanubian District Water Authority (Székesfehérvár). Meterorological data are from the Havi Időjárásjelentés (Monthly Weather Report) published by the Hungarian Meteorological Service. This work was supported by the Hungarian National Science Foundation (Project OTKA No. 6285).

References

[1]. Reynolds, C. S. (1990). Temporal scales of variablility in pelagic environments and the response of phytoplankton. *Freshwater Biology* **23**, 25-53.

[2]. Padisák, J. (1993). Species composition, spatial distribution, and the seasonal and interannual dynamics of phytoplankton in brown-water lakes enclosed with reed belts (Neusiedlersee/Fertő; Austria/Hungary). *BFB-Bericht* **79**, 13-29.

[3]. Padisák, J. (1993). Dynamics of phytoplankton enclosed with reed belts (Neusiedlersee/Fertő; Austria/Hungary). *Verh. Internat. Verein. Limnol.* **25**, 675-679.

[4]. Padisák, J. (1994). Relationships between short-term and long-term responses of phytoplankton to eutrophication of the largest shallow lake in Central Europe (Balaton, Hungary). In: *Environmental protection and lake ecosystems* (eds. H. Sund, H.-H. Geller, W. Xiaogan, Y. Kechang & S. Fengnind). Science and Technol. Press, Beijing, pp. 419-437.

[5]. Padisák, J. & Dokulil, M. (1994). Meroplankton dynamics in a saline, turbulent, turbid shallow lake (Neusiedlersee, Austria and Hungary). *Hydrobiologia* **289**, 23-42.

[6]. Padisák, J. & Istvánovics, V. (in press). Differential response of blue-green algal groups to phosphorus load reduction in a large shallow lake: Balaton, Hungary. *Verh. Internat. Verein. Limnol.* **26**.

[7]. Herodek, S. (1984). The eutrophication of Lake Balaton: Measurements, modelling and management. *Verh. Internat. Verein. Limnol.* **22**, 1087-1091.

[8]. Istvánovics, V. & Herodek, S. (1995). Estimation of net uptake and leakage rates of orthophosphate from ^{32}P uptake kinetics by a linear force-flow model. *Limnology and Oceanography* **40**, 17-32.

[9]. Herodek, S. & Istvánovics, V. (1986). Mobility of phosphorus fraction in the sediments of Lake Balaton. *Hydrobiologia* **135**, 149-154.

[10]. Bartha, Zs. (1974). The occurrence of *Aphanizomenon issatschenkoi* (Ussaczew) Proschkina-Lavrenko in Lake Balaton. *Annales Instituti Biologici (Tihany) Hungaricae Academiae Scientarium* **41**, 127-131.

[11]. Tamás, G. (1974). The occurrence of *Rhaphidiopsis mediterranea* Skuja in the plankton of Lake Balaton. *Annales Instituti Biologici (Tihany) Hungaricae Academiae Scientarium* **41**, 317-321.

[12]. Schmidt, A. (1977). Adatok Dél-magyarországi vizek algáinak ismeretéhez I. (Contributions to the knowledge about algae of waters of southern Hungary I.). *Botanikai Közlemények* **64**, 183-196.

[13]. Hamar, J. (1977). Data on the knowledge of the blue-green alga *Anabaenopsis raciborskii* Wolosz. *Tiscia (Szeged)* **12**, 17-20.

[14]. Horecka, M. & Komárek, J. (1979). Taxonomic position of three planktonic blue-green algae from genera *Aphanizomenon* and *Cylindrospermopsis*. *Preslia* **51**, 289-312.

[15]. Oláh, J., ElSamra, M. I., Abdel-Moneim, M. A., Tóth, L. & Vörös, L. (1981). Nitrogénkötés halhústermelô agroökoszisztémákban [Nitrogen fixation in fish producing agro-ecosystems]. *A Halhústermelés Fejlesztése* **10**, HAKI, Szarvas. [In Hungarian].

[16]. G.-Tóth, L. & Padisák, J. (1986). Meteorological factors affecting the bloom of *Anabaenopsis raciborskii* Wolosz. (Cyanophyta: Hormogonales) in the shallow Lake Balaton, Hungary. *J. Plankton Research* **8**, 353-363.

[17]. Istvánovics, V., Petterson, K., Rodrgio, M. A., Pierson, D., Padisák, J. & Colom, W. (1993). *Gloeotrichia echinulata*, a colonial cyanobacterium with a unique phosphorus uptake and life strategy. *J. Plankton Research* **15**, 531-552.

[18]. Gorzó, Gy. (1987). Fizikai és kémiai hatások a Balatonban elõforduló heterocisztás cyanobaktériumok spóráinak csírázására [The influence of physical and chemical factors on the germination of spores of heterocystic cyanobacteria in Lake Balaton]. *Hidrológiai Közlöny* **67**, 127-133. [In Hungarian with English summary].

[19]. Reynolds, C. S., Descy, J.-P. & Padisák, J. (1994). Are phytoplankton dynamics in rivers so different from those in shallow lakes? *Hydrobiologia* **289**, 1-7.

[20]. Kusel-Fetzmann, E. (1979). The algal vegetation of Neusiedlersee. In: *Neusiedlersee. Limnology of a shallow lake in Central Europe* (ed. H. Löffler), pp. 171-202.

[21]. Ruttner-Kolisko, A. & Ruttner, F. (1959). Zusammenfassung und allgemeine Limnologie. *Wiss. Arb. Burgenland* **23**, 195-201.

[22]. Padisák, J. (1983). A comparison between phytoplankton of some brown-water lakes enclosed with reed belt in the Hungarian part of Lake Fertõ. *BFB-Bericht* **47**, 133-155.

[23]. Hustedt, F. (1959). Die Diatomeenflora des Neusiedler Sees im österrreichishen Burgenland. *Öst. Bot. Z.* **106**, 390-430.

[24]. Bartalis, É. T. (1987-88). Egyes meteorológiai tényezôk hatása a Fertôrákosi öböl biológiai vízminôségére és fitoplankton összetételére [Influence of some meteorological factors on the water quality and phytoplankton compositions of the Fertörákos-Bay]. *Botanikai Közlemények* **74-75**, 153-192. [In Hungarian with English summary].

[25]. Padisák, J. (1984). The algal flora and phytoplankton biomass of the Hungarian part of Lake Fertô I.: Rákosi-bay. *BFB-Bericht* **51**, 17-29.

[26]. Ramberg, L. (1987). Phytoplankton succession in the Sanyati basin, Lake Kariba. *Hydrobiologia* **153**, 193-202.

[27]. Tilman, D. (1982). *Resource competition and community strucure.* Princeton University Press, Princeton.

[28]. Padisák, J., Reynolds, C. S. & Sommer, U. (1993). Intermediate Disturbance Hypothesis in Phytoplankton Ecology. *Developments in Hydrobiology* **81**, Kluwer Acad. Publ., Dordrecht, Boston, London. 199 pp. [Reprinted from *Hydrobiologia* **249**, 135-156].

[29]. Padisák, J. (1994). Identification of relevant time-scales in non-equilibrium population dynamics: conclusions from phytoplankton surveys. *New Zealand Journal of Ecology* **18**, 169-176.

[30]. Wilson, J.B. (1994). The intermediate disturbancc hypothesis of species co-existence is based on patch dynamics. *New Zealand Journal of Ecology* **18**, 176-181.

[31]. Reynolds, C. S. (1995). The intermediate disturbance hypothesis and its applicability to planktonic communities: comments on the views of Padisák and Wilson. *New Zealand Journal of Ecology* **19**, 219-225.

IMPACT OF CLIMATE-AFFECTED FLOWRATES ON PHYTOPLANKTON DYNAMICS IN DAM RESERVOIRS

BLANKA DESORTOVÁ AND PAVEL PUNČOCHÁŘ
T.G. Masaryk Water Research Institute
Podbabská 30, 16062 Prague 6
Czech Republic

Abstract

The influence of year-to-year fluctuations in rainfall on the seasonal development of phytoplankton is demonstrated using two examples from water-supply reservoirs in the Czech Republic. In both reservoirs, the occurrence of low flushing rates caused by reduced precipitation was accompanied by high phytoplankton densities, which significantly reduced the quality of the water. In the first reservoir (Římov), with a retention time of 92 days, this effect was apparent over the whole reservoir. In the second reservoir (Želivka), with a retention time of 430 days, this effect was only apparent in the upper (riverine) section. The possible decrease of flowrates as a consequence of climate change in the Czech Republic are discussed. Recent surveys suggest that reductions in the flushing rate, coupled with nutrient enrichment, could cause serious water quality problems in surface waterbodies that provide 58% of the current demand for drinking water.

1. Introduction

The geographical position of the Czech Republic and the hydrographic characteristics of its major rivers imply that its water resources are very sensitive to year-to-year variations in the weather and longer-term changes in the climate. In contrast to many other European countries, which receive drainage water fom neighbouring countries, atmospheric precipitation is the only source of water on the territory of the Czech Republic [1]. Water supply managers have therefore paid a great deal of attention to the problems of conserving water and increasing the volume of water stored within catchments. At present there are ca. 23000 fish ponds with a total volume of 625 × 10^6 m^3, 553 small reservoirs with a total volume of 36 × 10^6 m^3, and 195 large reservoirs

D.G. George et al.(eds.), Management of Lakes and Reservoirs during Global Climate Change, 127–139.

TABLE 1. Historical development of important dam reservoirs in the Czech Republic.

Year:-	1900	1920	1940	1960	1980	1995
Number of reservoirs	60	79	91	136	184	195

with a total volume of 3472×10^6 m^3; the last were built mainly in the 20th century. Table 1 shows how the number of reservoirs built for storing water has increased dramatically in the last fifty years.

These reservoirs have had strategic importance for the economic development of the country and 52 of them still provide 40% of the water required for drinking purposes. The water quality in most of the reservoirs is still very high but many have become enriched with nutrients in recent decades and produce excessive growths of algae that can cause serious problems for the preparation of drinking water. Recent estimates suggest that almost 28% of consumers are now supplied with water that periodically fails to meet the set quality criteria, and most of these problems are related to the presence of algal cells in the potable supply. However, very little is known about the frequency and intensity of such problems or how they are related to year-to-year changes in the weather. Compared with natural lakes, we would expect such man-made reservoirs to be more sensitive to changes in the climate, as their retention times are influenced by the regional hydrology and local demand for water [2].

The physical and chemical characteristics of reservoirs throughout the Czech Republic have been monitored for several decades, but very few of the records include information on the concentrations of total phosphorus or phytoplankton chlorophyll. In this paper, we collate some long-term records from three reservoirs that have been studied in rather more detail. Data on year-to-year changes in the total phosphorus and chlorophyll concentrations of a deep reservoir (Slapy) near Prague are used to illustrate a long-term trend in eutrophication, whilst similar data collated from two Bohemian reservoirs (Římov and Želivka) are used to illustrate the shorter-term effects of changes in the weather.

2. The progressive enrichment of Slapy Reservoir

Slapy, at an altitude of 271 m a.s.l., is located ca. 50 km south of Prague. It is a deep canyon-type reservoir with a relatively short retention time. It has a surface area of 13.1 $\times 10^6$ m^2, volume 270×10^6 m^3, maximum length 44.5 km, maximum depth 53 m, and a mean retention time of 38.5 days. Filled in 1954, Slapy belongs to the Vltava River Cascade of seven reservoirs, which were built primarily for power generation. Slapy has been monitored at regular intervals since 1958 by the research team at the Hydrobiological Institute of the Czech Academy of Sciences (CAS), and the results have been summarised in a number of publications [3–7].

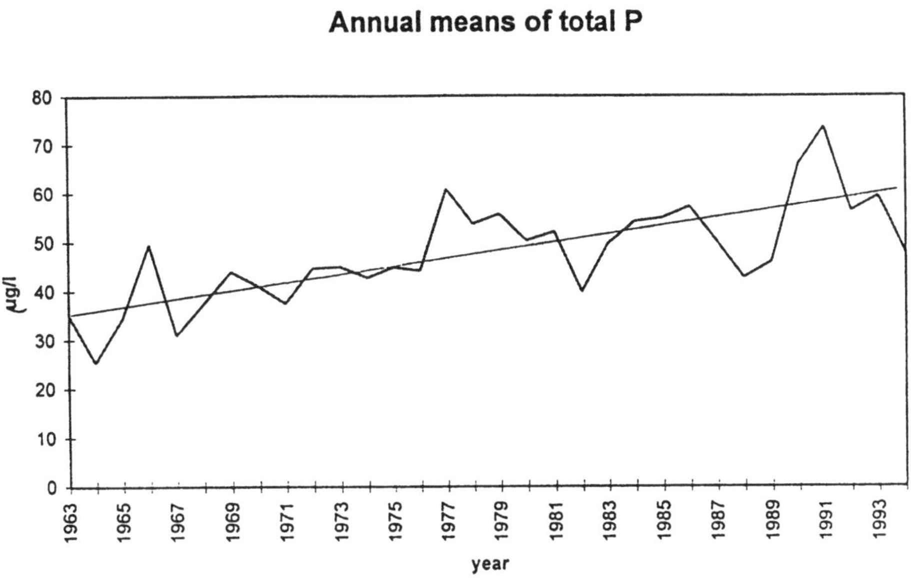

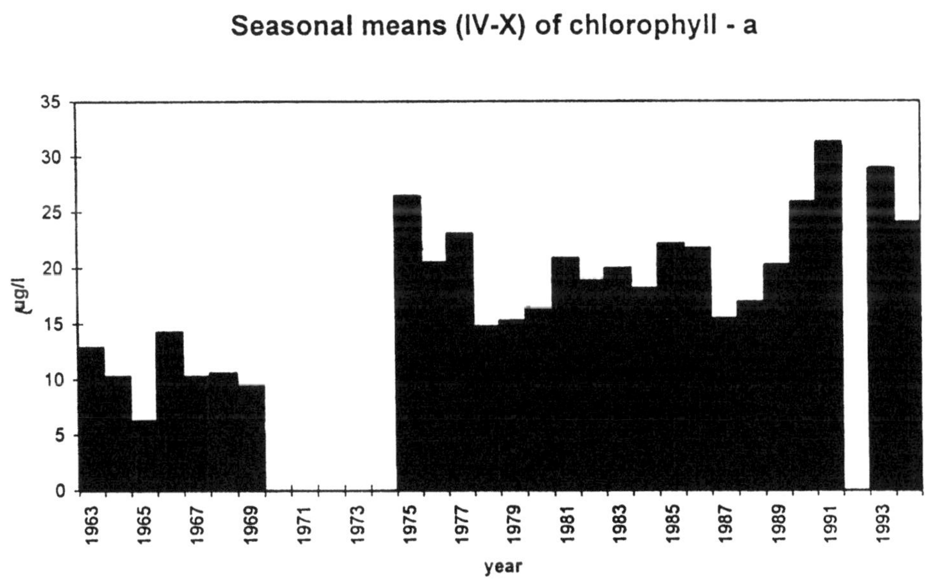

FIGURE 1. Year-to-year variations in the annual mean concentrations of total phosphorus (µg l^{-1}) and seasonal mean chlorophyll-*a* (April to October, µg l^{-1}) in Slapy Reservoir for the period 1963 to 1994. The superimposed line in the upper graph shows the calculated increasing trend for the annual mean concentration of total phosphorus. (Based on data from [6], [8] and [9]).

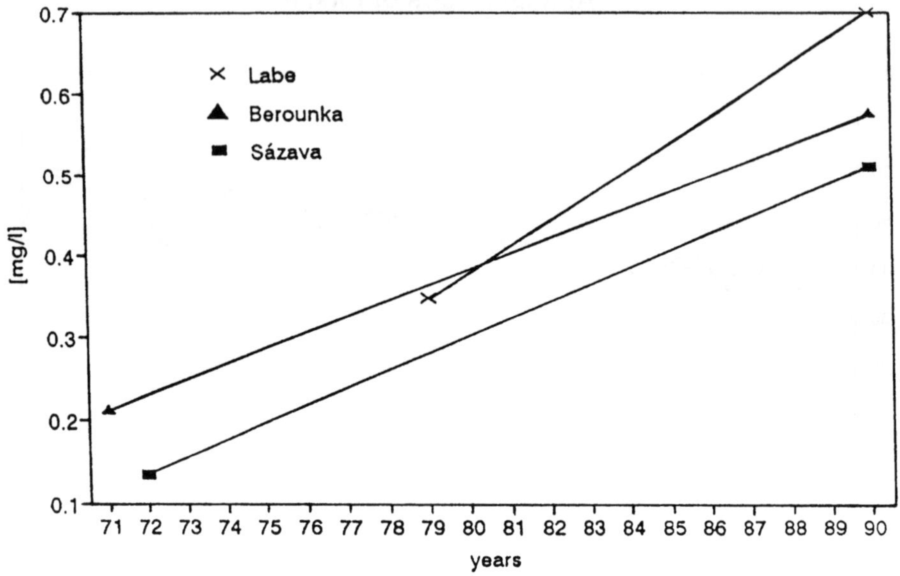
0.7
0.6
0.5
0.4
0.3
0.2
0.1
[mg/l]
Labe
Berounka
Sázava
71 72 73 74 75 76 77 78 79 80 81 82 83 84 85 86 87 88 89 90
years

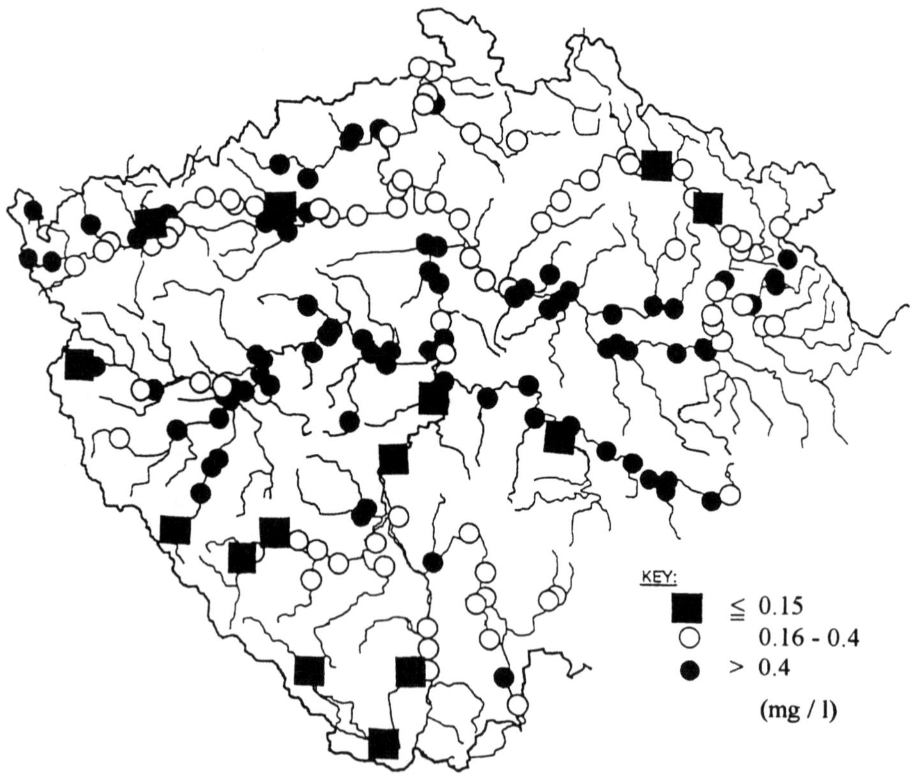
KEY:
≦ 0.15
0.16 - 0.4
> 0.4
(mg / l)

FIGURES 2 and 3 (*on the opposite page*). Concentrations of phosphorus (mg l^{-1}) in the Czech Republic. *Above:* Fig. 2. Long-term trends in mean phosphate at the mouths of three rivers. (Based on data in [10]). *Below:* Fig. 3. The distribution of total phosphorus mean values for 1991–1993 in Czech watercourses. (Based on data in [11]).

Fig. 1 shows year-to-year trends in the annual mean concentration of total phosphorus (TP) and the seasonal mean concentration of phytoplankton chlorophyll in the Slapy Reservoir between 1963 and 1994, inclusive. The concentration of TP has increased steadily over the 32-year period and is reflected in a parallel increase in the biomass of phytoplankton. Similar trends have been reported in other large reservoirs monitored at less frequent intervals and there have been comparable increases in the mean concentrations of phosphate found in three Czech rivers (Fig. 2). The distribution map in Fig. 3 shows the ranges of mean values recorded for TP during 1991–1993, in watercourses distributed throughout the Czech Republic. High concentrations of TP are currently being reported from a wide range of waterbodies.

3. The influence of year-to-year variations in flushing rate on the seasonal dynamics of phytoplankton in a reservoir with a short retention time (Římov)

The impact of year-to-year variations in flushing rate on the seasonal dynamics of phytoplankton in a reservoir with a short retention time can be demonstrated using data from the Římov Reservoir in the south of Bohemia. Římov is an important regional source of drinking water, situated on the River Malše; it is a relatively small waterbody, with a surface area of 2.1×10^6 m^2, volume 34.5×10^6 m^3, total length 13.5 km, maximum depth 43 m, and a mean retention time of 92 days. Římov has been intensively studied by the Hydrobiological Institute of CAS since it was filled in 1979 [12–16]. Fig. 4 shows changes in seasonal (April to October) mean concentrations of chlorophyll-*a* for the period 1979 to 1994. In spite of large year-to-year variations, there is a general trend for a gradual increase in phytoplankton biomass, indicating the moderately eutrophic state of the reservoir at the present time. Superimposed on this long-term trend, the year-to-year fluctuations in biomass are somewhat irregular, but the lowest concentrations of chlorophyll are typically recorded when the flushing rate (flowrate) of the reservoir is high (Fig. 4). This negative relationship between annual mean values of the flowrate and seasonal means of chlorophyll-*a* was first described for the period 1984 to 1991 [15]. The year-to-year changes in biomass are usually accompanied by shifts in the structure of the phytoplankton community and in the seasonal growth pattern. In the first three years after filling (1979–1981), maximum biomass each year was recorded during a spring peak of phytoplankton dominated by centric diatoms in the genus *Stephanodiscus* [12]. Later on, as the reservoir developed in age, the seasonal course of phytoplankton development changed and the annual chlorophyll maximum was recorded during summer. Phytoplankton structure changed too, as bloom-forming species became increasingly common during summer [16].

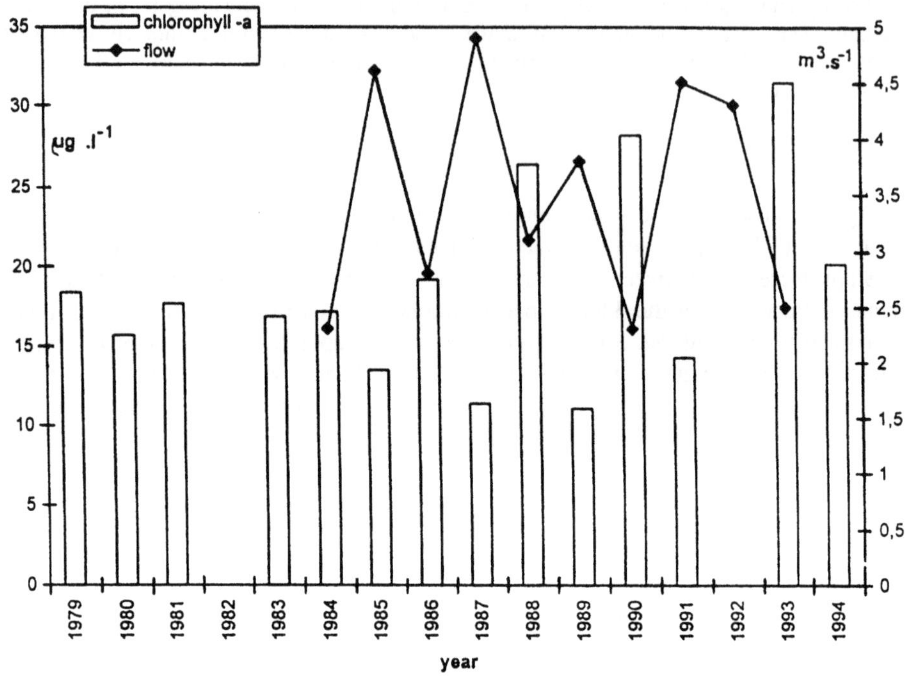

FIGURE 4. The relationship between year-to-year variations in the mean biomass of phytoplankton (April–October), measured as chlorophyll-*a* (μg l^{-1}), and annual mean flowrate (m^3 s^{-1}) in the Římov reservoir. (Based on data from [8], [12], [15] and [16]).

The bloom-forming cyanobacteria, mainly *Aphanizomenon flos-aquae*, were particularly common in the years with low flowrates (i.e. 1986, 1988, 1990). In the other years, the species composition of summer phytoplankton assemblages differed from year to year and was dominated by *Fragilaria crotonensis* in 1984, *Staurastrum planctonicum* in 1993 and 1994, and by a mixture of several algal species in 1991.

4. The influence of year-to-year variations in flushing rate on the seasonal dynamics of phytoplankton in a reservoir with a long retention time (Želivka)

The influence of year-to-year variations in flushing rate on the distribution of phytoplankton biomass in a reservoir with a long retention time is demonstrated using data for the Želivka drinking-water reservoir on the River Želivka. This reservoir is situated 70 km south-east of Prague and represents the most important source of drinking water for the city of Prague, and for some other municipalities in Central Bohemia. Želivka Reservoir was filled in 1977 and has a surface area of 14×10^6 m^2, volume 266.2×10^6 m^3, total length 38 km, maximum depth 53 m, and a mean retention time of 430 days. In the past, the concentrations of nutrients measured in the

TABLE 2. Mean values for concentrations of total phosphorus and chlorophyll-*a* (μg l^{-1}), and mean depth (m) for Secchi-disc transparency in Želivka Reservoir during 1980–1985 and 1992. (Based on data in [17] and [18]).

Year	Total Phosphorus	Chlorophyll-*a*	Secchi-disc depth
1980–1985	13.3	3.1	5.6
1992	26.7	6.8	3.8

River Želivka had been very low and this was the main reason for constructing a large drinking-water reservoir on its lower reaches. In recent years, however, concentrations of nutrients have increased substantially in the river and in the reservoir, and currently reach levels that periodically give rise to water treatment problems. Table 2 summarises the changes recorded in three water quality variables that have been measured in the euphotic layer near the reservoir dam during the last ten years. The mean values for total phosphorus and chlorophyll-*a* have increased by a factor of two and there has been a proportional decrease in the transparency of the water between the early 1980s and 1992.

The water quality situation is even worse in the upstream region of Želivka, well away from the point of abstraction. This riverine section of the reservoir now produces very dense blooms of algae in summer, and these are only dispersed slowly by the mass flow and wind-induced mixing. Fig. 5 shows the extent to which these conspicuous longitudinal gradients of phytoplankton biomass appeared in the reservoir in two successive years, 1991 and 1992. In both years, the gradual decrease of chlorophyll-*a* concentration along the reservoir was accompanied by an increase in water transparency. Fig. 6 shows the spatial variations in phytoplankton growth that characterise the lacustrine and riverine parts of the reservoir. At the downsteam end of the reservoir, the long-term pattern of phytoplankton development at the reservoir dam is characterised by a spring peak of phytoplankton biomass, dominated by the centric diatom *Aulacoseira subarctica*. From time to time, the spring growths of this diatom cause serious difficulties in the water treatment plant, which was not designed to process water containing high concentrations of phytoplankton. At the upstream (riverine) part of the reservoir, the seasonal development of phytoplankton is characterised by a summer peak of biomass (Fig. 6) composed of coccal cyanobacteria. Species in the genus *Microcystis*, mainly *M. aeruginosa*, *M. incerta* and *M. viridis*, dominate the phytoplankton assemblage. The chlorophyll-*a* concentrations recorded in this upstream part of the reservoir are particularly sensitive to year-to-year variations in the flushing rate, so a considerable proportion of the interannual variation in phytoplankton biomass can be explained by changes in the flow regime.

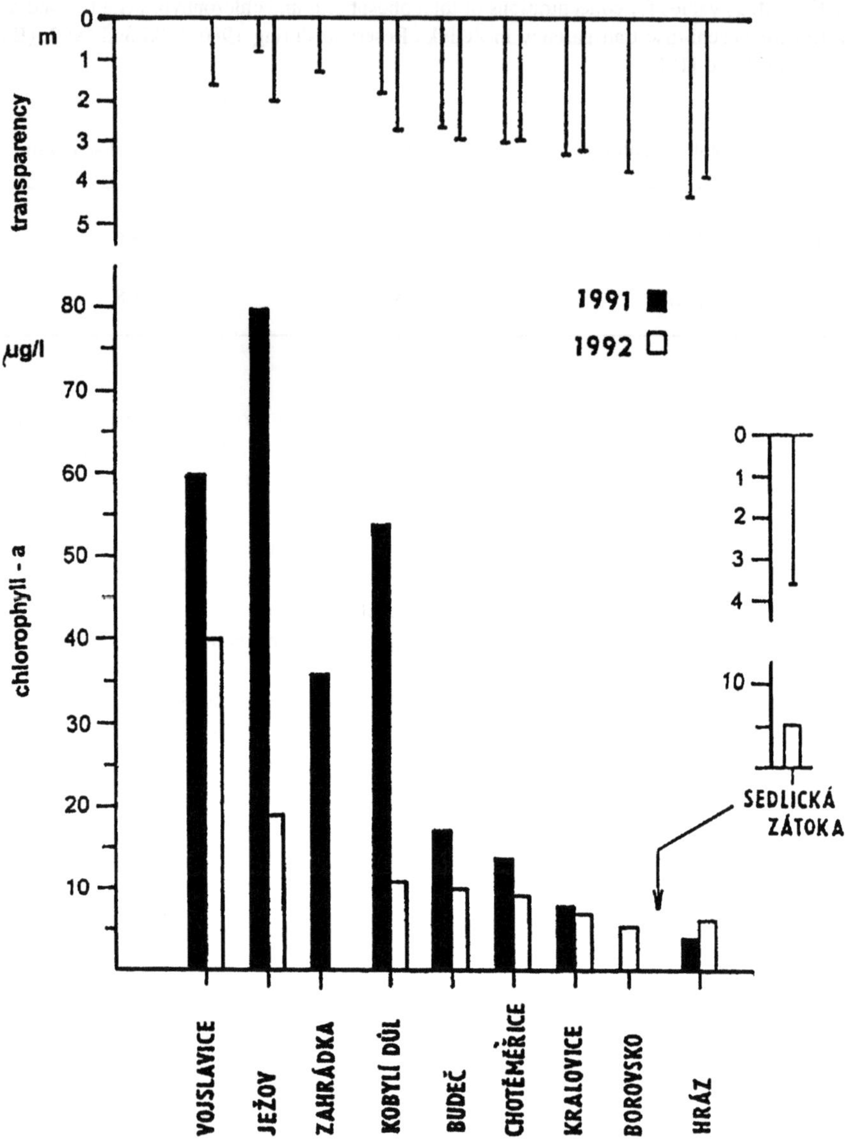

FIGURE 5. Annual mean concentrations of chlorophyll-*a* (µg l^{-1}) and depth (m) of Secchi-disc transparency along a longitudinal sampling profile of Želivka Reservoir in 1991 and 1992; see Fig. 6 for positions of the sampling sites. (Based on data in [18]).

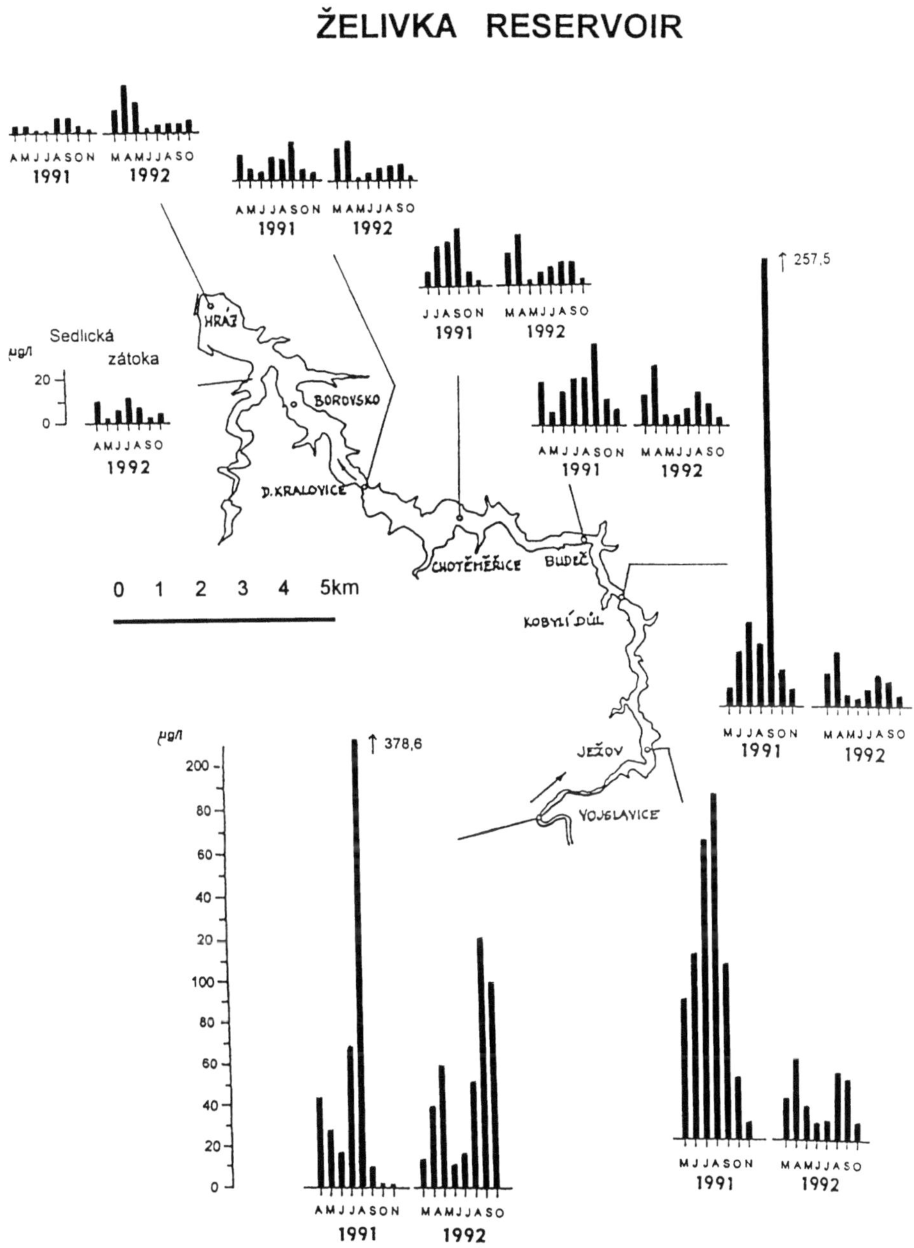

FIGURE 6. Seasonal changes in the concentration of chlorophyll-*a* ($\mu g\ l^{-1}$) during March–April to October–November along a longitudinal sampling profile of Želivka Reservoir in 1991 and 1992 (Desortová, unpublished).

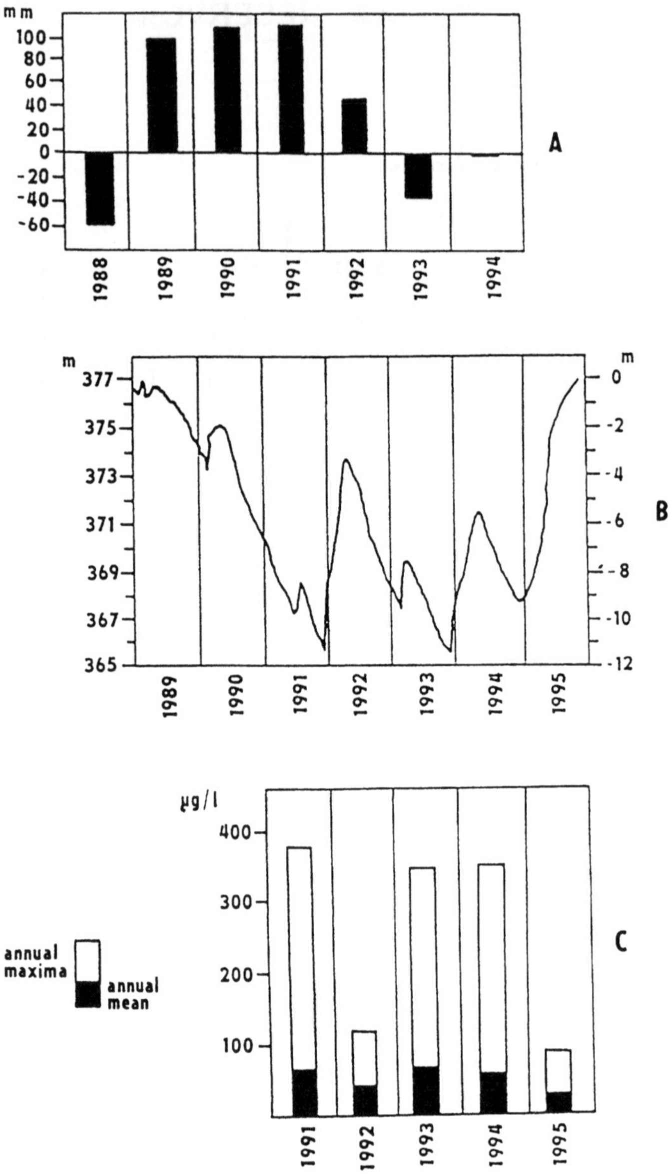

FIGURE 7. A, The development of precipitation deficiency (mm) on the territory of Bohemia between 1988 and 1994. B, Seasonal and interannual variations in water level (m below top level) in Želivka Reservoir between 1989 and 1995. C, Annual means and maximum chlorophyll concentrations (µg l^{-1}) in Želivka Reservoir during 1991 to 1995. (Based on data in [19] for A and B; Desortová (unpublished) for C).

Fig. 7 shows the relationship between year-to-year variations in rainfall, the recorded change in water level of Želivka Reservoir, and the annual biomass of phytoplankton. A decrease in the amount of water stored in the reservoir (Fig. 7B) was primarily caused by a deficiency of precipitation (Fig. 7A) and continuous abstraction of water for drinking, during unusually dry and hot summers. The high deficiency of precipitation recorded in the territory of Bohemia between 1989 and 1992 was primarily a function of the very dry summers. Fig. 7C shows that the year-to-year variations in phytoplankton biomass were negatively related to the rate of reservoir filling. In years with a small volume of water, 1991, 1993 and 1994, the chlorophyll-*a* maxima were 350 to 375 $\mu g\ l^{-1}$, whereas in a typical wet year (1995), when the reservoir was almost full, the chlorophyll maximum was below 100 $\mu g\ l^{-1}$ (Fig. 7C).

5. Possible impacts of climate change on reservoir phytoplankton dynamics

Results from our investigations on two morphologically different drinking-water reservoirs, Želivka and Římov, demonstrate that year-to-year variations in the flushing rate can have a major effect on the spatial and temporal patterns of phytoplankton development. In both reservoirs, high flows were accompanied by a decrease of phytoplankton biomass, whilst low flows resulted in relatively high concentrations of chlorophyll-*a*. In the smaller reservoir with the shorter retention time (Římov), this relationship was obvious for the entire reservoir. The changes in flushing rate were accompanied by changes in the composition of the phytoplankton; bloom-forming species of cyanobacteria became more abundant in years with low flows. In years where the flushing rate was higher the phytoplankton community tended to be dominated by diatoms or a mixed assemblage of several species. In contrast, in the large reservoir with a longer retention time (Želivka), the influence of flows on changes in the concentration of chlorophyll-*a* was observed only in the upper, riverine part of the reservoir, where extremely high concentrations were found when the low flows and low water levels occurred simultaneously. However, no changes in the composition of phytoplankton were observed and the phytoplankton community was commonly dominated by species of cyanobacteria. Year-to-year changes in the flowrate appeared to have very little effect on changes in biomass in the lacustrine part of the reservoir. The long-term changes recorded here were primarily related to the change in nutrient load, and the phytoplankton community was again dominated by the cyanobacteria.

These results suggest that any spatial or temporal changes in the rainfall that might occur in a warmer world would have a major impact on the qualitative and quantitative composition of the phytoplankton in reservoirs throughout the Czech Republic. The results from a national study on the impact of climate change [20] suggest that the hydrological regime will become increasingly unstable. Calculations suggest that a 10% decrement of precipitation might result in a reduction of 40 to 70% in the annual runoff in some regions of the Czech Republic. If we assume that the temperature will also increase and generate further losses by evapotranspiration, there could be a rapid increase in the number of water quality problems that are related to low flows. The deterioration of surface water quality due to increased growth of phytoplankton is likely to be particularly acute in dam reservoirs that are also subject to increased nutrient loads. Comprehensive research projects aimed at controlling eutrophication as well as identifying the sensitivity of specific waters to climate change need to be designed. Detailed monitoring programmes are also required and should be designed to include a wide range of physical, chemical and biological variables. In this respect, only a few of the 195 significant Czech reservoirs have been monitored in any detail (e.g. chlorophyll-*a* and total phosphorus were introduced into the schedule of routinely monitored variables only three years ago). For this reason, the development of target monitoring [21], and integrated water management based on reservoir catchments, is of critical importance to decision makers in the water industry.

References

[1]. Dvořák, V. & Hladný, J. (in press). Climate change hydrology and water resources impact and adaptation for selected river basins in the Czech Republic.

[2]. Straškraba, M. (1994). Vltava Cascade as teaching grounds for reservoir limnology. *Wat. Sci. Technol.* **30**, 289-297.

[3]. Hrbáček, J. (Ed.) (1966). *Hydrobiological Studies 1.* Academia, Praha, 407 pp.

[4]. Hrbáček, J. & Straškraba, M. (Eds.) (1973). *Hydrobiological Studies 2.* Academia, Praha, 348 pp.

[5]. Brandl, Z., Desortová, B., Hrbáček, J., Komárková, J., Vyhnálek, V., Sed'a, J. & Straškraba, M. (1989). Seasonal changes of zooplankton and phytoplankton and their mutual relations in some Czechoslovak reservoirs. *Arch. Hydrobiol. Beih. Ergebn. Limnol.* **33**, 597-604.

[6]. Desortová, B. (1989). Seasonal development of phytoplankton in Slapy Reservoir with special attention to the spring algal phase. *Arch. Hydrobiol. Beih. Ergebn. Limnol.* **33**, 409-417.

[7]. Straškrabová, V., Hejzlar J., Procházková, L. & Vyhnálek, V. (1994). Eutrophication in stratified deep reservoirs. *Wat. Sci. Technol.* **30**, 273-279.

[8]. *Annual report for the years 1991-1994.* Hydrobiological Institute, Academy of Sciences of the Czech Republic, České, Budějovice.

[9]. Javornický, P. & Komárková, J. (1973). The changes in several parameters of plankton primary productivity in Slapy Reservoir 1960-1967, their mutual correlations and correlations with the main ecological factors. In: *Hydrobiological Studies 2* (eds. J. Hrbáček & M. Straškraba), pp. 155-211.

[10]. *Year-books on water quality in the Czech watercourses, 1970-1990.* Czech Hydrometeorological Institute, Prague.

[11]. Elbe Project (1995). *Results and outcome.* T.G.M. Water Research Inst., Prague, 43 pp.

[12]. Desortová, B. (1982). Změny obsahu chlorofylu-a ve vodě Římovské, nádrže. In: *Kvalita vody a hydrobiologie údolní nádrže Římov.* Sb. semináře, České, Budějovice, pp. 59-61. [In Czech].

[13]. Vyhnálek, V., Komárková, J., Sed'a, J., Brandl, Z., Šimek, K. & Johanisová, N. (1991). Clear water phase in the Římov Reservoir (South Bohemia). Controlling factors. *Verh. Internat. Verein. Limnol.* **24**, 1336-1339.

[14]. Hejzlar, J., Balejová, M., Kafková, D. & Růžička, M. (1993). Importance of epilimnion phosphorus loading and wind-induced flow for phytoplankton growth in Římov Reservoir. *Wat. Sci. Tech.* **28**, 5-14.

[15]. Komárková, J. (1993). Cycles of phytoplankton during long-term monitoring of Římov and Slapy Reservoirs (Czech Republic). *Verh. Internat. Verein. Limnol.* **25**, 1187-1191.

[16]. Komárková, J. (1994). Phytoplankton cycles in the Římov Reservoir (South Bohemia). *Arch. Hydrobiol. Beih. Ergebn. Limnol.* **40**, 81-84.

[17]. Houk, V. (1988). Vývoj kvality vody ve vodárenské, nádrži na Želivce z biologického hlediska. *Vodní hosp.* **B38**, 123-130. [In Czech].

[18]. Desortová, B. (1994). Fytoplankton nádrže Želivka., In: *Ochrana jakosti vody vodárenského zdroje Želivka.* VUV T.G.M., Praha, pp. 84-94. [In Czech].

[19]. Vrabec, M., Ričicová, P. & Kessl, J. (1995). Hydrometeorologická charakteristika roku 1994 v České, republice. *Vodohosp. Techn.-Ekonom. Inform.* **7/8**, 241-248. [In Czech].

[20]. Hladný, J. & Dvořák, V. (1993). Hydrologický systém a vodní hospodářství. In: *Rizika změny klimatu a strategie jejich snížení.* Národní klimatický program ČR, Praha, pp. 111-125. [In Czech].

[21]. Adriaanse, M., van de Kraats J., Stoks, P. G. & Ward, R. C. (1995). *Proc. Internat. Workshop Monitoring Tailor-made, 20-23 September 1994.* Beekbergen, Netherlands, 356 pp.

MANAGING RESERVOIR WATER QUALITY IN THE VLTAVA RIVER BOARD AREA

JINDŘICH DURAS
Vltava River Board, Laboratory Plzeň
Denisovo n. 14
Czech Republic

1. Introduction

The Vltava River Board administers about 30 reservoirs of some importance in an area covering ca. 28000 km^2. Most of these reservoirs are still managed to optimise the quantity rather than the quality of water delivered. In many instances it is impossible to manage the reservoirs in any other way, as they were designed and built for satisfying quantitative rather than qualitative requirements. In recent years, however, more attention has been paid to the methods used to control water quality in drinking water reservoirs, as the price paid for the water often depends on its quality. In this short report, I describe some of the methods used by our specialists to improve the quality of the water abstracted from their storage reservoirs. Most of these methods are engineering solutions but they all require a sound understanding of the way lakes and other waterbodies respond to changes in the weather.

2. Changing the seasonal pattern of discharge

The simplest method of improving water quality is to decrease the residence time of water in a reservoir during periods when water quality is low. This method has been tested recently on Lučina Reservoir, where dense populations of the diatom *Asterionella formosa* may develop in April. Preliminary results seem promising and suggest that significant amounts of the diatoms could be washed out. In a similar way, large volumes of water containing high concentrations of nutrients can also be released in the autumn, thus minimising the risk of phytoplankton blooms in the following spring.

3. Improving the quality of water at the depth of abstraction

Some deep reservoirs have multiple outlets where water can be abstracted from different

D.G. George et al.(eds.), Management of Lakes and Reservoirs during Global Climate Change, 141–143.

depths. In others, the quality of the water being abstracted can be improved by lowering the water level so that water of better quality is located next to the outfall. This method is very useful, for instance, in situations where an inflow creates a lens of poor quality water during a short storm event.

4. Manipulating the seasonal pattern of discharge

The simplest method of improving water quality where there is plenty of water in the reservoir is to discharge a proportion of deep, poor-quality water before the reservoir is mixed in the autumn. This technique has been used successfully for the following purposes.

(1) To eliminate water of unsuitable quality before the autumn overturn occurs. Such water may, for example, contain very low concentrations of oxygen and high concentrations of managanese, although selective discharge may be most effective for stable substances such as humates.

(2) To improve the oxygen regime during summer or, in the case of meromictic reservoirs (e.g. Klíčava), also in spring. This reduces the hypolimnetic volume of anoxic water containing hydrogen sulfide, manganese, iron and phosphorus. Spring overturn in a meromictic reservoir may be significantly facilitated. The discharge technique is most useful in years when there is plenty of water.

(3) To lower the thermocline when the only available draw-off point is located in the hypolimnion. The purpose is to move water of better quality to the level at which it can be withdrawn.

(4) To decrease the overall residence time of water in a reservoir, as in Lučina Reservoir mentioned above in Section 2.

(5) To reduce the spring loading of nutrients by releasing relatively large quantities of water from the bottom of the reservoir after the autumn overturn or before the onset of spring stratification. The supposition that the supply of nutrients (which support the spring and summer phytoplankton blooms) can be diminished by this procedure, has never been confirmed.

5. Treating the waterbody with chemicals

The biological characteristics of many acid lakes can be improved temporarily by adding lime. Liming is not a permanent solution and it can be both labour and capital intensive. Liming is now practised on many small reservoirs in the Czech Republic and new methods are being devised which combine liming with selective discharge of water.

6. Biomanipulation

Biomanipulation techniques try to influence the structure of the zooplankton and phytoplankton communities by introducing predators (usually fish) that exercise a degree of top–down control. However, the idea that phytoplankton will be substantially suppressed by large zooplankters, such as members of the Daphniidae, has not been demonstrated very effectively in our medium-sized and large reservoirs. Moreover there is no conclusive evidence that these biomanipulations have had any significant effect on the critical issue – the quality of the raw water. Due to these facts and the relatively high costs of fish-stocking, this technique is not widely used in the Czech Republic, but young predatory fish are frequently released into reservoirs to reduce the population densities of planktivorous fish.

7. Discussion

Most of the methods currently being used to manage water quality depend on a plentiful supply of water. Such techniques cannot be employed readily if there is a marked reduction in the volume of water contained in the reservoirs in summer. The true systematic management of reservoir water quality is a relatively recent activity in our region and, so far, very little attention has been paid to the technical problems of managing reservoirs in a warmer world. More information is certainly required on the effect of year-to-year variations in the weather on water quality. In some systems, climatic effects can be very important, but it is sometimes difficult to identify and quantify these effects if the reservoir is also subject to large anthropogenic changes.

THE MASS INVASION OF SEVERAL BLUE-GREEN ALGAE IN TWO DRINKING-WATER SUPPLY RESERVOIRS IN SOUTHERN POLAND

HALINA BUCKA

Karol Starmach Institute of Freshwater Biology
Polish Academy of Sciences
ul. Sławkowska 17
31-016 Kraków, Poland

1. Introduction

The mass appearance of several planktonic blue-green algae (cyanobacteria), causing water-blooms in the dam reservoirs of southern Poland, has been observed more frequently in recent years. They have been noted in mid summer in the Goczałkowice Reservoir, and in late summer to early autumn in the Dobczyce Reservoir (Table 1). Goczałkowice (filled in 1955) is built on the upper course of the main river in Poland, the River Vistula (length 941.3 km). Dobczyce (filled in 1986) is built on the lower stretch of the middle course of the River Raba (length 137.4 km), the right-bank affluent of the Vistula. Their catchments cover 90% of the total area of Poland. I have studied the development of blooms caused by blue-green algae in both of these reservoirs, with particular emphasis on the shallower Goczałkowice.

There are contradictory theories about which environmental conditions specifically favour water-blooms containing blue-green algae. According to Kauppi et al. [7] the waterbodies concerned are shallow, rich in plant nutrients, only intermittently stratified, and have a relatively high pH. In their opinion the land-use on the catchment and related factors might have an influence on the mass development of algae. Reynolds & Walsby [8] suggest that water-blooms are caused by dense algal populations (most often by planktonic blue-green algae) following nutrient enrichment of a given waterbody, due to the inflow of different forms of pollution (e.g. treatment wastes and agricultural runoff) as the principal sources of phosphorus and nitrogen. Other hypotheses [9–11] link the mass development of blue-green algae in eutrophic waters with (for example) higher temperatures, low light levels, low N:P ratios, buoyancy mechanisms, and grazing by zooplankton, as well as the ratio of carbon dioxide to pH.

D.G. George et al.(eds.), Management of Lakes and Reservoirs during Global Climate Change, 145–151.

TABLE 1. Some morphometric features of Goczałkowice and Dobczyce reservoirs in southern Poland.

Variables	Goczałkowice	Dobcyze
Impounded river	Vistula	Raba
Year when filled	1955	1986
Altitude (m a.s.l.)	255.3	269.9
Length (km)	12	11
Width (km)	3.0–3.5	ca. 1
Mean depth (m)	5.2	11
Maximum depth (m)	13	28
Surface area (km^2)		
at max. water level	32	10.5
at min. water level	19.4	3.3
Capacity ($m^3 \times 10^6$)	16.3	99.0
Annual exchange of water	1.0–1.5	ca. 3.0

It must be stressed that, during water-blooms, results that are frequently taken to be the causes of the mass appearances of the algae, such as changes in pH values, are in fact simply a consequence of the rapid development of the algal population. The rise in pH results from increased photosynthesis by large numbers of algae, and is not the cause of water-bloom formation.

2. Physical–chemical aspects of Goczałkowice Reservoir

Continuous studies on Goczałkowice have been maintained since the reservoir was filled in 1955 [1–5]. Water taken from it by the local waterworks constitutes over 80% of the amount running into the reservoir. It is affected by municipal pollution and agriculture on the catchment, and is periodically affected by strongly eutrophicated pond waters in this region, at Gołysz.

The River Raba, together with other Carpathian tributaries and the upper Vistula above the Goczałkowice Reservoir, shows an excess of maximum water levels during the summer months, contrary to the situation in the lowland rivers where they occur during the spring [6].

The Goczałkowice Reservoir is shallow and pond-like (Table 1), with an east–west axis favouring strong penetration by winds (mean annual wind speed 3.0 m s^{-1}) that cause frequent and thorough mixing of the reservoir [12]. In the reservoir area, westerly winds from the Moravian Gate prevail [13]. The last author emphasised the importance of thermal water systems, also in the period of intensive movements of water masses, particularly during summer freshets.

In recent years there has been a decrease in the amount of oxygen in Goczałkowice, especially in the summer months, resulting from the decomposition of abundant algal growths [12]. However, wind-mixing in this shallow reservoir prevents the development of oxygen stratification.

A uniform phytoplankton distribution has been observed in the ice-free winter period as well as in the summer months, during the abundant development of algae. This is attributed to wind action which thoroughly mixes the entire water volume, along with a strong effect of waves [2].

In the period 1957 to 1982, small and short-lived falls in water temperature, with depth, were noted only in relatively long periods of calm weather. As is well known, water temperature depends on the air temperature. According to Augustyn [16], the mean annual air temperature for 1957 to 1982 varied by only ± 0.6°C, with an absolute maximum mean monthly air temperature of 25.2°C in July.

Winohradnik [13] emphasises a significant influence on the physical–chemical processes taking place in the water of the reservoir, due to changes in temperature, that are very important for chemical reactions in water treatment processes. Water-blooms are favoured by a constant rise in the content of nitrogen and phosphorus compounds observed in the reservoir. According to calculations given in [12], a phosphorus load of 0.6 g P m^{-2} $year^{-1}$ was carried into the reservoir in a period of 10 months, and this is 2–3 times higher than the permissible limit of 0.15 g P m^{-2} $year^{-1}$. Nutrients are carried into the reservoir by the rivers Vistula and Bajerka [17].

3. Algal blooms in Goczałkowice Reservoir

Recently, the reservoir has been dominated by the excessive development of blue-green algae (cyanobacteria), frequently leading to water-blooms. These always occurred in the summer months at high water temperatures (Table 2). In the opinion of various authors (e.g. [14, 15]), most algae increase their rates of photosynthesis and reproduction when water temperatures rise to 25 to 30°C.

In mid summer, at the begining of July 1992 during a severe drought [18, 19], there was a strong water-bloom in the pelagic and littoral parts of the reservoir. The bloom was duc to scvcral blue-green algae, mainly a dense population of *Aphanizomenon flos-aquae* (L.) Ralfs accompanicd by *Microcystis acruginosa* Kûtz., *Anabaena flos-aquae* (Lyngb.) Bréb., and *Gomphosphaeria naegeliana* (Unger) Lemm. (syn. *Woronichinia naegeliana* Elenk.). [In the Dobczye Reservoir a similar bloom in the autumn of 1995 (Table 2) was dominated by *Gomphosphaeria naegeliana*; other species were *A. flos-aquae, M. aeruginosa* and *M. incerta*.]

Generally speaking, in 1992 seasonal variations in the structure of phytoplankton communities were as follows [18]: in early spring the dominant group was Bacillariophyceae (*Asterionella formosa* Hass.), Chlorophyceae (mainly the species-rich genus *Scenedesmus*) were dominant at the begining of May, and in mid summer (July–August) cyanobacteria dominated, shared with large proportions of the species mentioned above. Their mass occurrence had not been noted before on such an enormous scale since the reservoir was filled in 1955. It ought to be added that in the period when the bloom developed, the water-level of the reservoir was lowered by more than 3 m, falling from 255.30 m a.s.l. in April to 252.05 m a.s.l. in mid October 1992. At the same time, 15 km^2 of the bottom area emerged above water [20].

TABLE 2. Some physical–chemical variables of the water in Goczałkowice and Dobczyce reservoirs during water-blooms in the summers of 1992 and 1995, respectively.

Variable	Goczałkowice	Dobczyce
Year	1992	1995
Temperature of surface layer (summer)	24.0–26.8	15.0
pH	ca. 9	9.1
Secchi-disc depth (m)	0.5–0.8	3.3
Oxygen saturation (%)	165	126
Nitrate-nitrogen (mg dm^{-3})	0.05–1.62	0.75
Phosphate (mg dm^{-3})	0–0.144	0.029
Chlorophyll-*a* in surface layer (μg dm^{-3})	234	21.3
Algal biomass (mg dm^{-3})	78.5	63.2
Numbers of algae per cm^3	49000 trichomes (*A. flos-aquae*)	320 colonies (*Gomphosphaeria*)

It is worth recording that during the period of the above water-bloom in Goczałkowice Reservoir, and during an earlier (smaller) bloom in 1988 (which also contained a large proportion of *Aphanizomenon*), the District Medical Officers of Health recorded a rise in the number of cases of spastic bronchitis, especially in children from the locality of Goczałkowice [4]. The mass appearance of algae (especially blue-green algae causing surface water-blooms) is dangerous for users of such waters [21–23].

4. Toxic effects of cyanobacterial blooms

Grazing by zooplankters may favour the development of those algae which are able to produce mucilage (e.g. *Aphanizomenon flos-aquae*); the surface of algal cells becomes stickier, leading them to aggregate. The aggregated cells are then too large to be consumed by small zooplankters, and the algae have a greater chance of survival [24]. Moreover, the intensive development of *A. flos-aquae* produces exotoxins which lead to the elimination of benthic animals that are potential consumers in the littoral zone; for example (among others) the Unionid bivalve molluscs *Unio tumidus* Phil. and *Anodonta piscinalis* L.). In 1992 the first animals to die from the extracellular toxic products of algae were in fact crayfish, *Orconectes limosus* Rafinesque [18]; 285 dead crayfish were noted on a 30-m stretch of shoreline. Also, nearly all chironomid larvae, oligochaetes and molluscs were killed.

Bivalve molluscs in the family Unionidae play a very important role as the largest filtering organisms; on average they filter up to 0.5 dm^3 of water per hour [20]. It has also been calculated that the molluscs collectively accumulate ca. 5 tonnes of phosphorus in their bodies and shells during their mass occurrence in the Goczałkowice Reservoir. In 1992 the fall in water-level, as well as metabolites from the water-bloom containing blue-green algae, caused the death of ca. 60% of the bivalves in the reservoir [20].

Cyanobacteria can produce both hepatotoxic peptides (e.g. some strains of *Microcystis* release microcystins into the water, which can cause liver damage [26]), and tumor promotors [23], as well as neurotoxic alkaloids. For example, strains of *Aphanizomenon* producing aphanotoxins, and *Anabaena* anatoxin-a, were found to be toxic by intraperitoneal mouse bioassay, causing convulsions and death by respiratory arrest [27].

5. Conclusions

Reasons for the sudden development of algal water-blooms are still controversial and require the consideration of every situation arising under different enviromental conditions. Generally, it is accepted that high concentrations of nitrogen compounds, linked with a surplus of phosphorus, as well as the retention time, type and age of the waterbody, are all responsible for blooms of algae.

Nevertheless, the intensity of water-blooms also depends, to a high degree, on water temperature related to the seasons of the year, and variation in the amount of rainfall on the catchment, which determines the frequency of water exchange in a given reservoir. In the 40-year period 1956 to 1995, the amount of rainfall in June and July combined has decreased locally, from ca. 300 mm to ca. 180 mm, whereas there has been a slight increase in September, from ca. 150 mm to 180 mm [28]. Long-term trends in rainfall and the associated weather patterns may affect the mean temperatures of both air and water, which in turn may favour the formation of water-blooms in dam reservoirs when water temperatures increase in the summer months.

The problem of algal blooms, and the potential toxic effects of blue-green algae (cyanobacteria), is of great importance for the Goczałkowice and Dobczyce reservoirs, as these supply drinking water for the Upper Silesian Industrial Region and the Cracow conurbation, respectively.

References

[1]. Krzyżanek, E., Kasza, H., Krzanowski, W., Kuflikowski, T. & Pajak, G. (1986). Succession of communities in Goczałkowice Dam Reservoir in the period 1955 - 1982. *Arch. Hydrobiol.* **106**, 21-43.

[2]. Pajak, G. (1986). Development and structure of the Goczałkowice reservoir ecosystem. 8. Phytoplankton. *Ekol. Pol.* **34**, 397-413.

[3]. Bucka, H. (1987). Ecological aspects of mass appearance of planktonic algae in dam reservoirs of southern Poland. *Acta Hydrobiol.* **29**, 149-191.

[4]. Bucka, H. & Żurek, R. (1992). Trophic relations between phyto- and zooplankton in a field experiment in the aspect of the formation and decline of water blooms. *Acta Hydrobiol.* **34**, 139-155.

[5]. Bucka, H., Żurek, R. & Kasza, H. (1993). The effect of physical and chemical parameters on the dynamics of phyto- and zooplankton development in the Goczałkowice Reservoir (southern Poland). *Acta Hydrobiol.* **35**, 133-151.

[6]. Wróbel, S. (1980). Zbiornik zaporowy w Dobczycach i jego ochrona - The Dobczyce dam water reservoir and its protection. *Zeszyty Problemówe Postepów Nauk Rolniczych* **235**, 205-215.

[7]. Kauppi, L. H., Knuuttila, S. T., Sandman, K. O., Eskonen, K., Luokkanen, S. & Liehu, A. (1990). Role of landuse in the occurrence of blue-green algal blooms. *Verh. Internat. Verein. Limnol.* **24**, 677-681.

[8]. Reynolds, C. S. & Walsby, A. E. (1975). Water-blooms. *Biol. Rev.* **50**, 437-481.

[9]. Sommer, U. (Editor) (1989). *Plankton ecology: succession in plankton communities.* Springer-Verlag, pp. 1-369.

[10]. Shapiro, J. (1990). Current beliefs regarding dominance by blue-greens: the case for the importance of CO_2 and pH. *Verh. Internat. Verein. Limnol.* **24**, 38-54.

[11]. Barica, J. (1994). How to keep green algae in eutrophic lakes. *Biologia, Bratislava* **49**, 611-614.

[12]. Kasza, H. (1992). Changes in the aquatic environment over many years in three dam reservoirs in Silesia (southern Poland) from the beginning of their existence - causes and effects. *Acta Hydrobiol.* **34**, 65-114.

[13]. Winohradnik, J. (1986) Development and structure of the Goczałkowice Reservoir ecosystem. 3. Characteristics of the reservoir and its usefulness to water supply systems. *Ekol. Pol.* **34**, 323-341.

[14]. Morduchaj-Boltovskoj, F.D. (1970). The effect of the heated water discharged from the cooling system of the Kanakovskaja thermal power station on the hydrology and biology of the Ivanskoe reservoir. In: *Productivity problems of freshwaters* (eds. Z. Kajak & A. Hillbricht-Ilkowska). Warszawa - Kraków, PWN, pp. 291-295.

[15]. Eloranta, P. (1982). Seasonal succession of phytoplankton in an ice-free pond warmed by a thermal plant. *Hydrobiologia* **86**, 87-91.

[16]. Augustyn, D. (1986). Development and structure of the Goczałkowice Reservoir ecosystem. 4. Meteorological conditions. *Ekol. Pol.* **34**, 343-350

[17]. Kasza, H. & Winohradnik, J. (1986). Development and structure of the Goczałkowice Reservoir ecosystem. 7. Hydrochemistry. *Ecol. Pol.* **34**, 365-395.

[18]. Krzyżanek, E., Kasza, H. & Pajak, G. (1993). The effect of water blooms caused by blue-green algae on the bottom macrofauna in the Goczałkowice reservoir (southern Poland) in 1992. *Acta Hydrobiol.* **35**, 221-230.

[19]. Kwandrans, J., Bucka, H. & Żurek, R. (1994). On the primary production and ecological characteristics of phytobenthos and phytoplankton in the littoral of the Goczałkowice Reservoir (southern Poland). *Acta Hydrobiol.* **36**, 335-355.

[20]. Krzyżanek, E. (1994). Changes in the bivalve groups (Bivalvia - Unionidae) in the Goczałkowice Reservoir (southern Poland) in the period 1983-1992. *Acta Hydrobiol.* **36**, 103-113.

[21]. Sivonen, K., Niemelä, S. J., Niemi, R. M., Lepisto, L., Luoma, T. H. & Räsänen, L. A. (1990). Toxic cyanobacteria (blue-green algae) in Finnish fresh and coastal waters. *Hydrobiologia* **190**, 267-275.

[22]. Reynolds, C. S. (1991). Toxic blue-green algae: the 'problem' in perspective. *Freshwater Forum* **1**, 29-38.

[23]. Lam, A. K.-Y., Prepas, E. E., Spink, D. & Hrudey, S. E. (1995). Chemical control of hepatotoxic phytoplankton blooms: implications for human health. *Wat. Res.* **29**, 1845-1854.

[24]. Kiorbøe, T. & Hansen, J. L. S. (1993). Phytoplankton aggregate formation: observations of patterns and mechanisms of cell sticking and the significance of exopolymeric material. *J. Plankton Res.* **15**, 993-1018.

[25]. Krzyżanek, E. (1989). Rola małży rodziny Unionidae w zbiorniku Goczałkowickim - Role of bivalves family Unionidae in the Goczałkowice dam reservoir. *Wszechśiat* **3**, 57-59.

[26]. Berg, K., Carmichael, W. W., Skulberg, O. M., Benestad, C. & Underdal, B. (1987). Investigation of a toxic water bloom of *Microcystis aeruginosa* (Cyanophyceae) in Lake Akersvatn, Norway. *Hydrobiologia* **144**, 97-103.

[27]. Vasconcelos, V. M. (1993). Toxicity of Cyanobacteria in lakes of North and Central Portugal. Ecological implications. *Verh. Internat. Verein. Limnol.* **25**, 694-697.

[28]. Szumiec, M. (1996). Ponds as the element of strategy applied in management of the country. In: *Inland conference of carp rearers concerning "Actual problems of pond farming in Poland"* held at Ustroñ, 11-13 March, 1996, pp. 9-17.

[18] Krzyzanek, E., Kasza, H. & Pająk, G. (1993). The effect of water blooms caused by blue-green algae on the bottom macrofauna in the Goczałkowice dam reservoir (Southern Poland) [illegible] Acta Hydrobiol. 35: [illegible]

[illegible] characteristics of [illegible] Goczałkowice Reservoir (southern Poland) [illegible] Acta Hydrobiol. [illegible]

[illegible] (1993) [illegible] the Goczałkowice Reservoir (southern Poland) in the period [illegible] Acta Hydrobiol. 35: 107-118.

[illegible]

WATER QUALITY IN THE LAKES AND RESERVOIRS OF ROMANIA

TEODOR L. CONSTANTINESCU
Romanian Water Authority
Str. Edgar Quintet 6
70106 Bucharest, Romania

1. Introduction

The territory managed by the Romanian Water Authority includes 122 natural lakes and 270 reservoirs which represent most of the country's 416 reservoirs. Some of the reservoirs managed are relatively large but others are temporary impoundments that are particularly susceptible to summer droughts. In 1995, a representative series of 95 lakes and reservoirs were sampled at regular intervals and their current physical, chemical and biological characteristics were assessed. The results of these surveys were then compared with the results of longer-term monitoring at a selected number of sites. In this paper, I summarise the results of the 1995 survey and present some physical and chemical data to show how the characteristics of the selected inland waters of Romania have changed over the last twenty to twenty-five years.

2. Methods

Water samples were collected from the lakes and reservoirs at intervals ranging from 2 to 6 months in 1995. Six key variables (temperature, transparency, oxygen concentration, nitrate concentration, phosphate concentration and phytoplankton biomass) were monitored at each site and the results used to produce a subjective classification of the waters into three categories that were closely related to their trophic status. Table 1 lists the various ways in which water from the three different categories is used. Only Category I waters are used for potable supplies but Category II and III waters can be used for a variety of purposes.

D.G. George et al.(eds.), Management of Lakes and Reservoirs during Global Climate Change, 153–157.

TABLE 1. The classification of Romanian lakes and reservoirs according to water quality and scope of use.

Water Quality Category	Scope of use
Category I	Drinking Animal husbandry Horticulture Salmonid culture Bathing Contact sports
Category II	Fish hatcheries (general) High quality industrial applications Urban and recreational use
Category III	Irrigation Hydro-electric power General industrial applications

3. The synoptic survey of lakes and reservoirs

Results from synoptic water quality surveys completed in 1995 are given in the Appendix. Most of the waters covered by the survey were reservoirs that formed part of a much larger hydrological network. Many of these reservoirs were relatively small, but the survey included one large reservoir (the Stanea-Costesti) in the Prut basin and two large reservoirs (Iron Gates I and II) in the Dunare (Danube) basin. The last two columns in the Appendix table show the "water quality" and "trophic status" of the lakes surveyed. Of the 79 waters in the survey, 56% are classified as Category I waters and a further 7% are in Category I/II. The results from the biologically-based "trophic status" evaluation were very similar; 26% of the waters sampled are classified as oligotrophic and a further 21% are identified as oligo/mesotrophic. Most of the water-supply lakes classed as eutrophic were relatively small, but one large reservoir in the Siret basin (the Cuib Vulturi) and one large reservoir in the Prut basin (the Halceni) contained high concentrations of phytoplankton.

4. Recent trends in water quality

The quality of the water in most Romanian reservoirs has always been good but recent reductions in the activity of several industrial concerns has brought about further improvements in water quality. This recent improvement can be illustrated by considering the long-term changes in the quality of the water and volume of discharge in the Arges basin between 1973 and 1995, as found in the Golesti Reservoir (Table 2).

TABLE 2. Mean concentrations of N and P (mg l^{-1}) in Golesti Reservoir (volume 86 x 10^6 m^3) between 1973 and 1995; values for discharge are based on monthly means (m^3 x 10^6).

Year	Discharge	Nitrogen	Phosphorus
1973	---	1.8	0.25
1980	60	1.2	0.15
1985	34	1.5	0.20
1988	41	2.3	0.33
1990	23	2.4	0.37
1991	48	2.5	0.22
1992	16	2.2	0.12
1993	21	2.3	0.13
1994	20	1.6	0.13
1995	33	1.7	0.13

In the Golesti Reservoir, concentrations of N increased in the early 1980s, remained steady in the early 1990s and then decreased rapidly in 1994. Concentrations of P also increased in the early 1980s but were already in decline by 1991. Most of this improvement in water quality was due to a decrease in the economic activity of the area, with a consequent reduction in the volume of waste-water. Some of these long-term variations, however, could be due to changes in the weather, particularly in the amount of summer rainfall. The volume of water discharged through the River Arges in the early 1990s is very much lower than that discharged in the early 1970s (Table 2) but the recent improvement in water quality has helped to counter the potentially damaging effects of these low flows.

5. Conclusion

The sampling programmes currently used to monitor regional changes in water quality will need to be expanded if we are to identify the more subtle physical effects of changes in the weather. In most areas, it is necessary to increase both the number of variables and the sampling frequency. The current monitoring programmes are planned by the Research Institute for Environmental Engineering (I.C.I.M) and the Romanian Water Authority (R.A). At present they have a database on nutrient concentrations covering more than ten years but have no information on heavy metals or the concentrations of organic compounds. The trends so far suggest that the impact of dry summers on water quality has not been very severe. However, the situation could become very much worse if dry summer conditions were combined with an increase in economic activity.

Appendix. Results from main lakes (asterisks) and reservoirs in Romania, sampled in 1995. Hydrographical basins are given in CAPITALS; total volume in millions of cubic metres; Uses: WS = water supply, En = energy, FC= flood control, F = fisheries, Re = recreation, Irr = irrigation, Nav = navigation, Ther = therapeutic waters; N = number of samples per year; WQC = water quality category; TS = trophic status: Olig = oligotrophic, Meso = mesotrophic, Eutr = eutrophic.

Waterbody	River	Volume	Uses	N	WQC	TS
TISA						
Calinesti	Tur	29	WS, En, FC	3	II	Meso
SOMES						
Colibita	Bistrita	84	WS, En, FC, F	3	I	Olig
Gilau	Somes	4	WS, En, F	4	I	Olig-Meso
Varsolt	Crasna	41	WS	3	II	Eutr
----	Firiza	17	WS, FC	4	I	Olig-Meso
MURES						
Bezid	Cusmed	15	WS, FC	3	--	Meso
Ighis	Ighis	13	WS, En	3	--	Meso
Petresti	Sebes	227	WS, En	3	I	Olig
Cincis	Cerna	41	WS	3	I	Olig
Hateg	Mare	15	WS, En	2	I	Olig
BEGA-TIMIS						
Surdue	Gladna	66	WS, En, Re, F	-	I	Meso-Eutr
Secu	Barzava	15	WS, En, Re, F	2	I	Olig-Meso
NERA-CERNA						
V. Iovan	Cerna	126	WS, En	2	I	Olig-Meso
Herculane	Cerna	16	WS	2	I	Olig-Meso
JIU						
V. Presti	V. Presti	5	WS	4	I	Olig-Meso
OLT						
Frumoasa	Frumoasa	11	WS	3	I	Olig
Tarlung	Tarlung	18	WS	4	I	Olig-Meso
G. Raului	Cibin	16	WS	4	I	Olig
R. Valcea	Olt	19	En	3	II	Meso
Babeni	Olt	78	En	2	II	Meso
Vidra	Lotru	340	WS, En	-	I	Olig
Bradisor	Lotru	38	WS, En	-	I	Olig
ARGES						
Vidraru	Arges	473	WS, En, FC, F	4	I	Olig
Zigoneni	Arges	13	En	2	I	Olig-Meso
Valcele	Arges	44	En	2	I	Olig-Meso
Budeasa	Arges	55	En, WS	2	I	Olig-Meso
Golesti	Arges	86	WS, En, FC, F	2	I	Olig-Meso
Rausor	Targului	68	En	2	I	Olig
Pecineagu	Dambovita	69	En	-	I	Olig
Vacaresti	Dambovita	54	En	-	I	Olig-Meso
Morii	Dambovita	20	WS, En, Re	4	I	Olig
Cernica	Colentina	9	WS, En, Re	4	I/II	Olig
Gradinari	Ilfovat	12	WS, En, F	2	I/II	Meso

IALOMITA						
Snagov*	------	17	---	-	--	Meso
Caldarusani*	------	5	Re	-	--	Eutr
Fundata*	------	10	Th	4	--	-----
Pucioasa	Ialomita	11	WS	4	I	Meso
Paltinu	Doflana	62	WS	2	I	Olig
Dridu	Ialomita	60	WS, En, Fc, F	2	II	Olig-Meso
SIRET						
Rogojesti	Siret	48	WS, En, Irr	3	I	Olig-Meso
Bucecea	Siret	25	WS, En, Irr	3	I	Meso
Dragomima	Dragomima	17	WS	4	I	Olig-Meso
Izv. Muntelui	Bistrita	1230	En, Re, WS	3	I	Olig-Meso
Batca Doamnei	Bistrita	10	En, Re	3	I	Olig-Meso
Poiana Uzului	Uz	90	WS, En	3	I	Olig
Galbeni	Siret	71	WS, En	3	IV	Eutr
Calimanest	Siret	44	En, Irr	4	II	Meso-Eutr
Solesti	Vaslui	47	WS, Irr	2	I	Eutr
Puscasi	Racova	21	---	-	II	Eutr
Cazanesti	Stavnic	21	WS, Irr, F	-	II	Meso
Cuib Vulturi	Tutova	55	WS, Irr	2	I	Eutr
Rapa Albastra	Simila	26	WS	-	II	Eutr
Pereschiv	Pereschiv	15	Irr, F	-	II	Eutr
Tungujei	Sacovat	25	WS	2	III	Meso
Jirlau*	Buzau	6	F	-	III	-----
Balta Alba*	------	5	Ther	-	--	-----
Siriu	Buzau	158	WS, Irr, En, FC	2	I	Olig
Candesti	Buzau	4	En, Irr	-	I	Olig
PRUT						
Stanca-Costesti	Prut	1400	WS, En, FC	4	I	Olig
Negreni	Baseu	20	WS, Irr, FC	-	I/II	Eutr
Catamaresti	Sitna	14	Irr, F	-	I/III	Eutr
Parcovaci	Bahlui	6	WS	-	I	Olig
Tansa	Bahlui	33	WS	-	I/II	Meso
Chirita	Chirita	7	WS	3	I	Olig
Halceni	Miletin	50	WS	-	III	Eutr
Prod Iloaici	Bahluiet	35	Irr, FC, F, Re	3	II	Eutr
DUNARE (DANUBE)						
Razelm	Danube Delta	909	---	-	I/II	-----
Galatui	Berza	9	---	2	I	-----
Iron Gates I & II	Danube	2900/1000	En, FC, Nav	6	I	-----
Sinoa	------	---	---	-	II	-----
Mariuta	Mostistea	14	Irr, En, F	-	I	-----
Iezer	Mostistea	280	Irr, En, F	-	I	-----
Tasaul	------	57	F	-	I/II	Meso-Eutr
Siutghiol	------	89	Re	2	II	Meso
Techirghiol	------	42	Ther	-	III	-----
Tatlageac	------	14	---	-	II	-----
Nuntasi	------	9	Ther	-	II	Eutr
Corbul	------	25	F	4	II	-----
Mangalia	------	--	F	-	II	Meso

LONG-TERM VARIATION OF CHLOROPHYLL CONTENT IN RYBINSK RESERVOIR (RUSSIA) IN RELATION TO ITS HYDROLOGICAL REGIME

N. M. MINEEVA AND A. S. LITVINOV
Institute for Biology of Inland Waters
Russian Academy of Sciences, Borok
152742 Yaroslavl, Russia

Abstract

Long-term changes in chlorophyll (CHL) during 1969 to 1984, in connection with hydrological conditions, were studied in two basins of the Rybinsk Reservoir (the third step of the Volga river system, northwest Russia). Year-to-year CHL fluctuations are quite similar in the Volga Reach and Main Part of the reservoir and depend upon hydrometeorological factors. Increases in the mean seasonal, most frequent and summer maximum CHL concentrations, demonstrate the accelerated eutrophication of the Main Part of the reservoir, which holds most of the total storage. Effects of water content, characterised in terms of total input, water level and water exchange intensity, are the most significant among hydrological factors. A clustering of years in fields of the first and second principal components corresponds to high-water periods (1969–1975) and low-water periods (since 1976). Strong influences of water content – controlled by human activity – on phytoplankton abundance (CHL) make it possible to provide ecosystem monitoring and management. Consideration is also given to the role of hydrological characteristics, including water balance, in producing undesirable changes that have occurred during recent years.

1. Introduction

Ecosystem development depends much on the influence of environmental conditions that include both natural abiotic factors and anthropogenic impacts. Complex hydrological structure is a distinguishing feature of large, shallow waterbodies with unstable water regimes. It is common knowledge that plankton indices make it possible to follow the course of ecosystem response to external influence; also that ecological

D.G. George et al.(eds.), Management of Lakes and Reservoirs during Global Climate Change, 159–183.

status and its variation can be evaluated in terms of trophic conditions. The common interrelated critera of phytoplankton abundance, photosynthesis and waterbody trophic level, including chlorophyll-*a* concentration (CHL), are of considerable current use in water ecology. Phytoplankton abundance determines many ecological relationships in natural waters and serves as a characteristic of important environmental events that, however, are not evident in a short time-period. For the purpose of ecological monitoring, long-term observations are of great scientific interest. Relatively few attempts to discuss the results of long-period observations on CHL are available [1–4]. The aim of the present work is to examine sixteen years of CHL fluctuations in the Rybinsk Reservoir in connection with some environmental parameters.

2. Materials and methods

Observations on phytoplankton pigments began in 1969 and have continued to the present. Original data are given in several publications [2, 5–10]. Chlorophyll data for 1983–1984 were obtained by Dr I. L. Pyrina (personal communication). Samples were taken every 2 or 3 weeks during the ice-free period from May to October, at the same six stations (Fig. 1) located within the Volga Reach (two stations) and the Main Part of Rybinsk Reservoir (four stations). Phytoplankton was sampled with a 1-m plastic tube (volume 4 litres), metre by metre, from the upper 0–2 m layer that may be considered as the euphotic zone. A special comparison made it possible to estimate the difference between 0 m to 2 m and 0 m (surface) to bottom samples, based on seasonal observations at Rybinsk Reservoir during May to October 1981–1982 (Table 1).

The results showed that there was not a strong difference between 0–2 m and surface–bottom chlorophyll content. Rare exceptions occurred during the blue-green summer bloom under calm and sunny weather when 0–2 m CHL was higher [2]. Subsamples (0.5 to 2 litres volume) from the mixed sample were concentrated by direct filtration through membrane filter No. 6 (pore size 3–5 μm) of Mytishi ultrafilters. Filtration of an adequate volume using membranes with smaller pore size was difficult, because of the high content of resuspended matter from bottom sediments. Suspensions of $CaCO_3$ and SiO_2 had been added previously to avoid CHL degradation under storage and enable its better extraction under grinding. Pigments were analysed in 90% acetone extract using a standard spectrophotometric procedure [11, 12]. This includes extraction under grinding, centrifugation, and measurement of absorbance in a 2-cm cell at 665 nm (until 1975) and 664 nm (since 1976), with subtraction of a turbidity correction at 750 nm.

The spectrophotometers used were the SP-4A and (since 1980) SP-26. CHL (together with pheopigments) was calculated using an equation recommended by UNESCO Working Group [12] and, from 1976, the Jeffrey & Humphrey (J&H) equation [13]. Comparison showed that there was no noticeable difference in CHL concentrations estimated with these two equations, as follows from regression equation (1) [14]:

$$CHL_{J\&H} = 0.04 + 1.05\ CHL_{UNESCO}\ ;\ n = 81,\ r = 0.99 \qquad (1)$$

TABLE 1. Chlorophyll concentrations (μg l^{-1}) in 0–2 m and surface–bottom samples in Rybinsk Reservoir for the period May–October, according to [2]. Means ± standard error; $t_{0.05} = 1.96$.

Basin	Year	CHL at 0–2 m depth	CHL at Surface–bottom	t criterion
Volga Reach	1981	14.4 ± 2.3	10.0 ± 1.3	1.66
	1982	10.9 ± 2.3	8.8 ± 1.4	0.78
Main Part	1981	18.2 ± 1.8	12.9 ± 1.0	2.57
	1982	11.6 ± 0.7	9.6 ± 0.5	2.32
Mologa Reach	1981	11.3 ± 2.2	8.8 ± 1.3	0.98
	1982	7.8 ± 0.4	7.5 ± 0.5	0.47
Sheksna Reach	1981	22.0 ± 4.4	18.0 ± 2.5	0.79
	1982	15.4 ± 1.3	14.8 ± 1.3	0.33

FIGURE 1. Map of Rybinsk Reservoir with locations (o) of six sampling stations. Lines indicate the borders of the basins.

For estimating trophic levels we followed the Vinberg scale [15] with its limits of < 1 μg l^{-1} (oligotrophy), 1–10 μg l^{-1} (mesotrophy) and > 10 μg l^{-1} (eutrophy). The latest scale published by Likens [16] makes those limits wider, with suggested limits of < 1–3 μg l^{-1}, 10–15 μg l^{-1} and > 15 μg l^{-1}, respectively. It seemed advantageous to combine the two scales and separate the moderate eutrophic type delimited by 10–15 μg l^{-1}.

Hydrometeorological data, including observations of the Hydrometeorological Service, are available for the entire period of reservoir existence. However, as we intend to discuss CHL data obtained during 1969 to 1984, only environmental data corresponding in time are represented below. Water temperature and Secchi-disc transparency were measured coincidentally with phytoplankton samples. Current and annual means of meteorological and hydrological parameters such as solar radiation, wind velocity and water fluxes were kindly presented by the Hydrometeorological Observation Service in the town of Rybinsk. The annual water exchange coefficient (K) was calculated using formula (2) [17]:

$$K = (V_1 + V_2) / 2V_0 \qquad (2)$$

where V_1 and V_2 are the total annual inflow and total water output volumes, and V_0 is the reservoir volume, all in km^3. All data were processed using standard statistical techniques, including regression and Principal Component analyses.

3. Rybinsk Reservoir and its hydrological characteristics

3.1. LOCATION

Rybinsk Reservoir at 58°00’–59°05’N, 37°28’–39°00’E (northwestern Russia) is the third step of the Volga River cascade system. Detailed descriptions, including phytoplankton, are given elsewhere [18, 19]. The reservoir was created in 1948 and its main characteristics are as follows: catchment area 150,500 km^2; normal level 101.8 m BS; total storage 25.42 km^3 ; surface area 4550 km^2; maximum depth 30.5 m; mean depth 5.6 m; maximum length 250 km; maximum width 56 km; water exchange 1.78 $year^{-1}$ and retention time 6.7 months (long-period averages).

The reservoir is located in the south taiga belt where precipitation (mean 616 mm $year^{-1}$ during 1947 to 1972, [18]) exceeds evaporation. Nevertheless, direct precipitation is not higher than 10% of the total water input. Inflow from three main tributaries – Volga, Sheksna and Mologa rivers – provides most of the total input of water into the reservoir. Four parts (basins) can be distinguished in the reservoir (Fig. 1): the central lake-like Main Part that covers ca. 60% of the area and contains ca. 80% of the total storage, and the more riverine Mologa, Sheksna and Volga reaches. Each tributary forms its own watermass differing in colour, transparency and conductivity. Such differences are usually most substantial during spring, and almost disappear in summer. The central Main Part is occupied by an inherent reservoir watermass.

3.2. ANNUAL CYCLE OF WATER LEVELS

The annual cycle of water level is divided into three periods: spring (progressive storage), summer–autumn (approximately constant level or small decrease of storage), and winter (decrease of storage). At the beginning of spring storage the daily increment of level increases from 1–3 to 20–30 cm and after that the level becomes more constant. Variable year-to-year periods of rising levels continue for 33 to 86 days starting from 5 March to 18 April (average date 5 April), and late May to early July is the time of highest levels. During summer and autumn the water level stays rather stable and is affected by tributary inputs and navigation (lock) flushes. By the winter, lateral input decreases to the minimal values, water loss through the cross-section of the Rybinsk electricity power-station becomes higher and, as a result, an intensive fall of level takes place. Decrease of storage is a slower process, with durations of 287 to 338 days (long-period average 306 days), ca. 5 times longer than that of progressive storage. The long-term monthly average amplitude of water level during 1947 to 1993 was 3.00 m and the highest value was 5.83 m. The lowest level usually occurs in March and the highest occurs in June. Long-period monthly levels and dependent morphometric characteristics are given in Tables 2 and 3.

3.3. WATER EXCHANGE

The intensity of water exchange depends on a series of natural and anthropogenic factors and has shown variable behaviour during the period of the reservoir's existence. Due to noticeable intra-annual (seasonal) variation in elements of the water budget, the respective intra-annual variations in water exchange are also large (Table 4). There is maximal water exchange in April or May, connected with decreased total storage (reservoir volume) and a strong increase in lateral input during spring floods. When input for storage is ended, water input and output decrease. Water exchange then becomes lower, being almost invariable during summer and falling to a minimum in August-September. A small increase in water exchange, starting in October, depends on rainy flood conditions and the next winter decrease of storage.

3.4. TEMPERATURE

Temperature conditions in the reservoir also undergo fluctuations from year to year. Ice-break (early dates 19 March to 3 April, late dates 27 April to 4 May) differs a little in different basins. The ice-free period, with full ice-melt, starts in the first half of May and continues for 180–190 days. During spring the reservoir Main Part contains winter water, with temperatures much lower than those of the riverine basins. Zones with high horizontal and vertical temperature gradients are formed as a result of mixing between warmer flood water and colder winter water of the central Main Part, and their subsequent warming. The vertical difference in temperature may be as much as 15°C, with a gradient in the thermocline region equal to 7°C per metre [20]. However, the temperature inhomogeneity is not long-lasting as it is destroyed by wind-mixing that provides heat transport within the water column. Warming of all the water-mass up to 15–20°C is over by the end of spring. Water temperature fluctuates within 20 to 23°C in the 0–2 m layer, 21 to 27°C on the surface and 18 to 22°C near the bottom. The highest temperature is observed in July in the surface layer and in August in all of the water column, as is shown by the long-term records (Table 5). Surface temperature

TABLE 2. Variations in monthly mean water level (m, BS) of Rybinsk Reservoir during 1948 to 1993.

Month	Mean	Maximum	Minimum
Jan	99.53	101.13	97.80
Feb	99.02	100.54	97.35
Mar	98.70	100.70	97.23
Apr	99.50	101.85	97.39
May	101.44	102.51	99.59
Jun	101.69	102.61	100.15
Jul	101.47	102.52	99.65
Aug	101.12	102.27	99.13
Sep	100.75	102.24	98.72
Oct	100.46	101.88	98.50
Nov	100.35	102.01	98.35
Dec	100.07	101.75	98.56

TABLE 3. Some morphometric characteristics of Rybinsk Reservoir during 1948 to 1993.

	Surface area (km^2)			Total storage (km^3)			Depth (m)		
Month	Mean	Max	Min	Mean	Max	Min	Mean	Max	Min
Jan	3388	4115	2632	15.71	21.68	10.50	4.6	5.2	3.9
Feb	3159	3843	2474	13.99	19.36	9.38	4.4	5.0	3.8
Mar	3016	3915	2431	13.04	19.97	9.08	4.3	5.1	3.7
Apr	3375	4475	2488	15.61	24.77	9.48	4.6	5.5	3.8
May	4270	4830	3415	23.01	27.88	15.91	5.4	5.8	4.6
Jun	4395	4885	3667	24.09	28.36	17.87	5.5	5.8	4.9
Jul	4285	4836	3442	23.14	27.93	16.12	5.4	5.8	4.7
Aug	4110	4698	3208	21.64	26.72	14.35	5.2	5.7	4.5
Sep	3937	4682	3025	20.16	26.58	13.10	5.1	5.8	4.3
Oct	3807	4490	2926	19.06	24.90	12.46	5.0	5.5	4.2
Nov	3757	4555	2859	18.64	25.47	12.02	5.0	5.6	4.2
Dec	3631	4425	2953	17.57	24.34	12.64	4.8	5.5	4.2

TABLE 4. Monthly water exchange coefficients ($year^{-1}$) in Rybinsk Reservoir. The three columns are long-term means for 1947 to 1990, a low-water hydrological year (1972–73) and a high-water hydrological year (1955–56).

Month	Long-term mean	Low-water year	High-water year
Jan	0.14	0.06	0.13
Feb	0.14	0.04	0.14
Mar	0.17	0.07	0.15
Apr	0.36	0.21	0.56
May	0.18	0.09	0.59
Jun	0.10	0.06	0.29
Jul	0.09	0.07	0.13
Aug	0.09	0.08	0.09
Sep	0.12	0.08	0.10
Oct	0.12	0.10	0.12
Nov	0.13	0.10	0.15
Dec	0.13	0.09	0.18
Year	1.78	1.05	2.63

TABLE 5. Long-term monthly mean water temperature (°C) in Rybinsk Reservoir. The values are long-term means at the surface, for 1951 to 1986, and in the water column (1978–1985 and 1988–1991). — indicates no measurements.

J	F	M	A	M	J	J	A	S	O	N	D
Surface temperatures (1951–1986):-											
—	—	—	0.9	8.1	15.9	19.3	18.7	12.7	6.1	1.5	—
Water column temperatures (1978–1985, 1988–1991):-											
0.6	0.9	0.8	1.0	5.9	14.1	18.9	19.1	13.9	8.2	2.8	0.6

is higher than the water column average during warming but the reverse tendency occurs during cooling. Intensive cooling marked by daily changes of 0.2 to 0.5°C starts in the second half of August, with a uniform temperature decrease in the water-mass. Values of 3 to 5°C are usually obtained by the end of October. The horizontal distribution of water temperature is quite uniform in early autumn.

Later, a weak temperature decrease takes place in the reaches of the Mologa and Sheksna rivers and also of small tributaries. Previous and present hydrometeorological conditions affect the spatial temperature distribution during freezing. Ice phenomena appear from 11 to 20 November (early dates) or 7 to 15 December (late dates). The lowest winter temperatures (< 1°C) are in the main tributaries, i.e. the Volga, Mologa and Sheksna rivers.

3.5. WATER TRANSPARENCY

Water transparency in the reservoir is low and never exceeds 2.0–2.5 m because of the frequent wind-mixing. Water colour is rather high due to the marshy catchment area. It corresponds to the mesohumic or mesopolyhumic type [21] and varies from 40 to 140 Pt-Co scale degrees in spring; the range is reduced to 40–60 Pt-Co degrees in summer and autumn. The lowest colour occurs in winter.

3.6. CHEMICAL CONTENT AND PHYTOPLANKTON

The oxygen regime is rather favourable, with the usual dissolved oxygen content near saturation. In its ionic composition, the reservoir water belongs to the Ca^{2+}/HCO_3^- type. Regarding nutrients, Rybinsk Reservoir is characterised by a rather high content of total phosphorus (TP) and total nitrogen (TN) [22], a feature of eutrophic waters [23]. The mean annual nutrient content was equal to 76–97 $\mu g\ l^{-1}$ TP and 1.06–1.31 mg l^{-1} TN in 1981–1982 [22], and 80 $\mu g\ l^{-1}$ TP and 0.95 mg l^{-1} TN in 1993 (Mineeva, unpublished). As the TN/TP ratio by mass varies around 12 to 15, nutrient content probably did not limit phytoplankton development.

The principal groups of planktonic algae consist of diatoms during most of the year, joined by blue-green algae in summer. The present trophic level of the reservoir is estimated as moderately eutrophic, according to the mean chlorophyll content during the ice-free period [1, 2, 8].

4. Results

4.1. PRINCIPAL HYDROLOGICAL EVENTS

Environmental conditions in the reservoir depend much on hydrometeorological characteristics. Long-period variations in some of them (as mean annual values) are shown in Fig. 2. Before outlining pecularities of the hydrological regime, brief mention should be made of the concept of water content. Conditions of water content in the reservoir can be characterised in terms of water level, total input and water exchange intensity. All these parameters are subject to the annual cycle, but each also depends on long-period variations. Intercentennial, 20- to 30-year rhythmic fluctuations (Bruckner cycles) were described by Shnitnikov [24]. A high-water period (transgressive phase) follows a low-water period (regressive phase) inside each cycle. Variations in parameters of the water regime during these periods result from variations in factors governing the total water input, i.e. atmospheric precipitation, air temperature, etc.

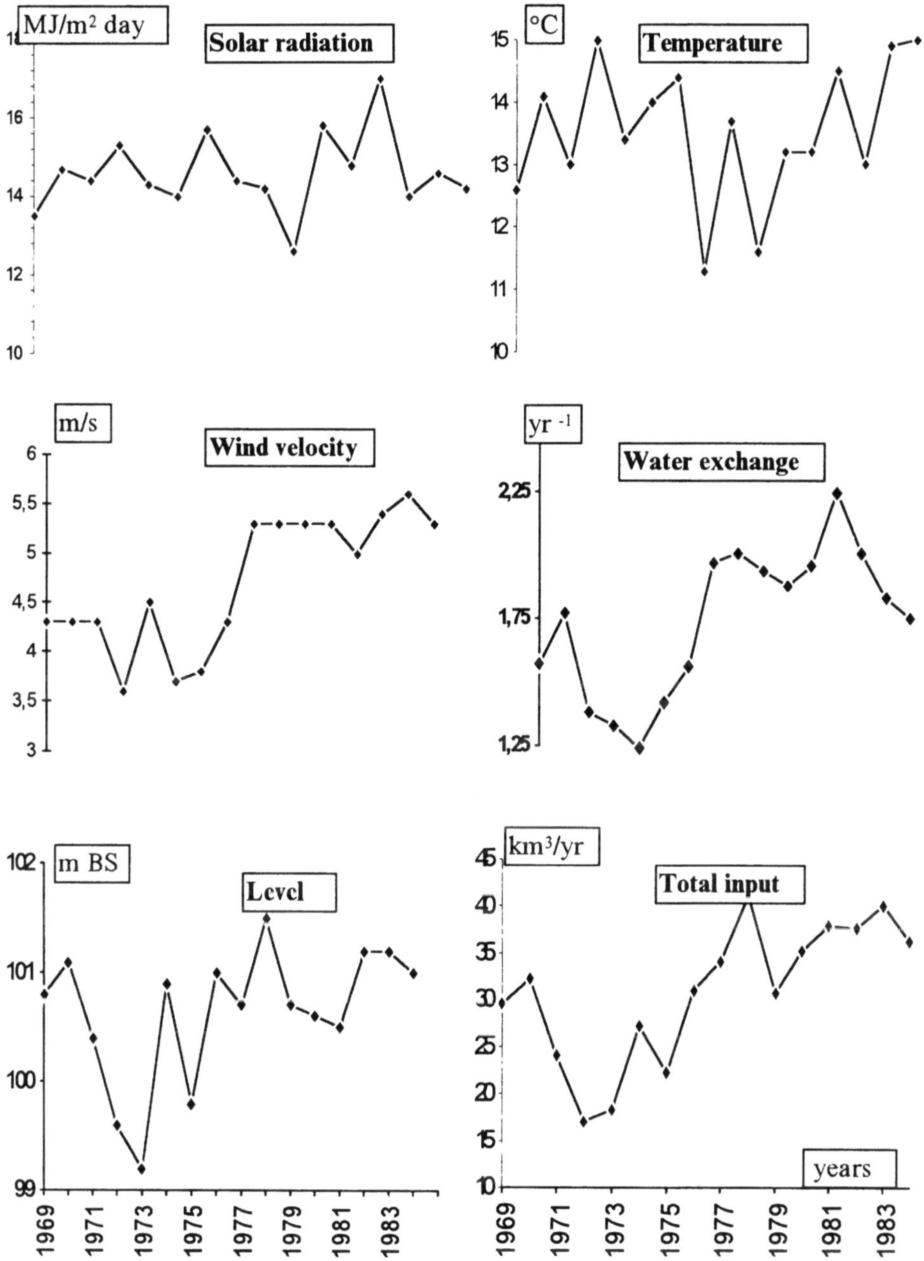

FIGURE. 2. Fluctuations (mean values) of some hydrological characteristics in Rybinsk Reservoir during the period 1969–1985.

TABLE 6. Hydrometeorological characteristics for Rybinsk Reservoir during high-water and low-water periods in 1969–1975 and 1976–1984. Means ± standard error; $t_{0.05}$ = 1.96. * Solar radiation total spectrum including PAR.

Variables	1969–1975	1976–1984	t-criterion
Surface area (km^2)	3789 ± 115	4073 ± 139	2.34
Water level (m, BS)	100.26 ± 0.28	100.93 ± 0.11	2.23
Total input (km^3 $year^{-1}$)	24.43 ± 2.14	36.03 ± 1.21	4.72
Total flow (km^3 $year^{-1}$)	24.09 ± 2.11	32.61 ± 1.08	3.59
Precipitation (km^3 $year^{-1}$)	2.00 ± 0.15	2.45 ± 0.12	2.34
Total input/precipitation	11.2 ± 0.5	13.8 ± 0.4	4.06
Water exchange ($year^{-1}$)	1.47 ± 0.07	1.95 ± 0.04	5.95
Wind velocity (m s^{-1})	4.1 ± 0.2	5.2 ± 0.1	4.92
Solar radiation (MJ m^{-2})*	14.6 ± 0.3	14.6 ± 0.4	0.0
Water temperature (°C)	13.8 ± 0.2	13.4 ± 0.4	0.89
Water transparency (cm)	148 ± 5	117 ± 2	5.76

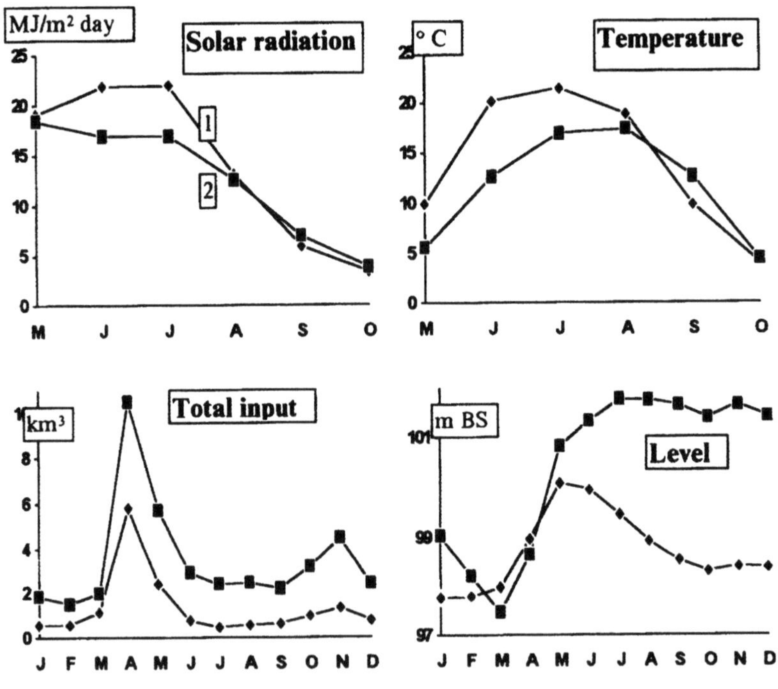

FIGURE 3. Seasonal cycle of hydrological characteristics in Rybinsk Reservoir during years with different water contents: 1 = 1973, extreme low water; 2 = 1978, extreme high water.

Three periods differing in water content can be distinguished over the time of the reservoir's existence: two high-water periods (1949–1962 and since 1976) and one low-water period (1963–1975) [17]). Our observations embrace two of them. Variations in certain characteristics during these periods are given in Table 6, which shows differences in reservoir characteristics between the two periods. Lower values of total input and total flow volume, precipitation, water exchange, water level and wind velocity are the features of the first period. These parameters became higher during the later years. The ratio of total input/precipitation also increases, indicating that lateral input provided most of the total input volume. A noticeable decrease in transparency occurred due to increased wind activity and total input. However, the two periods did not differ in solar energy income and water temperature that showed behaviour expected from features of reservoir geographical location ("zonal" characteristics). Hydrological parameters showed quite similar seasonal variations in the years that differed in water content. At the same time, most of them differ in their magnitude during different years (Fig. 3).

4.2. CHLOROPHYLL CONTENT AND ITS VARIATION

CHL values during all the period since 1969 varied within a wide range from 1–3 to 100–150 μg l^{-1}, following seasonal succession of biomass. Season variations in CHL are shown in Fig. 4. They did not change strongly from year to year as spring and summer peaks were the common features. The first maximum with CHL values of 18–52 μg l^{-1} (the limit values over all the period) occurred during May or the beginning of June. It may be observed at all the stations either simultaneously or with a short-term displacement. The spring maximum of phytoplankton develops under intensive input of solar radiation (daily sum 16–25 MJ m^{-2}) in the period of rapid warming, with water temperature about 10°C in the Main Part and 12°C in the Volga Reach (Table 7). This temperature difference corresponds with the differing regimes of warming in the two basins (see Section 3.4). Intensive water input and circulation during the spring flood provide high nutrient content and are propitious for diatom algae that then dominate the planktonic community.

TABLE 7. Chlorophyll content (μg l^{-1}) and water temperature (°C) in the Volga Reach and Main Part of Rybinsk Reservoir during the spring and summer phytoplankton peaks of 1969 to 1984. Means ± standard errors.

Variable	Volga Reach Spring	Summer	Main Part Spring	Summer
Chlorophyll	30.8 ± 3.0	33.1 ± 5.4	25.0 ± 7.4	32.5 ± 7.7
Temperature	12.5 ± 1.0	19.3 ± 2.3	10.6 ± 0.8	21.0 ± 0.4

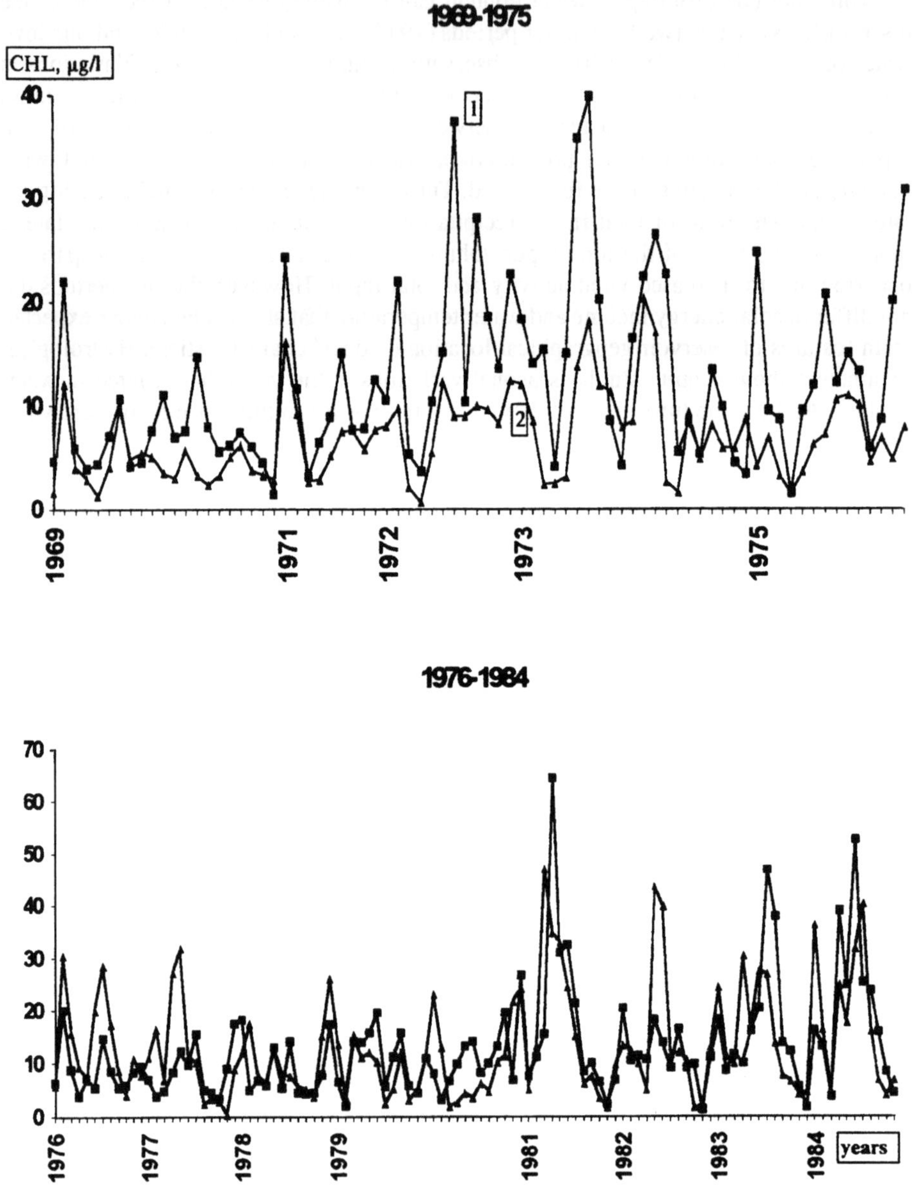

FIGURE 4. Seasonal fluctuations of chlorophyll (µg l^{-1}) in the Volga Reach and Main Part of Rybinsk Reservoir during 1969 to 1984.

The only weak correlation between CHL and abiotic factors – including total annual water input – has been found during this period (Table 7). (We calculated the correlation of CHL with total annual input volume because most of the latter comes during April–May).

The spring maximum is rather brief. Starting with the end of May or, more often, during June, the CHL content falls to 1–5 μg l^{-1} and the late spring (or early summer) phytoplankton depression occurs in the reservoir. With further warming a substitution of vernal species by the summer species takes place in the phytoplankton community. Zooplankton abundance also increases during this period [25], indicating an increased loss-rate by grazing. Starting 2 or 4 weeks later, the summer CHL peak begins its development. It corresponds to increased biomass that continues from July to September, with CHL concentrations of 10 to 77 μg l^{-1} in the Volga Reach and 8 to 127 μg l^{-1} in the Main Part. It develops during the maximum summer warming and seems to be determined by hydrometeorological conditions. Thus, the lowest CHL content was found in 1978 and 1980 under low solar radiation and water temperature. Additionally, a special pattern of water level with high-water up to autumn was observed during both these years. The highest CHL content, by contrast, is observed in years with the prevalence of hot and sunny summer weather (1972, 1973, 1981), including periods of extreme low total input volume and water level (1973). Such periods, characterised by low wind velocity and infrequent wind-mixing, provide favourable conditions for the development of blue-green algae. However, mixed algal communities that consist of diatoms and blue-greens [27] are most typical in summer. The CHL content during the summer peak, as well as during spring, does not depend strongly on the abiotic parameters mentioned above (Table 8). Control seems to be indirect due to a composite influence of environmental conditions on aquatic organisms. The single exception is the CHL summer maximum in the Volga Reach that correlates with solar radiation. This is not surprising for the algal community. Nevertheless, it had not been found either in a large, shallow and open-water basin or in the Volga Reach during spring, i.e. where and when stability of the water column was destroyed by the wind. The highest seasonal spring and summer values of CHL do not correlate with each other (Fig. 5).

TABLE 8. Correlation coefficients between chlorophyll and abiotic variables during the spring and summer phytoplankton peaks of 1969 to 1984 in Rybinsk Reservoir ($n = 16$, $r_{0.05} = 0.497$).

Variables	Volga Reach Spring	Volga Reach Summer	Main Part Spring	Main Part Summer
Water temperature	–0.42	–0.34	0.48	0.25
Solar radiation	0.35	0.75	0.14	0.32
Water level	0.10	–0.21	–0.10	0.13
Total input	–0.24	0.48	0.21	0.45

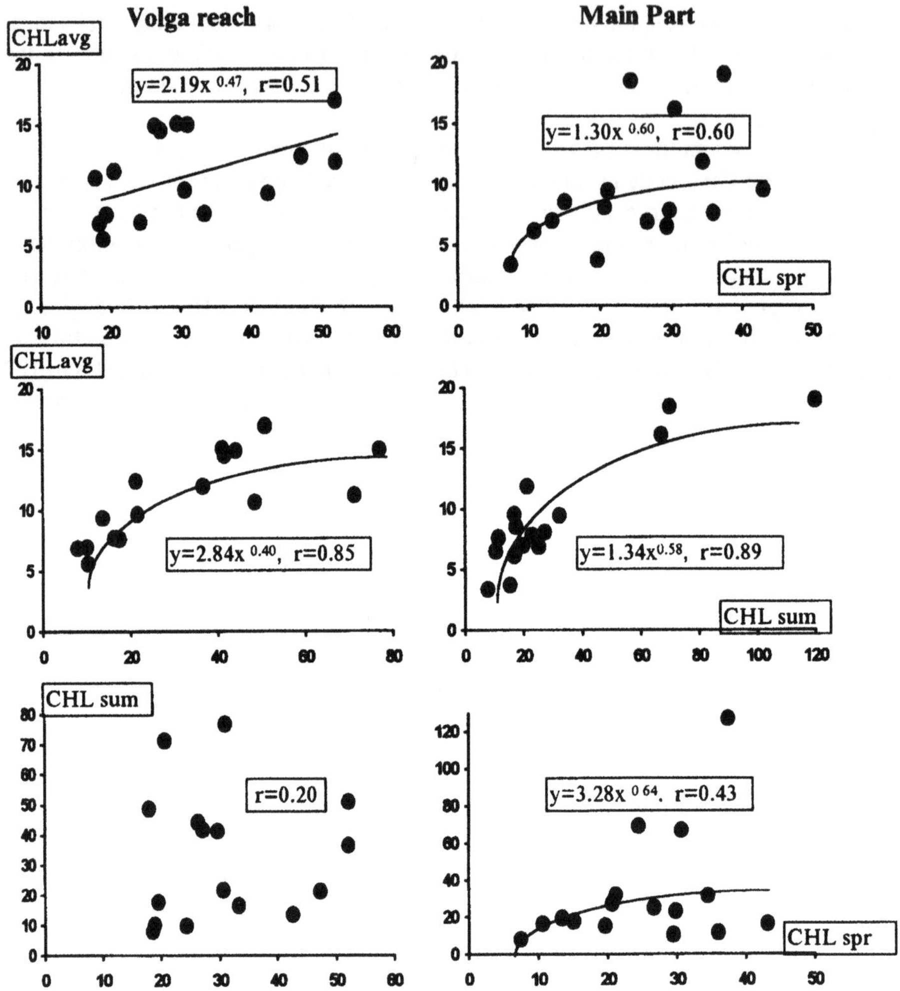

FIGURE 5. Relationship between the seasonal average and the seasonal maximum of chlorophyll (μg l^{-1}) in the Volga Reach and Main Part of Rybinsk Reservoir ($n = 16$, $r_{0.05} = 0.49$).

The survey of long-term CHL variations during 1969 to 1984 found some interesting points. First, it showed rather strong year-to-year fluctuations in CHL values. Thus, the highest spring CHL respectively varied by 2.8 and 5.8 times in the Volga Reach and the Main Part, the summer maximum CHL varied by 7.7 and 16 times, and the May to October average varied by 3.2 and 5.6 times (Fig. 6). On the whole, increased phytoplankton abundance (as chlorophyll) occurred during the years marked by hot and sunny (1972, 1973, 1975, 1981) or warm and cloudy (1983, 1984) weather, providing stable warming of the water masses or low water levels (1972, 1973) (see Fig. 2).

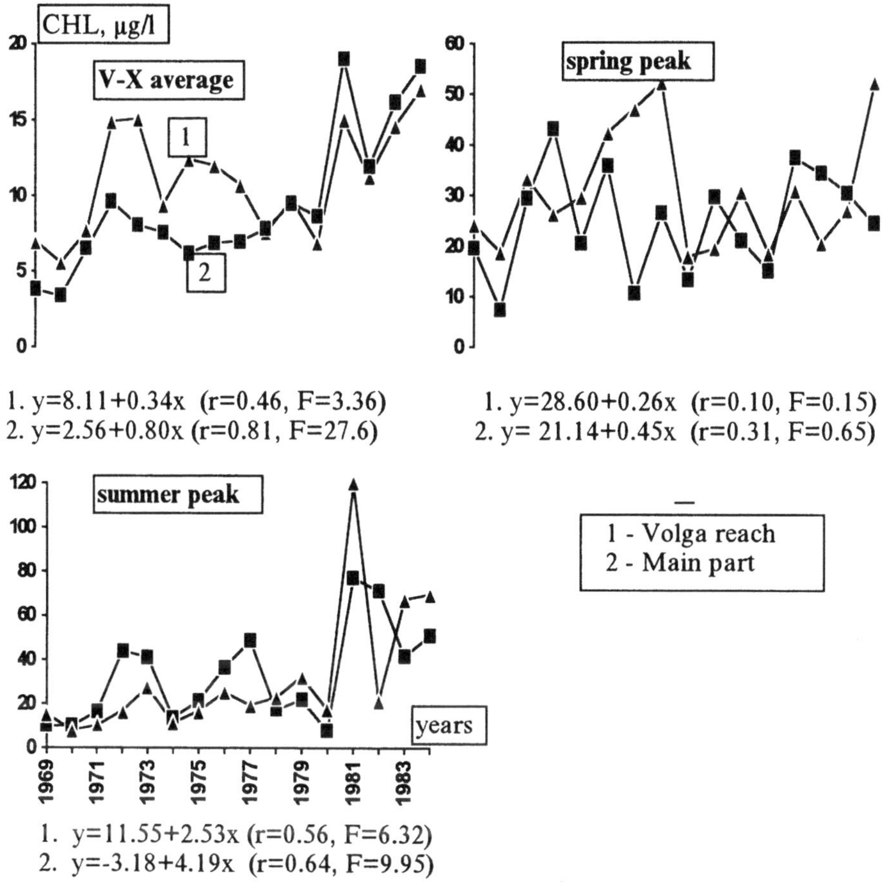

FIGURE 6. Variations in the seasonal average and the spring and summer maxima of chlorphyll (µg l^{-1}) in the Volga Reach (1) and Main Part (2) of Rybinsk Reservoir during 1969 to 1984 (n = 16, $r_{0.05}$ = 0.49; $F_{0.05}$ = 3.74).

A significant growth in average CHL has been found in the Main Part (Fig. 6), as estimated by the F-criterion. Empirical models showed a significant yearly chlorophyll increment that was equal to 0.87 µg l^{-1} in the Main Part, and a lack of it in the Volga Reach (see the equations given in Fig. 6). At the same time, summer maximum CHL became higher in both basins, in contrast to spring maximum CHL that did not change markedly (Fig. 6). A yearly increment in the CHL summer maximum, equal to 2.5 and 4.2 µg l^{-1} in the studied basins, was significant as estimated by the F-criterion. Growth of seasonal averages and summer maximum CHL in the Main Part was followed by simultaneous increases in the most frequent values, rising from 1–5 µg l^{-1} at the start and increasing to 5–10 µg l^{-1} in the 1970s and 10–20 µg l^{-1} in the 1980s (Fig. 7).

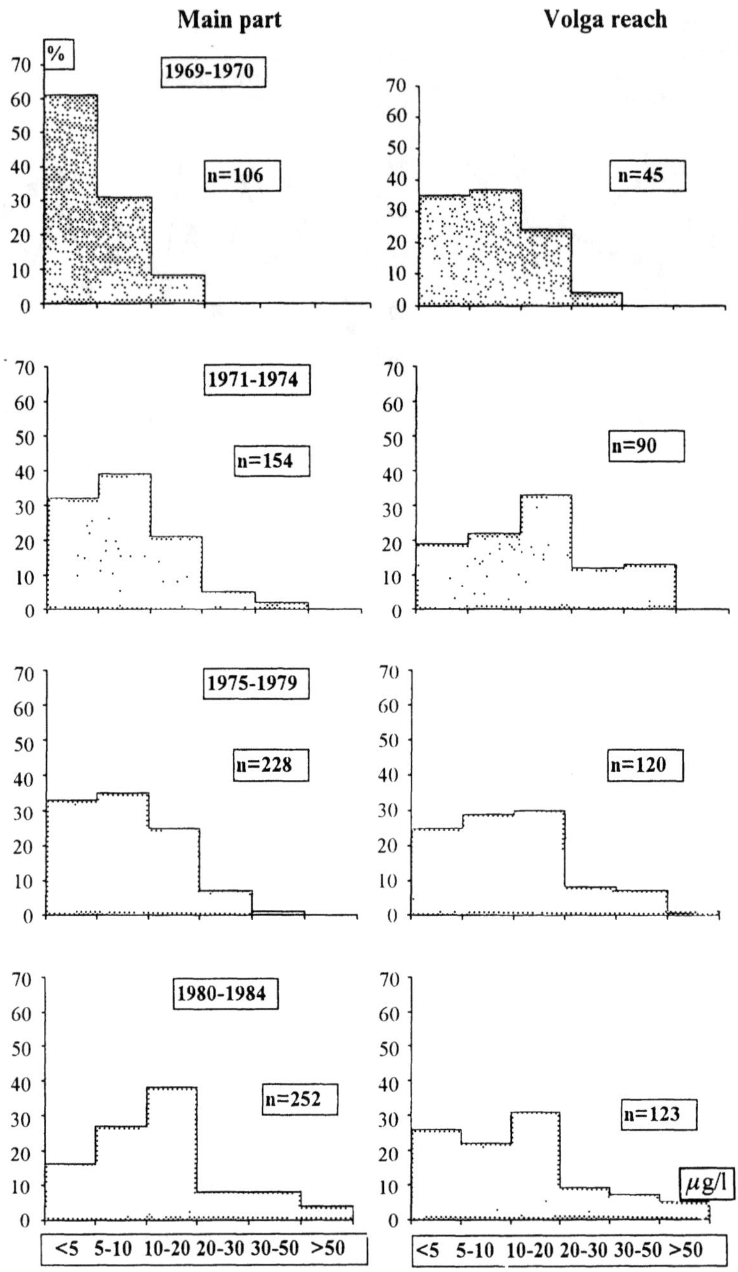

FIGURE 7. Relative frequency (%) of chlorophyll concentrations (μg l^{-1}) in the Main Part and Volga Reach of Rybinsk Reservoir during 1969 to 1984.

In the Volga Reach, CHL content was euqal to 10–20 μg l^{-1} over all the period. Both the mean and the most frequent CHL values in the Main Part were represented by magnitudes typical for mesotrophic waters (less than 10 μg l^{-1}, according to [15]) in the beginning of the observation period, and changed to more eutrophic values in the last years. If we consider CHL as a criterion of the trophic level, it became higher in the Main Part, as demonstrated by the increase in CHL values. The trophic status of the Volga Reach did not change to the same degree, where there was an increase in summer CHL only.

In accordance with results of Korneva [26] and Korneva & Mineeva [1] our conclusion about increased reservoir trophic state agrees with evidence of an increase in water saprobity connected with increased abundance of small-celled algae. These were *Stephanodiscus* spp. and small cryptomonads in spring-dominant complexes, and *Microcystis holsatica* accounting for up to 50–70% of total phytoplankton biomass in summer. Phytoplankton biomass (as fresh weight) varied in the range 0.84–3.41 mg l^{-1} (annual average) during 1964 to 1981 and did not reveal any trend. However, two periods were identified that differed significantly in mean biomass; these were 1954 to 1970 and 1971 to 1981, with means of 1.51 ± 0.52 and 2.34 ± 0.53 mg l^{-1}, respectively. The appearance of small-celled species (*r*-strategists) seems to restrain biomass from a significant increase. At the same time, an estimated increase in the chlorophyll/biomass ratio from 0.2–0.3% to 0.3–0.6% is evidence for a decline of phytoplankton mean cell volume under eutrophication [1].

It is common knowledge that phytoplankton development depends on the environmental conditions. However, each separate abiotic characteristic showed its immediate effect on CHL content only on rare occasions. A simple regression did not disclose any direct influence of abiotic factors on CHL values, either for the Volga Reach and Main Part of the Rybinsk Reservoir during 1969 to 1984, as estimated from annual means, or for the whole reservoir during 1981 to 1982, as demonstrated by our previous seasonal data then obtained [2] (Table 9). At the same time a *multiple* correlation of CHL with all those characteristics was rather high; R was equal to 0.57 for the 1981–1982 seasonal data.

TABLE 9. Correlation (r) of chlorophyll with hydrological variables in Rybinsk Reservoir, for 1969–1984 ($n = 16$, $r_{0.05} = 0.497$) compared with May–October 1981–1982 ($n = 108$, $r_{0.05} = 0.20$).

Variables	Average for 1969 to 1984 Main Part	Average for 1969 to 1984 Volga Reach	1981–1982 Main Part
Solar radiation	0.37	0.37	0.11
Water temperature	0.53	0.53	0.24
Water level	–0.38	0.13	–0.07
Wind velocity	0.07	0.52	0.03
Water exchange coefficient	–0.05	0.39	---
Water transparency	---	---	0.33
Water colour	---	---	0.07

Principal Component (PC) analysis revealed a strong influence of water content on phytoplankton development in both the Volga Reach and the Main Part (Fig. 8). Water level and water exchange ratio were closely related with the PC1 that included over 40% of the total variance and should be called the "component of water content". PC2 should be called the "component of temperature and solar radiation", as it was closely connected with those two parameters; it included ca. 30% of the total variance. Two periods were distinguished according to the projection of points (years) on the PC1 axis, corresponding with the low-water (1969–1975) and high-water (since 1976) periods. The same results were obtained in both basins, lending support for the effect of hydrological regime on phytoplankton abundance.

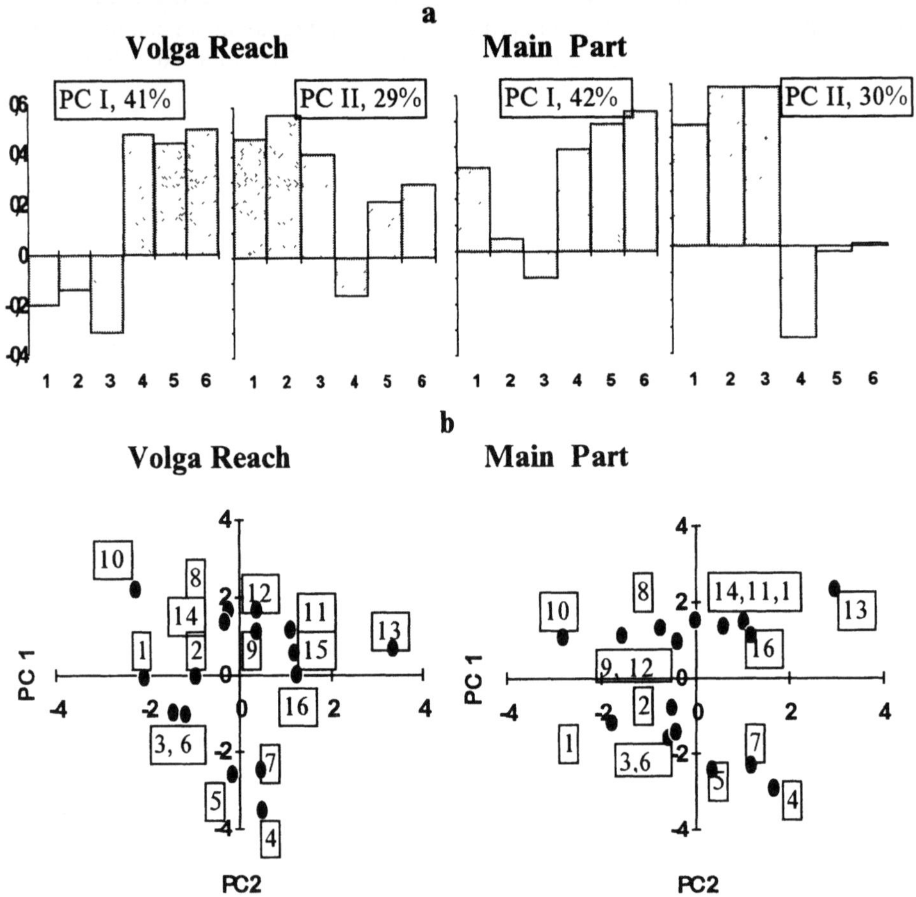

FIGURE 8. Relationships between annual mean chlorophyll content and hydrometeorological variables in the Volga Reach and Main Part of Rybinsk Reservoir, resulting from Principal Component analysis. (a), Relative weightings of variables (1 = chlorophyll, 2 = solar radiation, 3 = water temperature, 4 = water level, 5 = wind velocity, 6 = annual water exchange coefficients). (b), Ordination of years 1 to 16 (1969–1984) on the axes of PC1 and PC2.

CHL values calculated for the two distinguished periods, as well as trends of their variation, differed strongly in the two basins. Mean CHL, that varied within almost the same limits in the Volga Reach and the Main Part throughout 1969 to 1984 (5.6 to 17.0 and 3.4 to 18.5 μg l^{-1}, respectively, Fig. 6), was not the same regarding long-term trends. During the two distinguished periods it demonstrated a noticeable growth in the Main Part, as estimated from t-criteria (Table 10). There were no definable trends in mean CHL in the Volga Reach. Spring CHL did not show any long-term increases, but the summer maximum CHL rose markedly in both basins. Regarding the summer peak, there is a difference if one compares it with spring values of the two periods. Summer CHL values during the second period became higher relative to those of the spring peak during the later years, which may be considered as a feature of eutrophication. This CHL growth occurred during the later years that belong to the high-water period.

TABLE 10. Comparison of mean seasonal and spring and summer peaks of chlorophyll in the Volga Reach and Main Part of Rybinsk Reservoir during 1969–1975 and 1976–1984. Mean concentrations (μg l^{-1}) ± standard error; $t_{0.05} = 1.96$.

	Volga Reach			Main Part		
Chlorophyll	1969–1975	1976–1984	t	1969–1975	1976–1984	t
Seasonal mean	10.3 ± 1.4	11.6 ± 1.1	0.73	8.4 ± 1.1	11.7 ± 1.6	1.70
Spring peak	31.8 ± 3.8	30.0 ± 4.5	0.30	23.8 ± 4.9	25.9 ± 2.7	0.38
Summer peak	22.4 ± 5.4	41.8 ± 7.8	2.05	15.5 ± 2.4	45.9 ± 12.0	2.48

It might appear that all the conditions connected with increased water input have to restrict phytoplankton development. In reality it is possible to presume that increase in total input, because of increased lateral inflow during the high-water period, had led to nutrient enrichment and, as a result, to growth of CHL content. As reservoirs are artificial waterbodies, some features of their regime depend on human activity. Water level seems to be one of those characteristics that can be governed and determined at any instant. If we look for the influence of water level on CHL content, its effect on phytoplankton abundance does appear but during the low-water period only. The strong correlation of CHL with water level has been obtained from the example of the Main Part, which contains most of the total reservoir storage (Fig. 9). It justifies the conclusion that elements of the water regime strongly affect phytoplankton development. The regression model makes it possible to predict seasonal mean CHL under variations of water level (Table 11).

Such estimation seems to be quite realistic if one keeps in mind the alterations of high-water and low-water periods. At the same time, the recent CHL content is 1.4 times higher than in the early 1970s (see Table 9) and so a corresponding correction should be made (Table 11). An important point is that the reservoir trophic status should stay unchanged when the mean annual level is not lower than 98 m BS, so that calculated CHL does not exceed the real seasonal average values obtained over all the investigated period.

TABLE 11. Estimated values for mean chlorophyll concentrations (μg l^{-1}) at differing water levels (m, BS) in the Main Part of Rybinsk Reservoir, calculated from a regression equation for the low-water period, given in Fig. 9 (Estimate 1) and corrected by a factor of 1.4 (Estimate 2).

Water level	Mean chlorophyll concentrations Estimate 1	Estimate 2
95	19.3	27.0
96	16.9	23.7
97	14.5	20.3
98	12.2	17.1
99	9.7	13.6
100	7.3	10.2
101	4.9	6.9

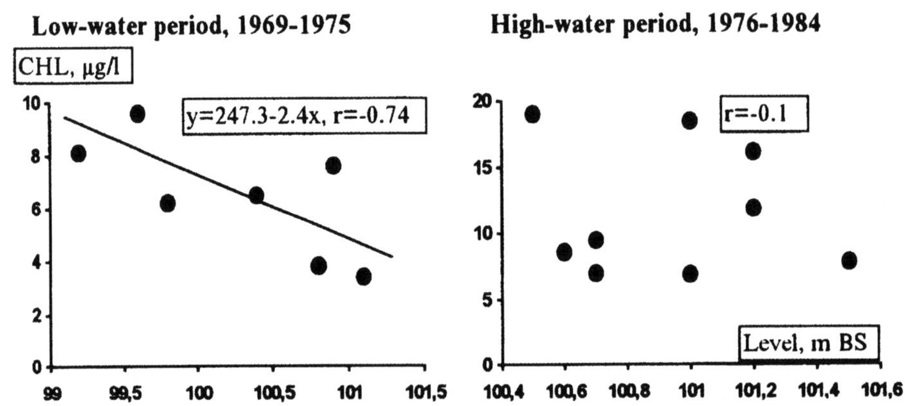

FIGURE 9. Relationship between mean chlorophyll concentrations (μg l^{-1}) and water levels (m, BS) in the Main Part of Rybinsk Reservoir during low-water and high-water periods.

5. Discussion

The following postulates are evident from the results obtained. CHL seasonal succession, that is intra-annual variability [27] with adaptation of community changes [28], is characterised by two or three peaks that seem to be usual for phytoplankton of temperate lakes and reservoirs. The same result has been obtained in English lakes [4], northwest Russian lakes [29], Volga cascade reservoirs (our data) and many others. Such two-peaked (diacmic [30]) or three-peaked (polyacmic) patterns with maxima in spring and summer or spring, summer and autumn maxima, is a feature of mesotrophic and moderately eutrophic waters.

Increase in trophic level within such limits does not affect the course of CHL seasonal changes in general. However, transformation of spring and summer maxima proceeds in such a way that summer values become much higher. Further progress in eutrophication has probably led to the development of one-peaked (monoacmic) seasonal dynamics with a prolonged summer maximum that has been shown for some lakes [29]. The situation observed in large reservoirs is quite similar and can be illustrated by the following examples (Fig. 10).

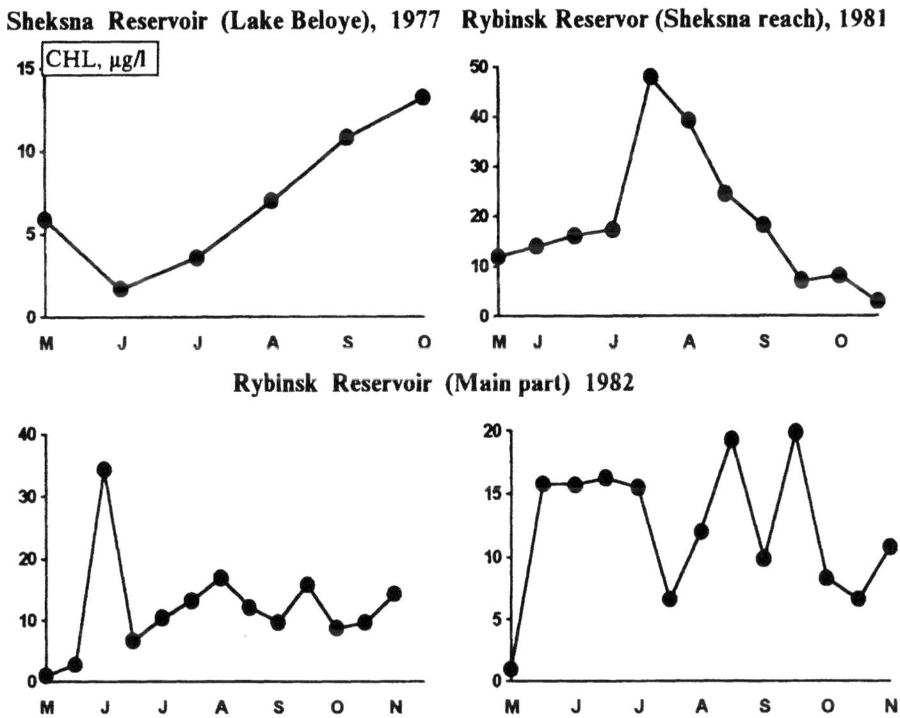

FIGURE 10. Seasonal variation in mean chlorophyll concentrations (µg l^{-1}) in reservoir basins with differing trophic levels.

Strong spring and autumn peaks have been found in the Belozyorsk Part of Sheksna Reservoir (northwestern Russia) with its highest CHL values of 15 μg l^{-1} and seasonal average ca. 5 μg l^{-1} [1], i.e. a weakly mesotrophic basin. In contrast, a single extended summer maximum was observed in the Sheksna Reach of Rybinsk Reservoir, the most eutrophicated area with higher CHL values of 50 to 100 μg l^{-1} and a May to October average of ca. 20 μg l^{-1} [2] (see Table 1). Year-to-year CHL fluctuations that are long-term in scale [27] show the increase in reservoir trophic level starting from the end of the 1970s. Increased volumes of total water input during the high-water period, and the influence of the Sheksna Reach, seem to be the principal reasons for trophic level changes in the Main Part of the reservoir. Eutrophication in the Volga Reach was insignificant, though its primary trophic level was higher in comparison with the Main Part because of the input of nutrient-enriched water by the River Volga [4].

Increase in trophic level is expressed in growth of seasonal averages, summer maxima and the most frequent CHL values that seem to be a result only of summer CHL growth. It is substantiated by a close correlation of summer peak and seasonal average CHL. In the case of Rybinsk Reservoir, the set of hydrological parameters exerts a primary control over algal abundance and productivity. Year-to-year CHL fluctuations are quite similar in both the Volga Reach and the Main Part, and depend upon hydrometeorological conditions. Water content is the most significant among them: clustering of years in the field of the first and second Principal Components correspond to high-water (1969–1975) and low-water (since 1976) periods. The strong effect of water content conditions – that can be expressed, for example, in terms of water level and is being controlled by human activity – was found during the low-water period. The regression model makes it possible to estimate phytoplankton development (CHL content) depending on variations in water content of the reservoir, and thus provides for ecosystem monitoring and management.

Acknowledgement

This work was supported by a grant from the Russian Foundation for Fundamental Research, Russian Academy of Sciences.

References

[1]. Korneva, L. G. & Mineeva, N. M. (1996). Estimation of trophic state changes in the Rybinsk Reservoir based on phytoplankton composition and pigment analysis. *Hydrobiologia* (in press).

[2]. Mineeva, N. M. (1986). Formation of phytoplankton primary production in water-bodies of different types. Dissertation manuscript, Institute for Biology of Inland Waters, Borok. [In Russian].

[3]. Pyrina, I. L. (1991). Long-period chlorophyll dynamics and productivity of plant plankton in the Rybinsk Reservoir. In: *Ecological aspects of regulation of growth and productivity of plants* (eds. O. V. Titova & V. I. Kefely). University Press, Yaroslavl, pp. 253-259.

[4]. Talling, J. F. (1993). Comparative seasonal changes, and inter-annual variability and stability, in a 26-year record of total phytoplankton biomass in four English lake basins. *Hydrobiologia* **286**, 65-98.

[5]. Mineeva, N. M. & Pyrina, I. L. (1986). Investigations of phytoplankton pigments in the Rybinsk reservoir in 1977-1979. In: *Biology and ecology of aquatic organisms* (ed. A. V. Monakov). Nauka Press, Leningrad, pp. 90-105. [In Russian].

[6]. Pyrina, I. L. & Mineeva, N. M. (1990). Content of phytoplankton pigments in water of the Rybinsk reservoir. In: *Flora and productivity of pelagic and littoral phytocenoses in the waters of the Volga basin* (ed. V. A. Ekzertsev). Nauka Press, Leningrad, pp. 178-188. [In Russian].

[7]. Pyrina, I. L., Mineeva, N. M., Korneva, L. G. & Letanskaya, G. I. (1981). Phytoplankton and its production. In: *Anthropogenic influence on large lakes of the north-west USSR. II. Hydrobiology and bottom sediments of Lake Beloye* (ed. D. N. Alexandrova). Nauka Press, Leningrad, pp. 15-64. [In Russian].

[8]. Pyrina, I. L., Mineeva, N. M. & Sigariova, L. E. (1986). Long-term investigation of phytoplankton pigments in the Rybinsk Reservoir. *Abstracts of the 5th USSR Hydrobiological Society Congress. Part 1, Tolyatty, September 15-19, 1986*, pp. 208-209. Kuibyshev. [In Russian].

[9]. Pyrina, I. L. & Sigariova, L. E. (1986). Phytoplankton pigments in the Rybinsk Reservoir during years with different hydrometeorological conditions. In: *Biology and ecology of aquatic organisms* (ed. A. V. Monakov). Nauka Press, Leningrad, pp. 65-89. [In Russian].

[10]. Yelizarova, V. A. (1973). Content and composition of plant pigments in waters of the Rybinsk reservoir. *Hydrobiological Journal* **2**, 23-33. [In Russian].

[11]. Lorenzen, C. J. & Jeffrey, S. W. (1980). Determination of chlorophyll in sea water. *UNESCO Technical Paper in Marine Science* **35**. UNESCO Paris.

[12]. UNESCO Working Group 17 (1966). Determination of photosynthetic pigments in sea water. In: *Monographs of oceanographic methodology*. UNESCO, Paris, pp. 9-18.

[13]. Jeffrey, S. W. & Humphrey, G. F. (1975). New spectrophotometric equations for determining chlorophylls *a, b,* c_1 and c_2 in higher plant, algae and natural phytoplankton. *Biochem. Physiol. Pflanzen* **167**, 191-194.

[14]. Mineeva, N. M. (1980). The method of calculating chlorophyll concentrations. *Biology of Inland Waters, Information Bulletin* **46**: 67-70. [In Russian].

[15]. Vinberg, G. G. (1960). *The primary production of the basins.* University Press, Minsk. [In Russian].

[16]. Likens, G. E. (1975). Primary production of inland aquatic ecosystems. In: *Primary production of the biosphere* (eds. H. Leith & R. H. Whittaker). Springer-Verlag, Berlin-Heidelberg-New York, pp. 185-202..

[17]. Fomichyov, I. Ph. & Litvinov, A. S. (1980). Long-period variations of water budget components in the Rybinsk Reservoir and their influence on water exchange and level. *Water Resources* **4**, 108-119. [In Russian].

[18]. Vikulina, Z. A. & Znamensky, V. A. (Eds.) (1975). *Hydrometeorological regimen of lakes and reservoirs of the USSR. The upper Volga reservoirs.* Hydrometeoizdat, Leningrad. [In Russian].

[19]. Kuzin, B. S. (Ed.) (1972). *The Rybinsk reservoir and its life.* Nauka Press, Leningrad. [In Russian; summary in English].

[20]. Litvinov, A. S. (1985). Temporal and spatial variability of the temperature fields in reservoirs. In: *Hydrophysical processes in rivers and reservoirs.* Nauka Press, Moscow, pp. 175-181. [In Russian].

[21]. Kitaev, S. P. (1984). Ecological fundamentals of lake bioproductivity in different natural zones. Nauka Press, Moscow. [In Russian].

[22]. Mineeva, N. M. & Razgulin, S. M. (1995). Influence of nutrients on chlorphyll content in the Rybinsk Reservoir. *Water Resources* **22**, 218-223. [In Russian].

[23]. Vollenweider, R. A. (1979). Das Nährstoffbelastungskonzept als Grundlage für den Eutrophierungsprozess stehender Gewässer und Talsperren. *Zeitschrift für Wasser und Abwasser Forshung* **12**, 46-56.

[24]. Shnitnikov, A. V. (1969). *Intercentury variability of the components of total moistening.* Nauka Press, Leningrad.

[25]. Rivier, I. K., Lebedeva, I. M. & Ovchinnikova, N. K. (1982). Long-term zooplankton dynamics in the Rybinsk Reservoir. In: *Ecology of aquatic organisms in the upper Volga reservoirs* (ed. A. V. Monakov). Nauka Press, Leningrad, pp. 69-87. [In Russian].

[26]. Korneva, L. G. (1993). Phytoplankton of the Rybinsk Reservoir: composition, peculiarity of spatial distribution, consequences of eutrophication. In: *Modern state of the Rybinsk Reservoir ecosystem* (ed. A. I. Kopylov), pp. 50-113. [In Russian].

[27]. Reynolds, C. S. (1990). Temporal scales of variability in pelagic environments and the response of phytoplankton. *Freshwater Biology* **23**, 25-54.

[28]. Harris, G. P., Haffner, G. D. & Piccinin, B. B. (1980). Physical variability and phytoplankton communities. II. Primary productivity by phytoplankton in a physical variable environment. *Archiv für Hydrobiologie* **88**, 393-425.

[29]. Trifonova, I. S. (1988). Oligotrophic-eutrophic succession of lake phytoplankton. In: *Algae and the aquatic environment* (ed. F. E. Round). Biopress Ltd, Bristol, pp. 107-124.

[30]. Hutchinson, G. E. (1967). *A treatise on limnology. Vol. II. Introduction to lake biology and the limnoplankton.* John Wiley & Sons, New York, London & Sydney.

LONG-TERM HYDROBIOLOGICAL INVESTIGATIONS ON LAKES IN NORTHWEST RUSSIA

V. G. DRABKOVA, V. V. SKVORTSOV,
T. D. SLEPOKHINA AND I. S. TRIFONOVA
Institute for Lake Research RAS
Sevastyanov, 9, St. Petersburg
196199 Russia

1. Introduction

Very often, monitoring of fresh water is regarded as monitoring eutrophication. Accelerated eutrophication of fresh waters evokes a serious degradation of water quality.

In spite of the fact that the effect of anthropogenic eutrophication was first marked at the end of the last century, intensive study of this process began only in the 1960s and 1970s. The problem of anthropogenic eutrophication of lakes on the territory of former Soviet Union (FSU) cropped up only recently, much later than in other countries.

Rates and intensity of lake eutrophication are most clearly detected in those lakes where a regime of study has been carried on for many years. This is the main reason why the interest in long-term ecological investigations increased markedly in our country during recent years. Study regimes on practically all large reservoir ecosystems have been organised since they were commissioned. However, there were not many lakes in the territory of the FSU on which long-term investigations had been carried out. Such lakes include the following:-

Beloe (Moscow Region); its study began in 1910–1912 [1].

The system of Naroch lakes: Naroch, Myastro and Batorin (Byeloruss); their study began in 1946 and has continued up to the present [2].

Trakai Lakes (Lithuania) situated in a cultural landscape; the study of these lakes has been carried out intermittently since 1952 [3].

Lake Sevan; studies began in 1923 [4].

Lake Pletsheevo; studied intermittently since 1921, and since 1978 a wide range of studies have been carried out on this lake [5].

D.G. George et al.(eds.), Management of Lakes and Reservoirs during Global Climate Change, 185–204.

With the exception of the above lakes, two in the northwest of Russia have been studied by the Institute for Lake Research (RAS). These are Ladoga, the largest lake in Europe, and the small Lake Krasnoye in the Karelian Isthmus, within the Ladoga watershed. Studies on Ladoga have been continuing since 1956; the Krasnoye studies have been carried out since 1964.

2. Lake Ladoga

2.1. GENERAL FEATURES

Ladoga is the largest lake in Europe and one of the most northern among the largest waterbodies in the world. Catchment conditions have provided the ecosystem of the lake with high quality water; however, since the early 1960s there have been signs of deterioration.

The surface area of Ladoga is ca. 17700 km², ca. 14.6 times smaller than its total catchment area. By comparison, this ratio in the Laurentian Great Lakes in North America ranges only from 1.4 to 3.4 [6]. Therefore, the influence of the catchment area on all processes inside Ladoga is likely to be considerable in spite of the lake's very large size and water volume.

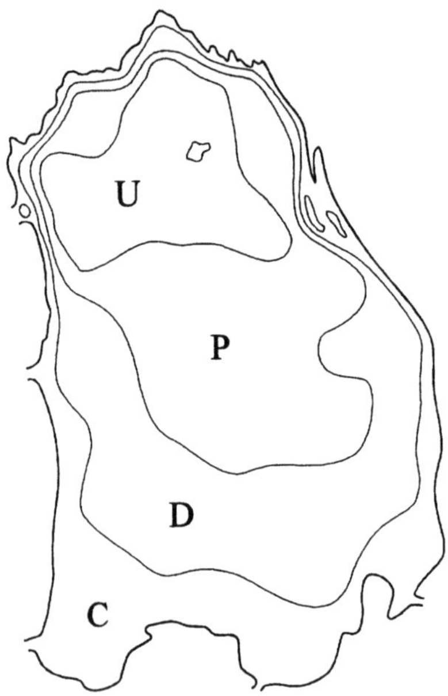

FIGURE 1. Scheme of basic morphometric zones in Lake Ladoga: C = coastal; D = declinal; P = profundal; U = ultraprofundal.

With reference to the total territory of Russia, the Lake Ladoga region is characterised by a high level of economic development, with a concentration of producer industries greater than elsewhere in Russia. This has considerably hindered the task of protecting this unique lake. One must also remember that the quality of the drinking water of 5 million inhabitants of St. Petersburg depends on its ecology. Therefore it is necessary to preserve the natural high quality of Ladoga water and its fishery resources, although the water resources of the lake are intensively exploited both by industry and agriculture. The lake receives effluent waters from 594 industrial and 680 agricultural enterprises. The total volume of polluted sewage water, flowing into Ladoga, is 400 x 10 m^{-3} $year^{-1}$ [7].

The enormous surface area of Ladoga and its morphometry have determined the main characteristics of the natural conditions for its biota. The northern part is the deepest, with an average depth of 82 m and a maximum of 230 m. The southern part is shallow with an average depth of 15 m. Glacial forms of relief such as skerries and fjords are numerous at the north whereas there are many stones, gravel and sandy areas in the west and southern part, with vast sandy beaches on the east shore. Different parts of the lake are characterised by different hydrochemical characteristics, resulting partly from large differences in productivity of the phytoplankton and zooplankton, as well as the benthos. Human impact now affects the whole lake, but this is patchy, depending on which part of the waterbody is being considered.

For investigating internal processes the lake area effectively can be divided into four zones [8]. These are (1) the coastal zone, with depths less than 15 m (average depth 9 m); (2) the declinal zone, at depths from 15 to 52 m (average 30 m); (3) the profundal zone, at depths from 52 to 89 m (average 66 m); (4) the ultraprofundal zone, at depths greater than 89 m (average 113 m) (Fig. 1). It is also necessary to analyse bays and fjords as separate individual entities.

2.2. PLANT NUTRIENTS

Long-term investigations of Ladoga since 1956 have shown that the concentration of phosphorus in the water is the main factor which determined the primary production, the species composition of phytoplankton and, ultimately, the trophic status of the lake and its component parts.

The development of the Ladoga ecosystem may be divided into four time periods: (1) until 1976, when the lake was oligotrophic with a mean total phosphorus concentration of ca. 10 μg l^{-1}; (2) 1976 to 1983, a period of intensive anthropogenic eutrophication when the mean total phosphorus concentration rose to 27 μg l^{-1}; (3) 1983 to 1986, a period of stable, high total phosphorus concentrations (mean 22 μg l^{-1}) during which primary production was high; (4) since 1986, a period of relative stabilisation.

The period from 1976 to 1983 is characterised by great changes in the lake ecosystem. During this time the process of anthropogenic eutrophication was largely attributable to an increasing nutrient load. In the period from 1956 to 1962 the average phosphorus load was 0.14 g m^{-2} $year^{-1}$, whereas during the period from 1976 to 1979 the load reached 0.39 g m^{-2} $year^{-1}$.

A sharp rise in phosphorus input to Lake Ladoga occurred in the early 1970s following the conversion of the Volkhov aluminium plant to accommodate new raw materials (apatite-nepheline ore). The average concentration of phosphorus in the Volkhov River subsequently increased from 46 mg l^{-1} (1959–1962) to 230 mg l^{-1} (1976–1979). It was calculated that whilst it was subjected to a loading of 4000 t $year^{-1}$ the lake might remain oligotrophic or slightly mesotrophic (a permissible level). However, a loading of 7000 t $year^{-1}$ exceeded the critical level and under this regime the lake would become eutrophic [8] (Fig. 2).

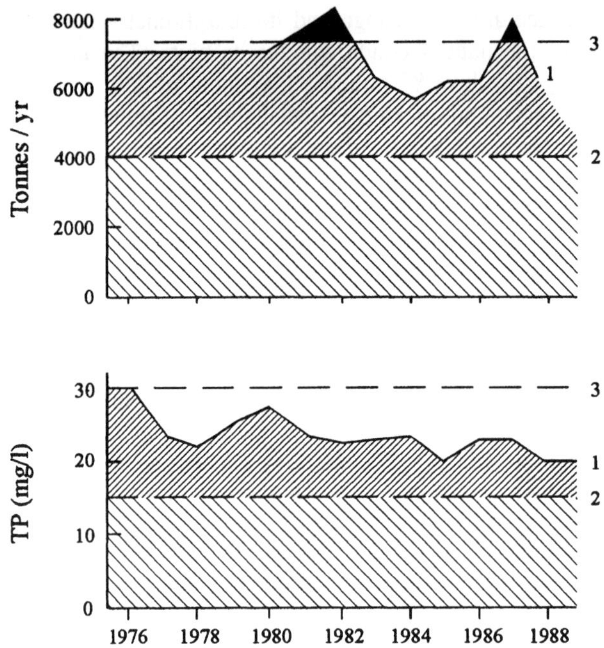

FIGURE 2. (*Above*) Input of total phosphorus into Lake Ladoga (tonnes per year) and (*below*) mean concentration (mg l^{-1}) in the lakewater. Concentration levels: 1 = observed; 2 = permissible; 3 = critical.

An increase in the phosphorus loading of the lake occurred due to the high water levels experienced in 1981–1982, 1987 and 1991. Thus both factors (anthropogenic load and natural cycling) were responsible for the fact that the load exceeded the critical level. The concentration of total phosphorus in the lake water was also very high between 1976 and 1983 (Table 1); a decrease in the oxygen concentration of the hypolimnion was also registered in the profundal and ultraprofundal zones during this period (Fig. 3).

TABLE 1. Mean phosphorus concentrations (μg l⁻¹) in Lake Ladoga [8].

Year	Total phosphorus Spring	Summer	Autumn	Phosphate phosphorus Spring	0–10 m	10 m–bottom
1959–1962	---	10	---	3	2	2
1976–1980	25	26	27	14	7	10
1981–1983	24	25	22	15	9	12
1984–1986	21	23	22	14	6	7
1987–1992	20	20	21	9	4	8

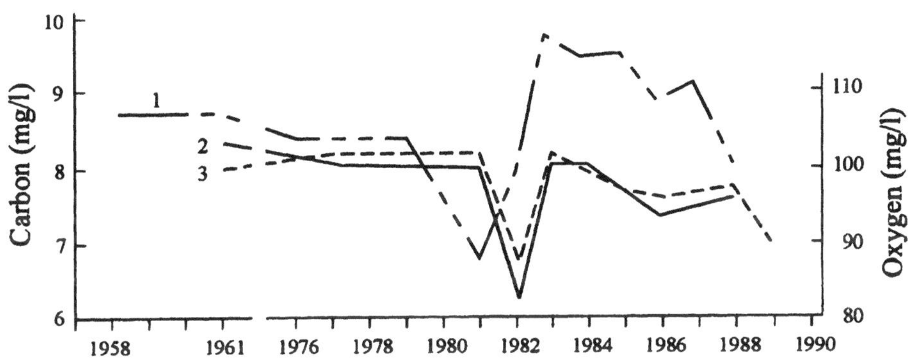

FIGURE 3. The average concentration (mg l⁻¹) of organic carbon (1), and the oxygen concentrations (% saturation) in the hypolimnium of the ultraprofundal (2) and profundal (3) zones of Lake Ladoga during the study period (1958–1989).

2.3. PHYTOPLANKTON

During the period of intensive anthropogenic eutrophication of the lake (1976–1983), seasonal community composition of the phytoplankton changed; species characteristic of oligotrophic lakes were replaced by those normally found in eutrophic waters. This was most clearly seen in the summer plankton community when blue-green algae (Cyanobacteria) dominated the general community, but diatomic species were almost completely absent. After 1984 a replacement of eutrophic species by mesotrophic diatoms was observed and by 1989 the composition of the dominant summer phytoplankton became practically identical to that observed during the oligotrophic period of 1956–1962 [8]. At present, Cryptophyte species dominate in summer; such algae are often associated with the eutrophication of standing waters.

The period of intensive anthropogenic eutrophication of the lake was characterised by maximum concentrations of chlorophyll-*a*. During this period the greatest oscillations in concentration were observed, the maximum exceeding the annual mean by 15–17 times, whilst in 1984 this value was only 7–9 times. During the initial stages of anthropogenic eutrophication the higher values were marked in coastal zones and particularly in the declinal zones (Fig. 4).

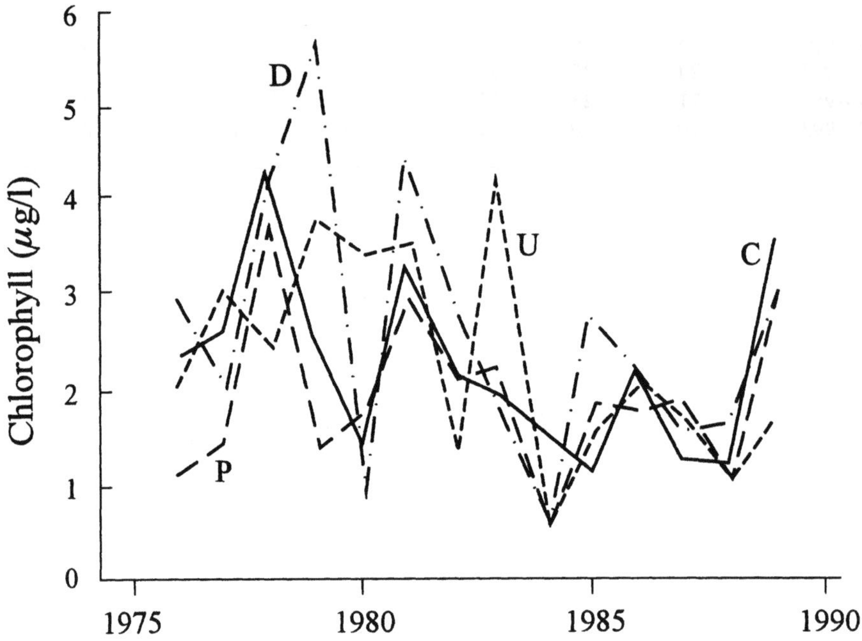

FIGURE 4. Mean annual concentrations of chlorophyll-*a* (μg l^{-1}) in phytoplankton of coastal (C), declinal (D), profundal (P) and ultraprofundal (U) zones of Lake Ladoga.

The overall production of phytoplankton in Ladoga during the period of vegetative growth increased from 1976 to 1989, but this increase did not correlate directly with the changes in phosphorus load. This was due partly to the inertia of limnological processes in the lake and partly to variability of natural conditions in various parts of it (Fig. 5). For visible mass development of algae to occur, periods of calm weather were necessary. If stormy conditions prevailed then the productivity of the algae changed dramatically. For surface blooms caused by blue-green algae to manifest themselves in Ladoga, a period of more than 3 days of calm weather with temperatures higher than 16°C was required [9]. Such conditions and events were observed every year in the coastal and declinal zones, but in profundal and ultraprofundal zones of the lake they occurred only during the warmer seasons.

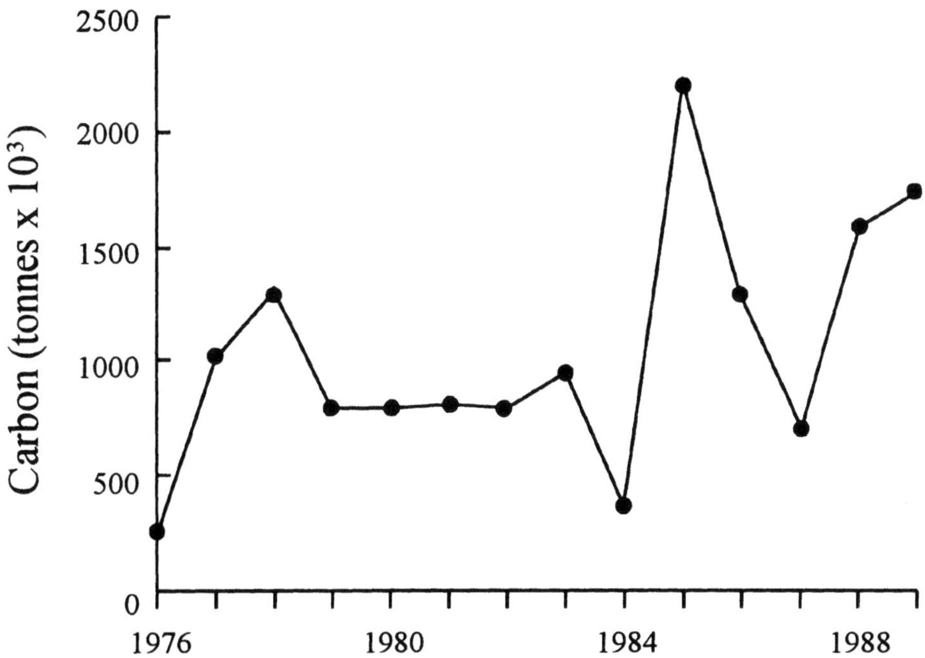

FIGURE 5. The overall production of phytoplankton in Lake Ladoga during the growing seasons of 1976 to 1989.

2.4. BACTERIA

The counts of bacteria recorded in the plankton show considerable spatial variability. Numbers in the profundal and ultraprofundal zones vary between those characteristic of mesotrophic lakes and those found in eutrophic waterbodies. In the latter stages of this study some stabilisation of bacteria population estimates was noted, but only in the epilimnion. In the hypolimnion, on the other hand, there has been a tendency for the average count of bacteria to increase steadily.

2.5. ZOOPLANKTON

Interestingly, the species composition of the pelagic zooplankton, as well as its density and biomass, have not changed significantly during the last 40 years (Table 2). However, in the coastal regions of the lake, the species composition and biomass of the zooplankton during the 1990s were characteristic of mesotrophic and sometimes even eutrophic lakes, whereas in the pelagic areas the composition was that of oligo-mesotrophic waters with only a slight trend of eutrophication in the uppermost layers [10].

TABLE 2. Numbers and biomass of summer plankton during long-term investigations in the pelagic zone of Lake Ladoga. The values are weighted means for the water column 0 to 70 m (Mean ± SE) at stations 55 and 82 [10].

Zooplankton	1950–1960	1970s	1980s	1990s
Numbers, Ind. 10^3 m^{-3}	11.1 ± 0.61	12.6 ± 3.6	16.4 ± 3.1	13.3 ± 2.2
Biomass, mg m^{-3}	214 ± 76	291 ± 58	224 ± 36	268 ± 88

2.6. ZOOBENTHOS

The distribution of the zoobenthos in Lake Ladoga depends mainly on the type of bottom deposit and depth of the biotope, but eutrophication has also influenced these communities. In the southern part of the coastal zone the density, biomass and productivity of bottom invertebrates has increased. There has also been a slight increase in the populations of the profundal and declinal zones, but at maximal depths (ultraprofundal) there was practically no change (Table 3).

Some changes in species composition were also noted during the long-term investigations of benthic communities. The most rapid shifts have occurred in the coastal zone, where 85% of the complete list of 385 species and forms of Ladoga benthos [11] were encountered. Different natural conditions, and especially human activities, have formed various biocenoses ranging from oligtrophic to hypereutrophic and even strongly polluted areas where bottom fauna was absent. With the changes in human impact, one can observe significant alterations in the benthic biocenoses.

In the deeper zones of the lake the species composition of the benthos was more stable, but during the last two decades some of the most sensitive species of invertebrates have disappeared, including the glacial relicts *Gammaracanthus loricatus* Sars and *Pisidium conventus* Clessin.

The trophic state of the declinal zone, as evaluated by the species composition of the zoobenthos during the 1970s and 1990s, indicated that it was on the border between an oligotrophic and mesotrophic condition. The profundal and ultraprofundal zones, on the other hand, were at different stages of oligotrophy.

TABLE 3. The average density (N, Ind. m^{-2}), biomass (B, wet weight g m^{-2}) and net production (P, g C m^{-2}) of macrozoobenthos in four zones of Lake Ladoga during two periods.

Zones	Productivity	1976–1984	1992–1994
Ultraprofundal	N	233	216
	B	0.50	0.44
	P	0.15	0.12
Profundal	N	807	1125
	B	2.27	3.26
	P	0.40	0.57
Declinal	N	1130	1335
	B	3.02	4.52
	P	0.58	0.21
Nearshore (southern part)	N	1100	2920
	B	3.02	8.94
	P	0.44	1.75

3. Lake Krasnoye

3.1. GENERAL FEATURES

The Institute for Lake Research, RAS, has been carrying out investigations of Krasnoye since 1964. This lake is situated on the Karelian Isthmus, which lies between the Gulf of Finland and Lake Ladoga, an area of 15000 km². Forests occupy about 65% of the territory and bogs occupy from 5 to 6%. The Karelian Isthmus is rightly called a lakeland; there are more than 700 lakes in this region.

The nature of the Karelian Isthmus is affected by human economic activity. In spite of the high proportions of forests it has long been a region of intensive agriculture. In the last few decades the human impact on the whole territory surrounding the lakes has become noticeably greater. The scale of development has increased steadily; large settlements have appeared and the application of mineral fertilizers has increased. The role of the Karelian Isthmus as a recreational area has been growing steadily.

Krasnoye is located in the central part of the Karelian Isthmus. Morphological data for the lake are as follows: area 9.1 km², specific drainage basin 18.4 km², maximum depth 14.6 m, average depth 6.6 m. The water-mass of the lake is fully renewed approximately every fourteen months.

Long-term studies of this lake have shown a close correlation between meteorological conditions during the year and the hydrological structure of the lake, and the resulting biological regime. The main components controlling the biological regime of Krasnoye are meteorological and hydrological factors, and to a lesser degree anthropogenically-controlled variables.

The period from 1960 to 1970 was characterised as a phase of decreasing humidity and the following decade was a phase of the lowest water levels (a dry period). The highest water temperatures and lowest values for water level were recorded in 1972 and 1973. During these years there was also minimal flow of water through, and exchange with, the lake system (Fig. 6b).

When the 20-year dataset is examined, 1964, 1972, 1975 and 1983 were noted as the most warm, with the highest values of positive temperatures, heat storage and total solar radiation. The most stable stratification was observed during the summers of 1966 and 1972, and the coldest years were 1965, 1977 and 1982.

During the stable summer stratification of 1966 and 1972 (Fig. 7), very low oxygen concentrations were observed (Table 4), although this was also true of some other years when stratification was apparently less stable.

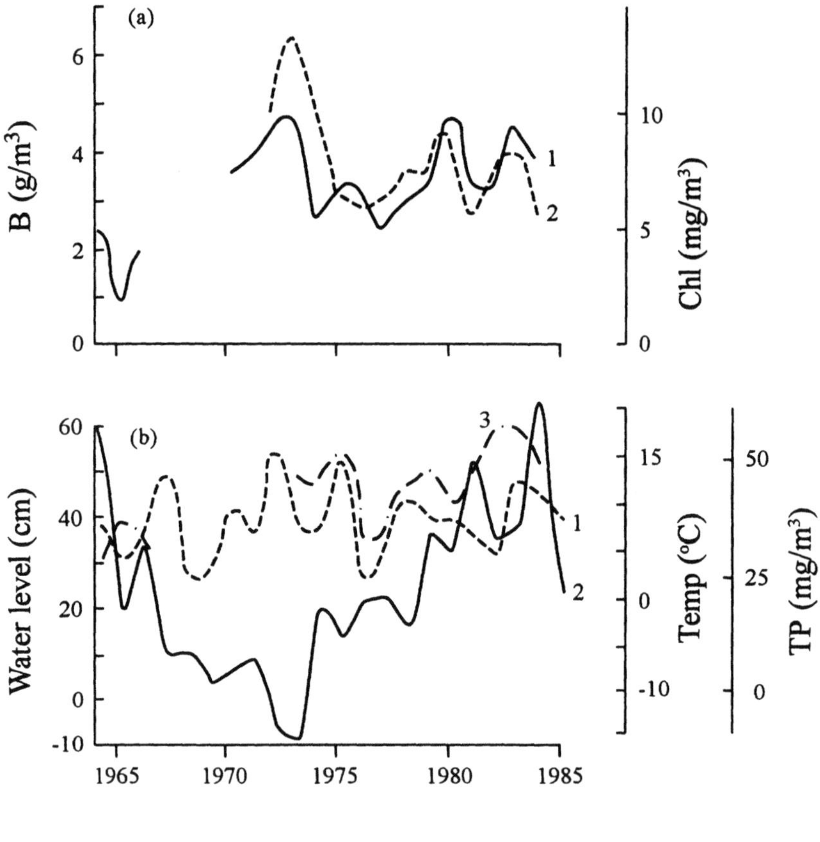

FIGURE 6. Long-term changes in phytoplankton productivity of Lake Krasnoye: (a), mean biomass (1, g m^{-3}) and chlorophyll-*a* (2, mg m^{-3}) for May–October; (b), mean water temperature (1, °C) for May–October, and mean annual water level (2, cm) and mean annual total phosphorus concentration (3, mg m^{-3}) for 1964 to 1984–85.

TABLE 4. Minimum values of oxygen concentrations in the hypolimnion of Lake Krasnoye from 1963 to 1984 [11].

Year	Late winter Date	O_2 mg l^{-1}	%	Summer Date	O_2 mg l^{-1}	%
1963	19 Apr	6.6	49	13 Aug	0.3	3
1964	22 Apr	4.6	36	17 Jul	2	20
1965	22 Apr	3.1	24	18 May	5.5	48
1966	27 Apr	2.2	17	26 Jul–9 Aug	0.3	2
1967	24 Apr	8.8	63	10 Aug	3	30
1970	---	---	---	21 Jul	1	9
1971	16 Apr	7.4	54	5 Aug	0.8	8
1972	25 Mar	2.2	17	27 Jul–5 Aug	0	0
1973	27 Apr	3.3	25	25 Jul–5 Aug	0	0
1974	14 Mar	1	8	13 Aug	2	19
1975	25 Mar	7.1	53	31 Jul	0.7	7
1976	13 Apr	4.8	36	16 Aug	0.9	9
1977	8 Apr	6.8	51	15 Aug	0.5	5
1978	10 Apr	3.1	24	14–29 Aug	0.8	8
1979	10 Apr	2.7	20	27 Jul	0.4	4
1980	14 Apr	3.2	24	14–25 Aug	0	0
1981	11 Apr	4.5	34	5–18 Aug	0.3	3
1982	6 Apr	6.3	48	13 Aug	0.4	3
1983	6 Apr	5.1	40	13 Aug	0.9	9
1984	27 Mar	3.3	26	30 Jul	1	10

3.2. NUTRIENTS

From 1980 to 1990, total phosphorus concentrations in tributary streams increased 2–5-fold, especially small streams receiving waste-water from poultry farms. But concentrations hardly altered in the main inflow (Strannica), which provides 60% of the total inflow to Krasnoye. Hence changes in the lake were not significant, although a small increase in the 1980s is apparent (Table 5).

TABLE 5. Mean concentrations (mg m^{-3}) of total phosphorus and nitrogen in Lake Krasnoye from 1973 to 1984 [11].

Nutrient	Years	Annual means	May to October	June to August
Phosphorus	1973–1977	43	45	39
	1981–1984	51	49	45
Nitrogen	1977–1979	57	52	51
	1981–1984	73	70	66

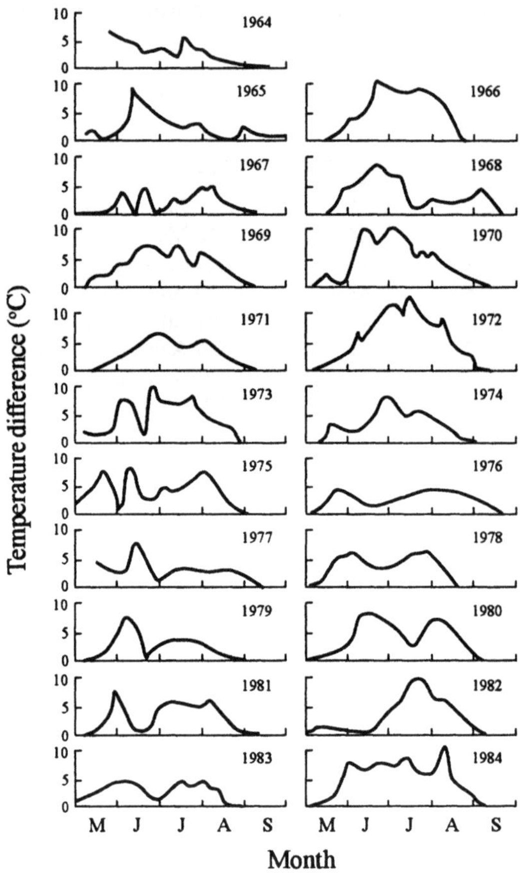

FIGURE 7. Differences in temperature between the surface and bottom waters of Lake Krasnoye during the summers of 1964 to 1984.

3.3. PHYTOPLANKTON

Analyses of long-term changes in phytoplankton biomass show considerable stability in the productivity of this community between 1970 and 1990. The average phytoplankton biomass (4.8 g m^{-3}) during the growth season between 1980 and 1990 was very close to the mean value obtained for the previous decade (3.98 g m^{-3}), as was the chlorophyll-*a* content of the phytoplankton. However, the biomass of phytoplankton during these two decades was much higher than the mean biomass recorded between 1960 and 1970 (2.0 g m^{-3}) during the high water-level phase (Fig. 6a).

It is obvious that the increase in phytoplankton productivity was much slower than in the late 1960s and at the beginning of the 1970s, when there was a drastic change in the total phosphorus content of the lakewater and phytoplankton biomass (productivity). Changes in the structure of the phytoplankton community were also observed. Between 1970 and 1980 the highest phytoplankton biomass was usually recorded during the spring pulse of diatoms in May (mainly *Aulacoseira islandica*). In the following decade the peak of biomass was as a rule observed in July, with the peaks in these years being rather small. The reason for the poor development of *A. islandica* might be found in the decrease of the duration of homothermy in spring, as well as the increase in organic matter and total phosphorus content of the water (a result of amelioration and eutrophication).

3.4. BACTERIA

Changes in the bacterioplankton between 1970 and 1980 included an increase in bacterial numbers and productivity in the lake. The numbers of bacteria in the water increased even further in the decade from 1980. The highest values of both numbers and biomass of bacteria were recorded in 1984 (Fig. 8), a year in which the highest water levels were recorded during this long-term investigation. The increased flow and high water levels may well have resulted in a greater input of organic matter which, in turn, may have stimulated bacterial growth.

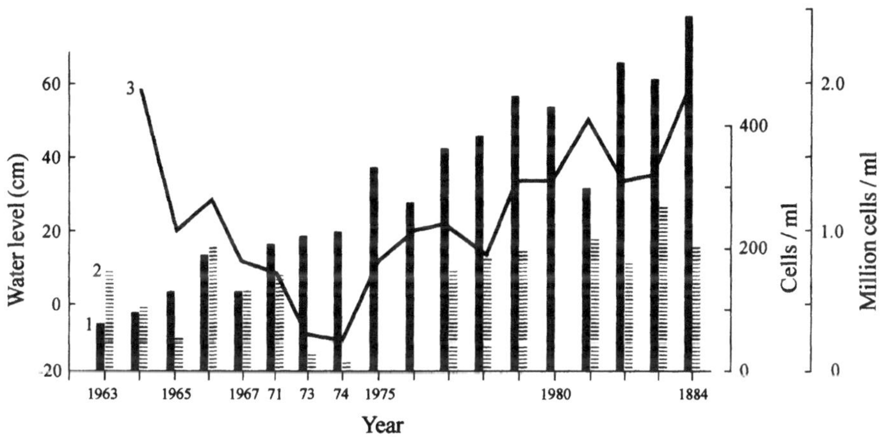

FIGURE 8. Total numbers of bacteria (1, millions ml^{-1}), numbers of bacteria (2, ml^{-1}) growing on BPA medium, and water level (3, cm) in Lake Krasnoye. The values for bacteria (variables 1 and 2) are means for the period May to October; values for water level (variable 3) are annual means.

3.5. ZOOPLANKTON

The structure and abundance of the zooplankton community was also studied in this investigation and changes were observed over the period from 1964 to 1975. The structural transformations are expressed as changes in the proportions of taxonomic groups, shifts in predominant species and decreasing values of species diversity indices. Mean values for abundance and biomass of the zooplankton community (during the ice-free period) have more than doubled during this period (from 75,000 to 178,000 ind. m^{-3} and from 1.4 to 2.5 g m^{-3}). Further investigations showed that some stabilisation of the zooplankton community had occurred in the mid-1970s (Table 6). But this was accompanied by changes in community structure, particularly a decrease in numbers of the dominant species of the genus *Bosmina* and an increase in members of the genus *Daphnia*. The proportion of Rotifera in the total zooplankton (both numbers and biomass) also increased. Although the values for zoplankton biomass between 1980 and 1990 were very close to those recorded during the previous decade, the highest values recorded for zooplankton numbers and biomass during the whole of this investigation were observed in 1982 and 1983.

TABLE 6. The density (thousands of individuals m^{-3}) and biomas (g m^{-3}) of zooplankton in Lake Krasnoye. Mean values for May to October are shown for each year [11].

Years	Numbers	Biomass
1964	75	1.4
1966	99	1.4
1970	147	1.6
1972	167	1.8
1973	136	1.4
1975	178	2.5
1977	216	1.6
1978	216	1.9
1979	149	1.2
1980	148	1.3
1981	256	1.7
1982	206	2.5
1983	304	3.0
1984	151	1.6
Means for the 1960s	87	1.4
Means for 1970–1980	190	1.8

A comparison of the density and biomass of planktonic communities during the warmer years indicated an increase in the values obtained for these variables and, therefore, a slow eutrophication of the lake (Table 7).

TABLE 7. Water temperature (sum of positive temperatures each year (t, °C)), maximal heat storage (mJ m^{-2}) and water levels (cm) related to the numbers (N) and biomass (B) of planktonic communities in Lake Krasnoye during the warmest years in three decades of investigations [11]. Estimates for planktonic communities are: phytoplankton (B, g m^{-3}); bacterioplankton, (N, 10^6 cells ml^{-1} and B, g C m^{-2}); Zooplankton (N, 10^3 Ind. m^{-2} and B, g m^{-3}).

Year	Temp.	Heat storage	Water level	Phytoplankton	Bacterioplankton		Zooplankton	
				B	N	B	N	B
1964	2307	557	59	2.6	0.3	0.2	75	1.4
1972	2449	557	– 9	4.1	1.4	0.4	167	1.8
1983	2323	535	34	4.7	1.7	0.5	304	3.0

3.6. ZOOBENTHOS

Investigations of the macrobenthos of Lake Krasnoye have been conducted from 1963 to the present day, excluding an interval from 1986 to 1990. The meiobenthic community has been studied since 1975. Samples have been taken from two of the main biotopes of the lake (i.e. the littoral zone at 1.5 to 2.0 m depth and the profundal zone at 10.5 m depth), twice during each month from May to September and once in March, April and November.

Structural changes observed in the littoral benthic community relate to an increase in the organic matter content of sandy sediments. During the study period the abundance and biomass of the psammophylous populations decreased whereas those of the pelophylous population increased. It has been established that the abundance of the littoral macrobenthos is inversely related to the water level in the lake [11]. Long-term studies of the meiobenthic community show that its density and biomass increased significantly from the period 1975–1980 to the period 1991–1995, with changes in the ratio of the predominant taxonomic groups.

The species composition and abundance of the profundal macrobenthic community were controlled largely by the oxygen concentration in the hypolimnion. As a consequence of frequent periods of oxygen deficiency in the near-bottom layer of water, in the early 1970s, species diversity of the macrobenthos decreased significantly. Unfortunately, populations of relict species such *Pontoporea affinis* and *Mysis oculata* var. *relicta* have completely disappeared, whereas *Chaoborus flavicans* became a normal component of the benthic community [11].

Year-to-year fluctuations in the quantity and biomass of the benthic community in Krasnoye are determined, on the one hand, by competition between two species (*Chironomus anthracinus* Zett and *C. plumosus* L.) and, on the other hand, by the oxygen concentration in the water of the hypolimnion. During periods of oxygen

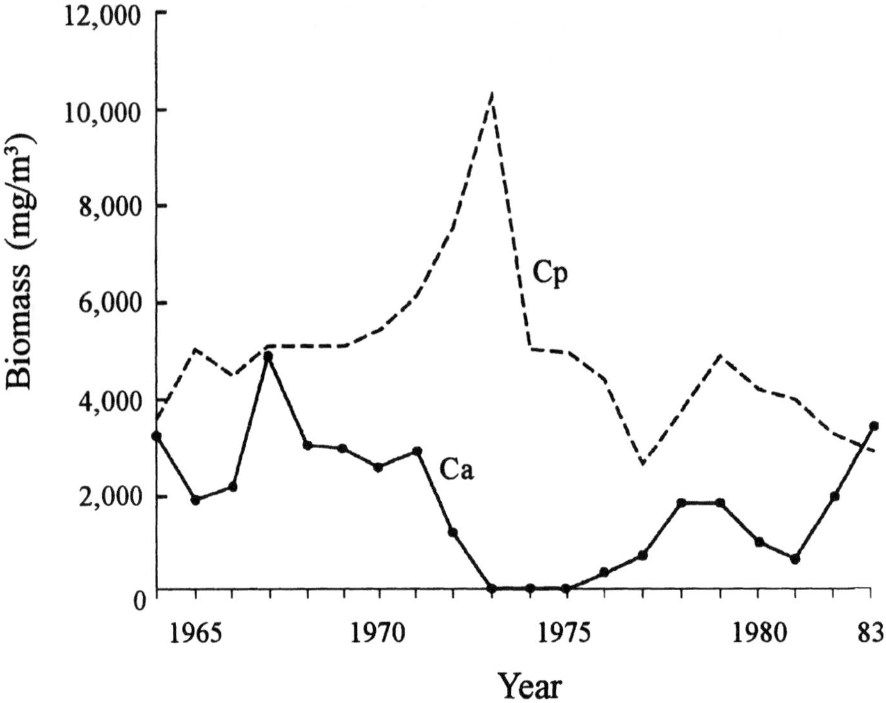

FIGURE 9. Changes in the annual biomass of *Chironomus anthracinus* (Ca, solid line) and *C. plumosus* (Cp, broken line) in Lake Krasnoye from 1964 to 1983.

deficiency the population density of *C. anthracinus* decreased (and completely disappeared in 1972) and, as a result, the population density of *C. plumosus* increased. When the population density of *C. anthracinus* returned to normal levels, the abundance of *C. plumosus* declined (Fig. 9). The mean annual values for the biomass of *C. anthracinus* correlate with water levels in Krasnoye, whereas those of *C. plumosus* are inversely related to this parameter (Fig. 10).

FIGURE 10 (*on the opposite page*). Relationships between the relative water levels (open circles and broken lines) and the annual biomass (solid squares and solid lines) of *Chironomus anthracinus* (upper figure) and *C. plumosus* (lower figure) in Lake Krasnoye during 1964 to 1983. The curves represent best-fit muliple regression lines (see the text).

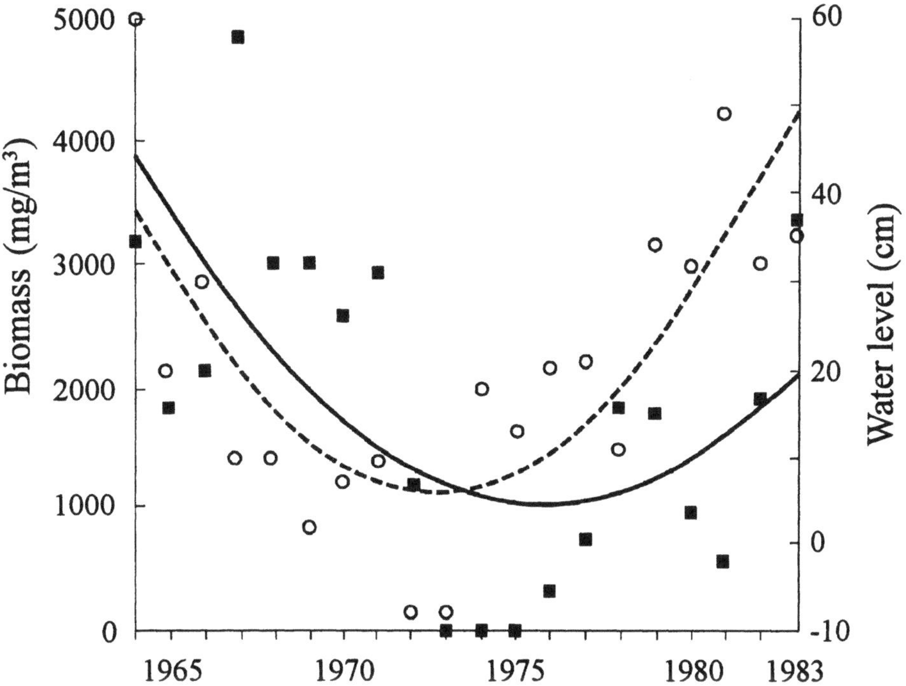

5000
4000
3000
2000
1000
0
Biomass (mg/m³)
60
40
20
0
-10
Water level (cm)
1965
1970
1975
1980
1983

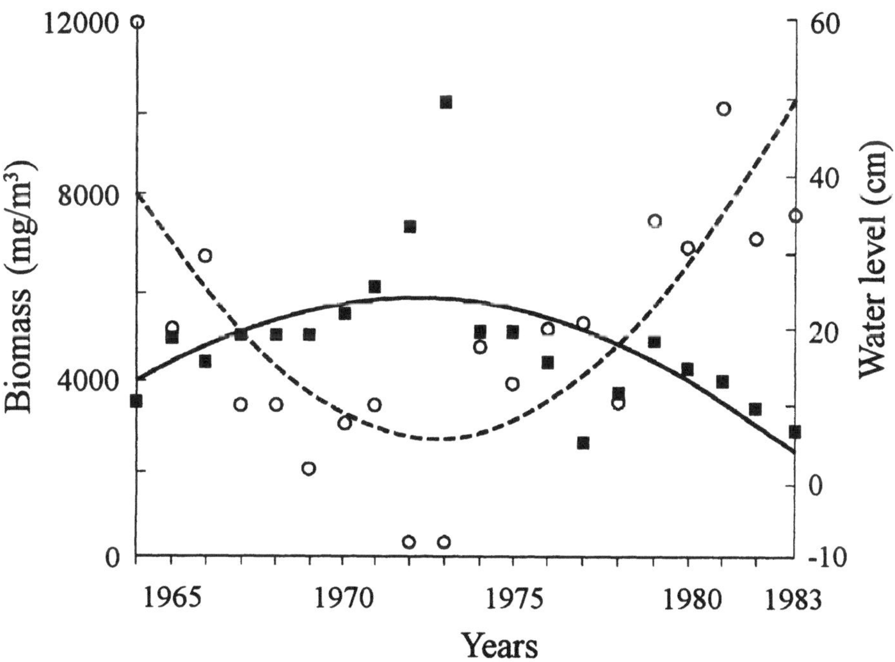

12000
8000
4000
0
Biomass (mg/m³)
60
40
20
0
-10
Water level (cm)
1965
1970
1975
1980
1983
Years

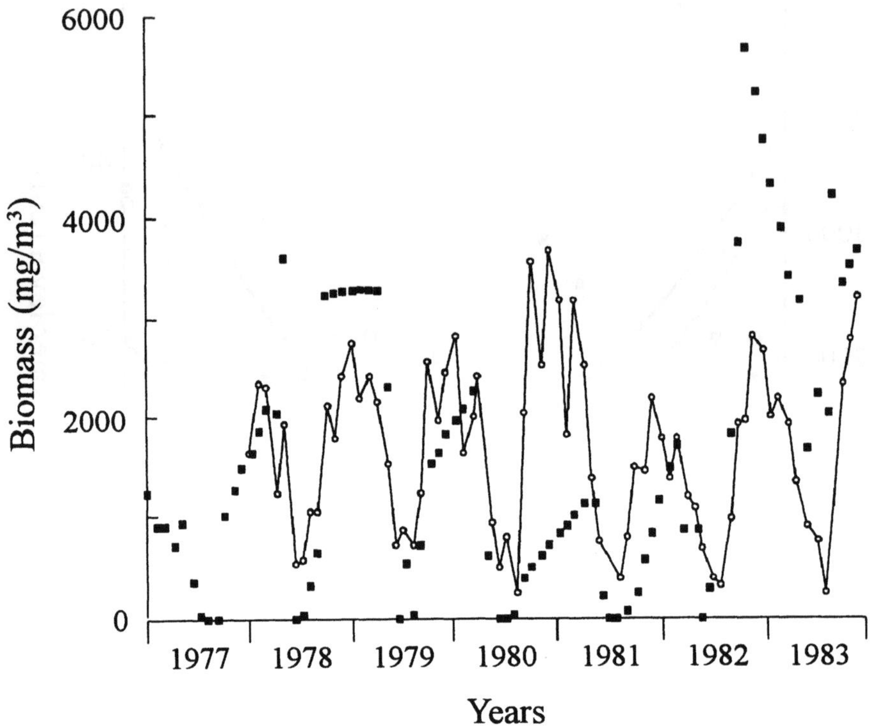

FIGURE 11. Observed (solid squares) and modelled (open circles and solid line) values for the biomass of *Chironomus anthracinus* in Lake Krasnoye.

The long-term series of data from 1964 to 1983 on the population of *C. anthracinus* was analysed using both time-series analysis and multiple regression. Two types of tentative predictive models were constructed. The results obtained for the prediction of chironomid density and biomass using an autocorrelation model with 2-year forward stepwise mode are in fairly good agreement with the observed data. The multiple regression model shows significant correlation between the population biomass of *C. anthracinus* and chlorophyll-*a* concentrations in the lakewater during preceding months. The best-fit formula was as follows:

$$Y = 112\ CHL_{t-4} + 99\ CHL_{t-5} + 126\ CHL_{t-7} + 118\ CHL_{t-10}\ ; \qquad R^2 = 0.657$$

where Y is *C. anthracinus* (mean monthly biomass), CHL is the average monthly concentration of chlorophyll-*a* and *t* is the time-period (Fig. 11). This correlation appears to owe more to trends rather than absolute values and therefore further refinement is required.

On the whole, and bearing in mind comparisons with changes in many other lakes during the last decade, the condition of Krasnoye, based on hydrochemical and hydrobiological indicators, can be considered as stable with an underlying process of slow eutrophication.

3.7. CONCLUSION

The experience gained in monitoring Lake Krasnoye indicates that evaluation of water level and physical conditions will be of paramount importance in assessing its evolution in relation to anthropogenic influences such as the input of limiting nutrients. This is particularly important in a waterbody such as Krasnoye which is at the early stages of eutrophication.

References

[1]. Stalmakova, G. (1968). Zoobentos Ladozskogo ozera. (Zoobenthos of Lake Ladoga). In: *Biologicheskie resursy Ladozhskogo ozera.* Nauka, Leningrad, pp. 4-70. [In Russian].

[2]. Vinberg, E. (ed.) (1985). *Ekologicheckaja sistema Narochanskich ozer.* (Ecosystem of Naroch Lakes). Minsk. [In Russian].

[3]. Klimkaite, I. N. (1977). Osobennosti evtrotfirovanija ozer Litovskoij SSR (na primere Trakajskich ozer). (Peculiarities of eutrophication of Trakai lakes, Lithuanian SSR). In: *Anthropogenic eutrophication of natural waters* (Abstracts of conference), 139-143. Chernogolovka. [In Russian].

[4]. Parparov, A. S., Parparova, R. M. & Simonjan, A. A. (1977). Nekotorye pokasateli intensivnosti protsessa evtrofirovanija ozer Sevan. (Some indexes of the intensity of Lake Sevan eutrophication). In: *Anthropogenic eutrophication of natural waters* (Abstracts of conference), 160-162. Chernogolovka. [In Russian].

[5]. Butorin, N. V. & Skljarenko, V. L. (eds.) (1989). *Ecosistema ozera Pleschcheevo* (Ecosystem of Lake Pleshcheevo). Nauka, Leningrad.

[6]. Ragotzkie, R. (1988). Great Lakes ecosystem experiment. *Verh. Internat. Verein. Limnol.* **23**: 359-365.

[7]. Voropaeva, G. & Rymyantzev, V. (1991). *Kontseptsia sovershenstvovania prirodopolzovania v basseine Ladozhskogo ozero.* (Conception for the perfection of nature, making use of the Lake Lagoga watershed). Gidrometeoizdat, Leningrad, pp. 1-28.

[8]. Petrova, N. & Terzhevik, A. (eds) (1992). *Ladozhskogo ozera. Kriterii sostoijinia ekosistemy*. (Lake Ladoga. Criteria of ecosystem state). Nauka, St Petersburg. [In Russian].

[9]. Petrova, N. (ed.) (1982). *Antropogennoe evtrofirovanie Ladozhskogo ozera.* (Anthropogenic eutrophication of Lake Ladoga). Nauka, St Petersburg. [In Russian].

[10]. Andronikova, I. (1996). Zooplankton in monitoring of lake eutrophication, with special reference to Lake Ladoga. In: ?? (eds. H. Simola, M. Viljanen, T. Slepukhina & R. Murthy; in press).

[11]. Trifonova, I. (ed.) (1988). *Metodologicheskie aspekty limnologicheskogo monitoringa*. (Methodological aspects of limnological monitoring). Nauka, Leningrad. [In Russian].

THE TIMING OF WARMING - AN IMPORTANT REGULATOR OF CHANGE IN LAKE PLANKTON COMMUNITIES

RITA ADRIAN
Institut für Gewässerökologie und Binnenfischerei
im Forschungsverbund Berlin e.V.
Müggelseedamm 260
D-12587 Berlin, Germany

1. Introduction

Since successional changes in the plankton occur on time-scales that range from a few days to several months, plankton communities are likely to be good indicators of regional changes in the weather. Some lake plankton communities have already been studied for several decades [1, 2]; others have been studied for shorter periods (Gaedke et al. this volume, pp. 39–5; Padisák et al. this volume, pp. 111–125) and comparisons have been drawn between a series of lakes in the same area (Kratz et al., this volume, pp. 273–287). For ecosystems that exhibit distinct seasonal patterns of succession, the timing of climatic change will be crucial. An increase in air temperature or precipitation during winter/spring, as forecasted for the northern European area (Davies et al. this volume, pp. 1–13), will certainly affect lakes early in the year but could also influence changes later in the season or in other years if they influence the concentration of the algal innoculumn or the pool of resting stages in the zooplankton. In this paper the effects of moderate winter air temperatures on the bosminid and cyclopoid copepod community in the Heiligensee will be discussed. The role of direct versus indirect temperature effects, such as alterations in food quantity and quality, predation pressure and life-history traits, are evaluated. The study shows that species-specific life-history traits may be key factors for the competitive potential of zooplankton species in a changing environment.

2. Study site

The Heiligensee, a small lake located in Berlin, Germany (52°36'N, 13°1'E), has a maximum depth of 9.5 m and covers an area of 0.3 km^2. During the summer it has a

D.G. George et al.(eds.), Management of Lakes and Reservoirs during Global Climate Change, 205–222.

well-developed thermocline. Annual means of chlorophyll-*a* concentrations typically range between 23 and 67 $\mu g l^{-1}$, and total phosphorus concentrations reach from 161 to 321 $\mu g l^{-1}$; both are well above the limits defined for eutrophic lakes [3]. More detailed descriptions of the physical, chemical and biological characteristics of the site can be found in a paper by Adrian et al. [4]. Within the last 21 years (1975–1994), severe changes in water temperatures, nutrient dynamics and the plankton communities, have been recorded. During this period, the area experienced several unusually mild winters and many of the changes recorded appear to be connected with this change in the weather. In this study, I analyse some of the climatic factors that influence the seasonal development of the crustacean zoooplankton in the Heiligensea.

Three *Daphnia* species coexist in the Heiligensee (*D. galeata, D. cucullata, D. hyalina*), but their relative abundances have changed during the course of this study. In recent years *D. galeata,* which used to be dominant during the spring, declined to minor densities, while the smaller *D. cucullata* increased in abundance. For *D. hyalina*, which is generally present in low numbers, there has been no obvious change. Differences in sensitivity to the presence of cyanobacteria (*galeata* > *cucullata*), early invertebrate predation pressure by *Cyclops vicinus* (*galeata* > *cucullata*), and differences in the importance of direct temperature effects (*cucullata* > *galeata* > *hyalina*) acted at the same time and in the same direction, viz. favouring the smaller *D. cucullata* (details are discussed elsewhere).

3. Methods

The zooplankton was sampled at 1-m intervals from the surface to 8 m depth using a 7-litre plexiglass cylinder trap. Sampling intervals were usually monthly, except for intervals of 2 to 7 days in the spring of 1980, 1984 and 1986. Data for the bosminids (1980, 1984) and the cyclopoid copepods (1975–76) from earlier years were taken from Schumpelick [5], Müller [6] and Schubert [7] respectively. Water temperatures and samples for chlorophyll-*a* analyses [8] were taken in vertical profiles (at 1-m intervals) over the deepest portion of the lake at biweekly (1975–1984) to mostly monthly intervals thereafter. The nonparametric Mann-Kendal test [9] was used for time-series analyses. Significant trends are inferred if the test statistic is > +1.96 or < –1.96 ($p<0.05$). Positive values of this test statistic indicate a rising trend whilst negative values indicate a falling trend. More details of this statistical procedure are given in [4].

4. Results

4.1. AIR TEMPERATURES

Between 1975 and 1995, local air temperatures in Berlin [10] showed a significant rising trend (test statistic 2.34*, $p<0.05$), which largely consisted of an increase in air temperatures during winter and spring (Fig. 1). Within the last eight years (1987–88 to 1994–95) mean winter air temperatures (December–February) have always remained above the long-term mean (1975–1995). Relative to the period 1975–1994 a significant rising trend was found in April (test statistic 3.01*, $p<0.05$). This local change in air

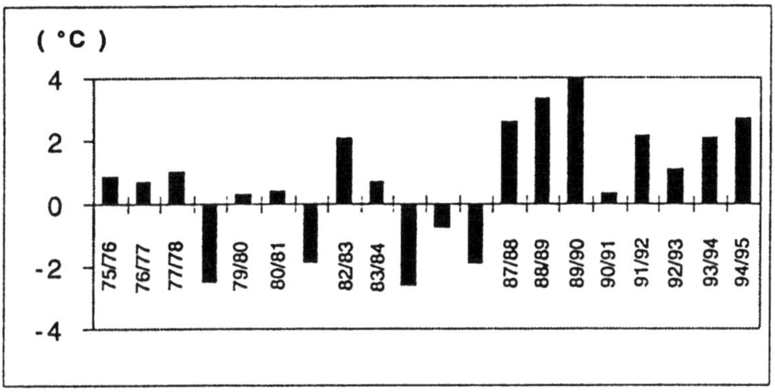

FIGURE 1. Deviation of mean winter air temperatures (December–February) from the long-term mean, relative to 1975–1994. Air temperatures derived from hourly measurements were recorded in Berlin-Dahlem [10].

temperatures is consistent with general findings of climatic change over northern Europe (see Davies et al. this volume).

4.2. WATER TEMPERATURES

The year-to-year changes in water temperature were closely correlated with changes in air temperature, especially during spring. Mean water temperatures in March–May increased by 1.7 °C between 1975 to 1987 (mean of 8.48 °C) and 1988 to 1995 (mean of 10.17 °C) (Fig. 2).

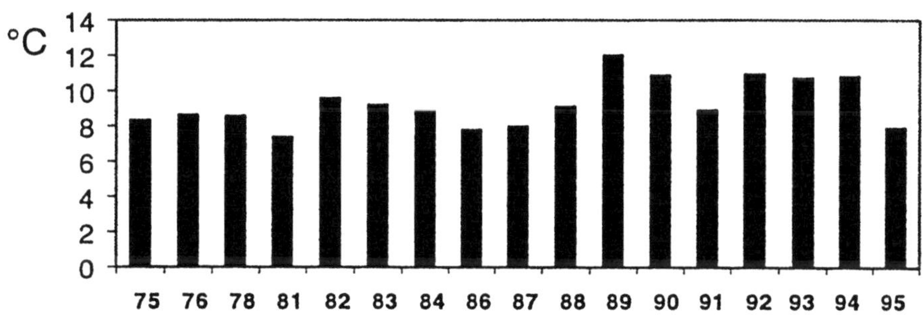

FIGURE 2. Mean spring water temperatures (March–May) of the upper 0–4 m water stratum in the Heiligensee between 1975 and 1995. Years with missing values have been excluded.

The highest increases were recorded in April, where a linear regression produces an average increase of 2.6°C over the 21-year period. Mean winter water temperatures (December–February) differed by only 0.32°C between 1975–76 (mean of 2.90°C) and 1986–87 (mean of 3.22°C). Direct temperature effects are likely to be of minor importance during winter, but the duration of ice-cover can have a major effect on circulation, penetration of light and the growth of phytoplankton. Furthermore, mild winters and enhanced air temperatures in early spring may lead to an early thermal stratification in spring, causing early chemical (P, Si) and turbulence gradients which have an impact on the phytoplankton species· composition. Note that, hitherto, no significant trends in water temperatures have been recorded during the summer.

4.3. SEASONAL DISTRIBUTION OF ALGAL BIOMASS

Between 1978 and 1994, there was a significant increase in the biomass of algae recorded in the Heiligensee during winter (test statistic 6.05*). Mean winter algal biomass (December–February) increased by 72% between the earlier years (1978–1987) and the mild winter period of 1987–88 to 1994–95 (Fig. 3). However, there was no direct match between this increase in biomass and winter temperature; the change was gradual and the full response was recorded only after a 2-year delay.

In the mild winters of 1987–88 and 1988–89 algal biomass remained within the range found in earlier years. During spring (March–May), summer (June–August) and autumn (September–November) the increase was 30%, 21% and 58% respectively (Fig. 3). Considering single months, significant rising trends were found in January, February, October and November (test statistics were 2.85*, 2.41*, 2.24* and 2.85* respectively). The increase in algal mass during winter and spring (December–June) was accompanied by an increase in the proportion of cyanobacteria and a decrease in the numbers of diatoms and cryptophytes.

4.4. ROTIFERS

The rotifer populations found in the Heiligensea frequently serve as an important source of food for cyclopoid copepods, but no obvious changes were recorded in this community between 1986 and 1994 (Fig. 4a). However, monthly sampling intervals are inappropriate to follow the population development of rotifers, as abundances are most probably underestimated. The spring peak in 1986 can be explained by weekly sampling intervals, which clearly demonstrates the high likelihood of underestimations when using monthly sampling intervals. There is, however, some evidence for an increase in abundance during winter in recent years (Fig. 4b), parallel to the observed increase in algal mass (Fig. 3).

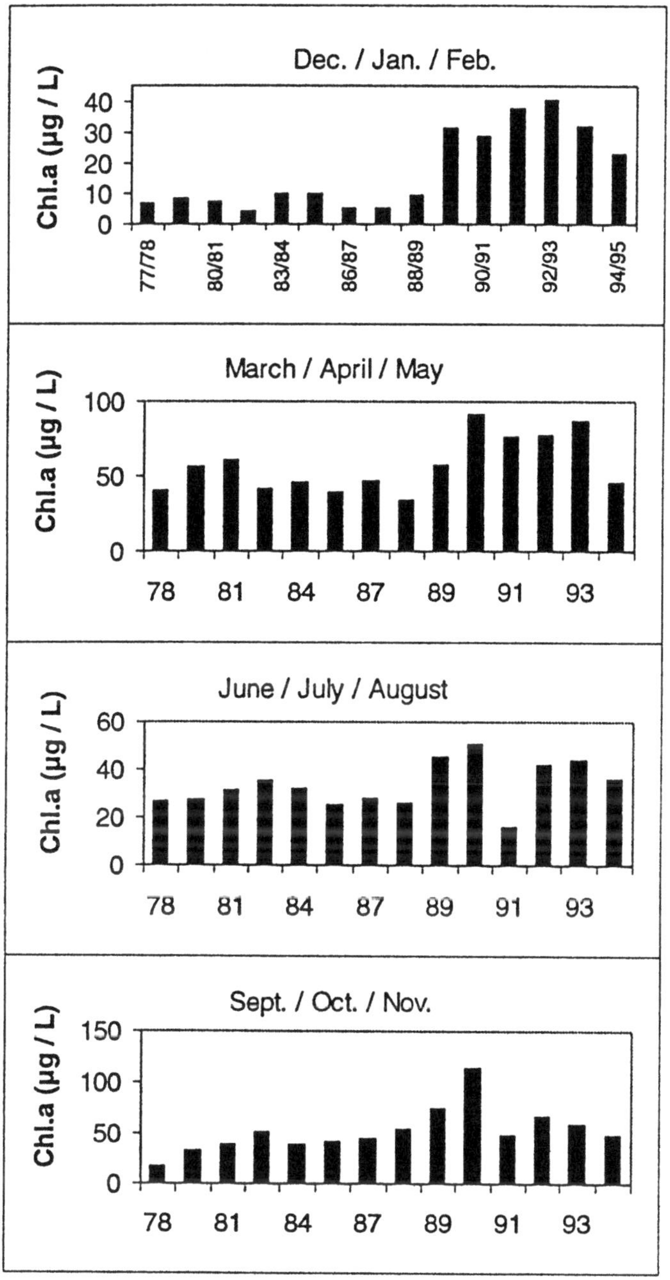

FIGURE 3. Mean chlorophyll-*a* concentrations (µg l^{-1}) in the upper 0–4 m of the Heiligensee during winter, spring , summer and autumn, between 1978 and 1994–95.

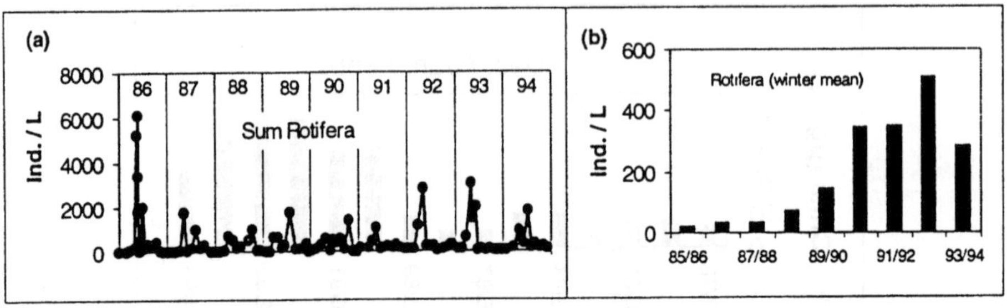

FIGURE 4. The numbers of rotifers (l^{-1}) in the Heiligensee between 1986 and 1994. (a) Total numbers and (b) mean numbers during each winter (Dec–Feb).

4.5. OVERALL CHANGES IN CRUSTACEAN ZOOPLANKTON

Parallel to the observed changes in water temperatures and phytoplankton mass there have been strong responses in the population numbers of individual crustacean zooplankton species. Four (*D. cucullata, C. vicinus, Thermocyclops oithonoides, Bosmina longirostris*) of the 19 numerically important crustacean species showed an increase, whereas another four (*D. galeata, C. kolensis, Eudiaptomus graciloides, Eudiaptomus gracilis*) exhibited marked declines in numbers.

4.6. BOSMINIDS

In the Heiligensee we found three bosminid species (*Bosmina longirostris, B. coregoni coregoni, B. coregoni thersites*) (Fig. 5). *B. longirostris*, the dominating species, developed its population maximum during spring. In recent years (1988–1994) the population peak numbers generally lie above those recorded in 1980, 1984 and 1986. Because of short sampling intervals (2–4 days) in 1984 and 1986, an underestimation of population numbers can be excluded. Given these years as a baseline calibration for the early and mid 1980s, there is some evidence for an increase of *B. longirostris* in recent years. Between 1980 and 1988, peak abundances were generally found in May–June, whereas in 1989 through 1994 high numbers were already present in April–May. The two summer–autumn species (*B. c. coregoni, B. c. thersites*) showed decreasing tendencies in recent years. *B. c. thersites* totally disappeared in 1994, following minor numbers in 1993 (Fig. 5).

4.7. CYCLOPOID COPEPODS

Between 1975–76 and 1986 through 1994, we found that several cyclopoid species (*Cyclops vicinus* and *Thermocyclops oithonoides*) became more abundant, *C. kolensis* became less abundant, and populations of some species remained more-or-less

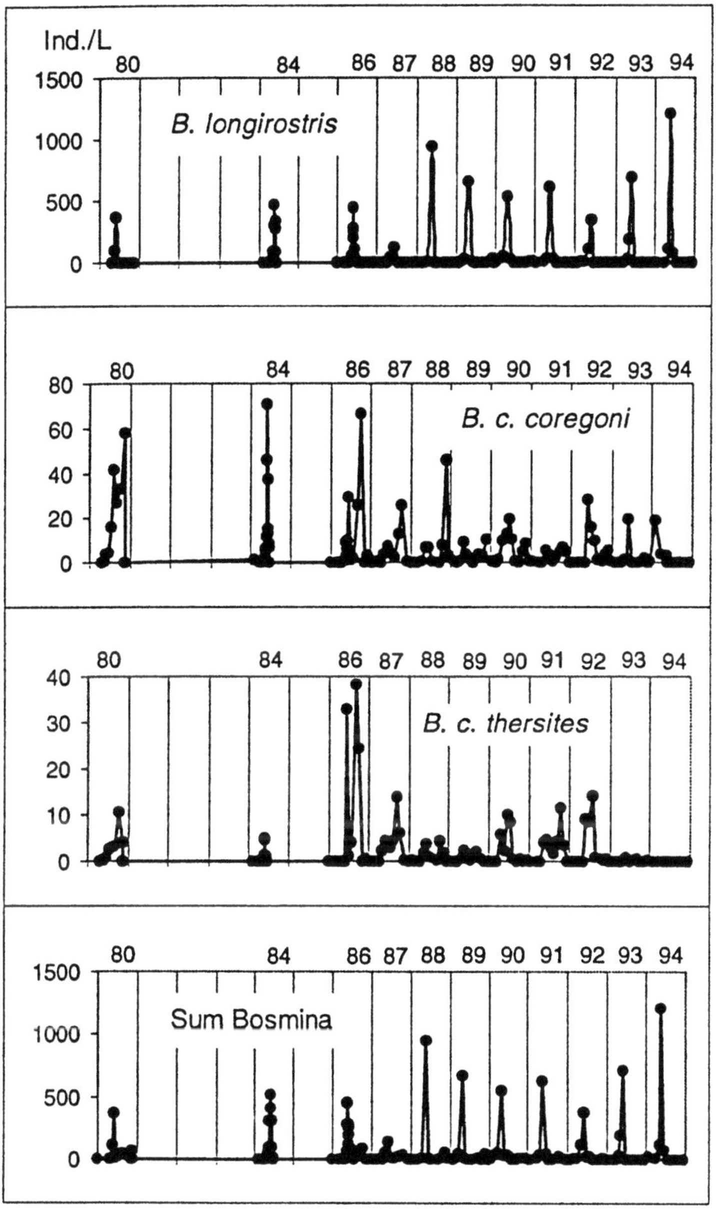

FIGURE 5. Abundance patterns of three *Bosmina* species (*B. longirostris, B. c. coregoni, B. c. thersites*) in the Heiligensee, in the period 1980 to 1994. Total numbers (l^{-1}) of the three species are summarised as "Sum Bosmina".

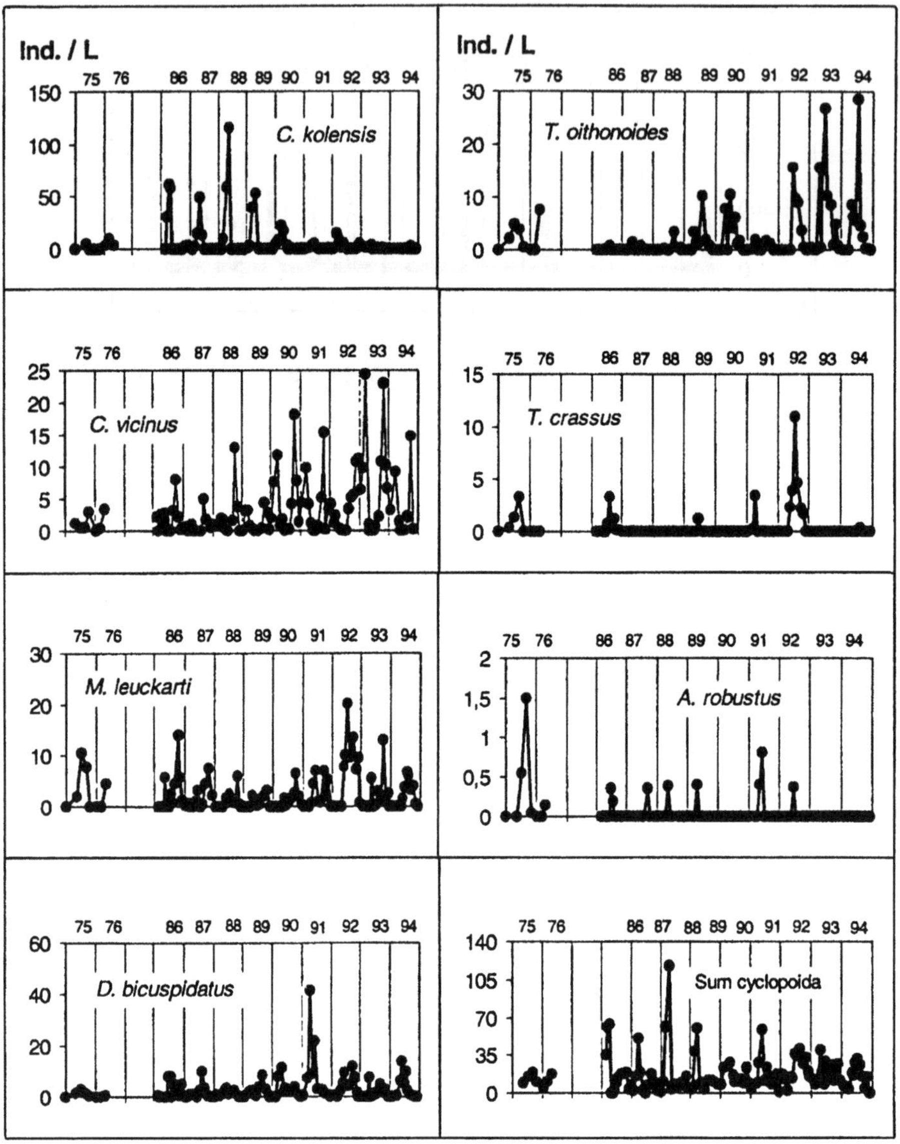

FIGURE 6. Abundance patterns of the seven cyclopoid copepod species in the Heiligensee, during the period 1975 to 1994. Total numbers (l^{-1}) of all species are summarised as "Sum cyclopoida".

constant (*Diacyclops bicuspidatus, Mesocyclops leuckarti*) (Fig. 6). *Thermocyclops crassus* and *Acanthocyclops robustus,* both summer species, were not found in 1993 and 1994 (with one exception in August 1994 for *T. crassus*).

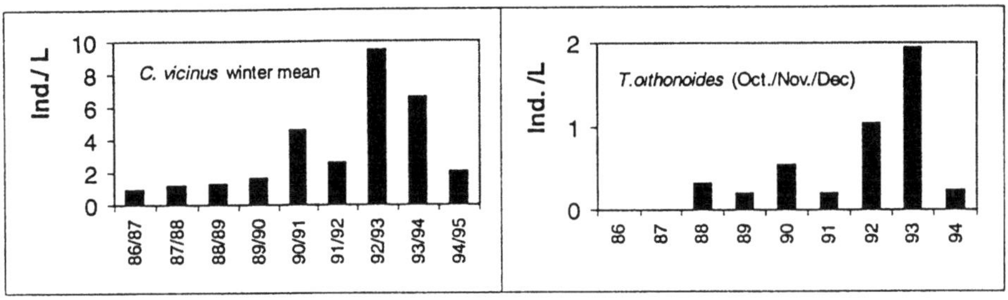

FIGURE 7. Mean numbers (l^{-1}) of *Cyclops vicinus* during winter (Dec–Feb) and *Thermocyclops oithonoides* during autumn (Oct–Dec) in the Heiligensee, from 1986–87 to 1994–95.

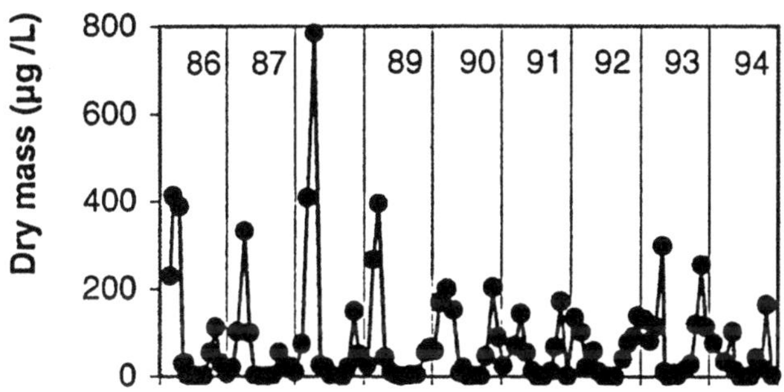

FIGURE 8. Dry mass (µg l^{-1}) of *Cyclops kolensis* plus *C. vicinus* in the Heiligensee, during the period 1986 to 1994.

Cyclops kolensis typically developed a single maximum during winter and early spring. During summer, late copepodites entered diapause [11] and the species totally disappeared from open water. *C. vicinus* usually developed a first maximum during spring and a second maximum during autumn (maximum in November). During summer, late copepodite stages diapause in the sediment [11]. Overlaps of *C. kolensis* and *C. vicinus* populations used to be basically restricted to early spring. This pattern, which is often found for these species [12, 13] was common in the Heiligensee until 1988–89. Thereafter *C. kolensis*, the dominant species during winter, has decreased to minor numbers in recent years (1990–91 to 1994). In contrast, *C. vicinus* has increased during spring and autumn (Fig. 6) and was found in unusually high numbers during winter, starting in 1990–91 (Fig. 7). Thus the mean winter densities (December–February) between 1986–87 and 1989–90 were 1.29 Ind. l^{-1}, compared with 5.07 Ind. l^{-1} between 1990–91 and 1994–95. Based on mean individual dry weights for adult *C. vicinus* (11.05 µg Ind^{-1}) and *C. kolensis* (6.77 µg Ind^{-1}) [14], we found there was an overall decrease in dry mass in the period 1990 to 1994 (Fig. 8).

During summer, five cyclopoid species coexisted in the Heiligensee (*D. bicuspidatus, M. leuckarti, T. oithonoides, T. crassus, A. robustus*) (Fig. 6), which differed in the timing of their peak abundances. Because of the generally low numbers of *A. robustus* and *T. crassus*, possible changes or extinction are hard to evaluate so far. Thus only *D. bicuspidatus, M. leuckarti* and *T. oithonoides* are considered in the following paragraph.

While densities of *D. bicuspidatus* and *M. leuckarti* remained at similar levels over the entire investigation period, there was an increase in peak numbers of *T. oithonoides* during the summers of 1992 and 1993 (Fig. 7). Overall, *T. oithonoides* extended its pelagic phase. A tendency for an early population development was found in 1989 to 1994 when individuals were first found in April–June compared with May–July in 1986 to 1988. Furthermore, an extension of occurrence in October through December has been recorded. *T. oithonoides* used to be absent during autumn–early winter in 1986 and 1987, but was found in low numbers thereafter (1988–1994) (Fig. 7).

Although monthly sampling intervals are critical for recognising peak numbers, even for copepods, there is some evidence that the timing of maximal densities has changed for the three species. Between 1986 and 1988, peak abundances of *T. oithonoides* were found in August–September, but thereafter in June–July. *M. leuckarti* used to develop maximal numbers in September (1986–1990), but between 1991 and 1994 peak numbers were found in June or in August. In 1986 to 1989, peak abundances of *D. bicuspidatus* were found in June to September, whereas in 1990 to 1994 maximal numbers were already present in April–May (except for 1992 when maximal numbers were recorded in October).

Considering the total numbers of cyclopoid copepods, no obvious change became apparent (see Fig. 6, "Sum of cyclopoida"). Cyclopoid naupliar stages (N1–N5) and copepodite C1–C5 stages were present in similar numbers throughout the period 1986 to 1994 (Fig. 9).

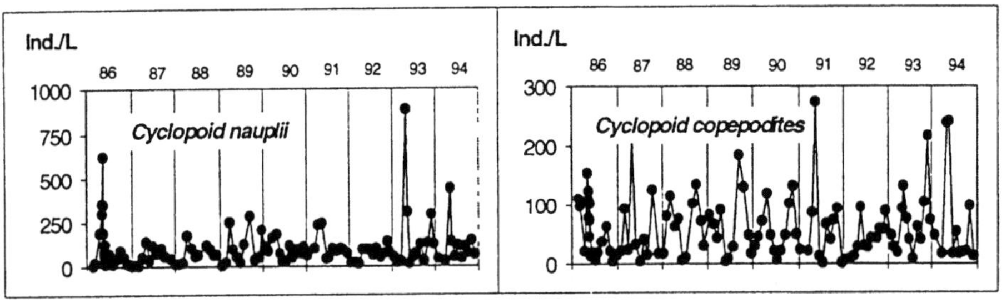

FIGURE 9. Numbers (l^{-1}) of cyclopoid nauplii (N1–N5) and copepodites (C1–C5) in the Heiligensee during 1986 to 1994.

5. Discussion

The results of this study, in conjunction with known aspects of the ecology of zooplankters, suggest that global warming could have some important effects on the dynamics of zooplankton communities in eutrophic, stratified, temperate lakes. The results for the Heiligensee suggest that the most important effects will be those associated with a change in water temperature (e.g. the effect of mild winters). This study also demonstrates the difficulty of identifying causal mechanisms in complex plankton communities and suggests that different groups (e.g. the cladocerans and the cyclopoids) could respond in different ways to changes in temperature. Cladocerans, which mature quickly and usually reproduce parthenogenetically, are more likely to respond to direct temperature effects rather than to indirect effects like altered food resources. Responses by copepods are even more complex, given that they develop through a number of larval stages, exhibit different diapausing strategies, and may even become cannibalistic. Moreover, as different species overlap in their population development, competition and predation may alter substantially as abundances and/or species composition may change after thermal stress.

In recent years the spring increase in temperature has reached levels (>10°C) where the reproductive rates of zooplankters are strongly influenced by temperature [15, 16]. Such direct temperature effects could have a particularly important effect on the population development of *B. longirostris*, a species that inhabits the upper water layers and typically increases rapidly during spring. As *B. longirostris* showed little change in filtration and rejection rates when fed with *Lyngbya*-enriched food assemblages [17], the observed changes in food quality are likely to have rather low negative effects on *B. longirostris*, especially since *Lyngbya limnetica* was one of the species that increased substantially during winter and spring in the Heiligensee (for more details see [4]). However, *B. longirostris* probably experienced a decline in available edible algal mass. Based on a conversion factor of 0.2 [18] for the ratio of carbon to biovolume, the edible algal fraction (sum of diatoms, cryptophytes and chlorophytes: these usually comprised most of the non-cyanobacterial mass) between January and June showed a decreasing tendency between 1986 (0.13–2.66 mg C l^{-1}) and 1993 and 1994 (0.10–0.35 mg C l^{-1} and 0.02–1.16 mg C l^{-1} respectively). Given that *B. longirostris* is also able to feed over a broad size range [19, 20], the question of whether *B. longirostris* has been limited by food, and whether this has changed during the investigation period, remains open. The ability to feed more efficiently at lower food concentrations than daphniids [21] may have provided a competitive advantage over *D. galeata*, which has decreased in abundance in recent years (discussed elsewhere). Following this decline, *B. longirostris* may have profited from a release in competition for the same food resource with *D. galatea*, which is a more efficient filter-feeder. In the Heiligensee, both species overlap during spring.

Underlying parameters for changes in the abundance of the two summer species (*B. c. coregoni*, *B. c. thersites*) remain unclear. Unfortunately we do not have long-term records of the monthly phytoplankton composition for the entire investigation period. In terms of chlorophyll-*a* concentrations, algal mass availability during summer remained on similar levels in all years under examination (Fig. 3). However, as seen for the noted changes during spring, alterations in food quality were probably decisive underlying parameters for changes in abundance of the two summer species as well. Since we have no records of the long-term development of *Chaoborus* and fish populations, questions related to predation remain open.

The two cold-water species, *C. kolensis* – which decreased to minor mumbers, and *C. vicinus* – which increased in abundance (Fig. 6), differ in their diapausing strategy. Diapause in *C. kolensis* is assumed to be primarily influenced by physiological readiness, i.e. diapausing copepodites are rather unaffected by external cues over relatively long time-periods [22]. In contrast, *C. vicinus* is able to respond directly to external cues such as photoperiod, oxygen and turbulence, and it is better adapted to unpredictable environments [12, 22, 23]. Artificial aeration at low and high temperatures resulted in increased emergence rates of diapausing *C. vicinus* copepodites [22]. It can be hypothesised that the absence of long ice-cover periods, in conjunction with the enhanced food resources during winter, may have contributed to the increase of *C. vicinus* during winter and spring, as considered below.

(i) The absence of long ice-cover periods prevented inverse thermal stratification, leading to mostly well-oxygenated and turbulent conditions in the sediment near-water layers, causing an early termination of the diapause of *C. vicinus* copepodites. Oxygen depletion during winter in the sediment near-water layers was generally restricted to years with long ice-cover periods (data not shown here). (ii) Due to enhanced algal mass in autumn and winter (Fig. 3) and the enhanced densities of rotifer prey during winter (Fig. 4b), only a reduced portion of late copepodite stages of *C. vicinus* may have entered diapause in autumn, while an unusually high portion developed to the adult stage, forming overwintering populations (Fig. 7). Since adult *C. vicinus* are omnivorous with a high herbivory component (>50% [24–26]) they should have profited from the increased availability of algal and rotifer prey during winter. The densities of both food components lay well beyond levels where ingestion rates of *C. vicinus* became independent of prey density [27]. This is confirmed by the positive correlation between maximum numbers of *C. vicinus* in spring and mean algal mass during winter (Fig. 10a), and by the positive correlation between mean densities of *C. vicinus* and rotifers during winter (Fig. 10b). Our results support earlier observations, where seasonal fluctuations in *C. vicinus* were also positively correlated with favourable food conditions [12, 23, 28]. The change in algal composition towards a dominance of cyanobacteria during winter and spring was not necessarily disadvantageous for adult cyclopoid copepods exhibiting a raptorial feeding mode. Large cyanobacteria such as *Microcystis aeruginosa* were found in faecal pellets of adult *C. vicinus* [27]. After the decline of *C. kolensis*, *C. vicinus* may also have been favoured by a higher ratio of available food per unit of cyclopoid dry mass, since an overall decrease in cyclopoid dry mass has been recorded (Fig. 8).

For species that develop two separate peaks of abundance, like *C. vicinus* (spring–autumn), effects of changes in one season may propagate into the following seasons. It can be hypothesised that an accentuated pool of diapausing copepodites during summer, resulting from the enhanced population during the previous spring, contributed to the population increase in the autumn. More than 70% of the variation

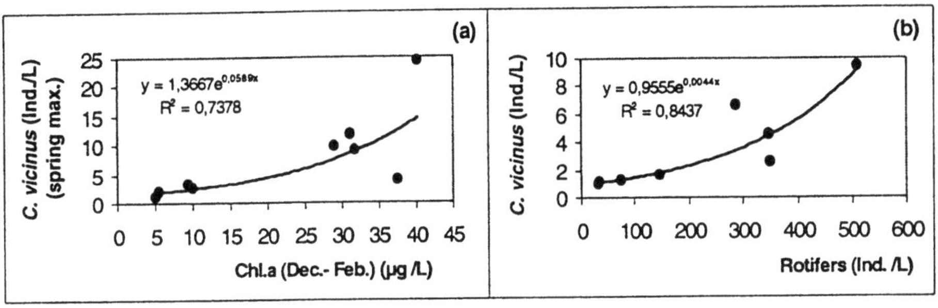

FIGURE 10. (a) Correlation between maximum numbers (l^{-1}) of *C. vicinus* in the Heiligensee during spring (Mar–May) versus mean chlorophyll-*a* concentrations ($\mu g\ l^{-1}$) in the upper 0–4 m water stratum during winter (Dec–Feb). (b) Correlation between mean numbers of *C. vicinus* versus mean numbers of rotifers during winter.

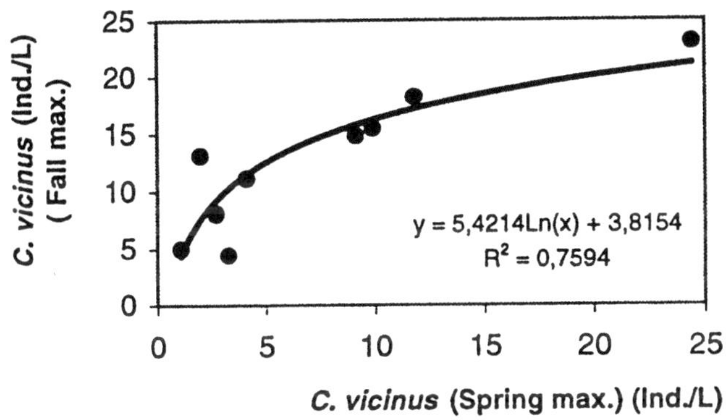

FIGURE 11. Correlation between maximum numbers of adult *C. vicinus* in the Heiligensee during spring and autumn.

in peak numbers during autumn could be explained by the peak numbers during spring of the same year (Fig. 11). Changes in the timing of autumn overturn, affecting the emergence rates of *C. vicinus* copepodites, can be excluded. Holomixis was generally achieved by mid October in all of the studied years [4].

Omnivorous *C. kolensis*, which also exhibits a high dietory component [29] but has a rather inflexible diapausing behaviour, was not able to profit from the enhanced availability of both algal and rotifer prey during winter. Early predation by the larger *C. vicinus*, which is known to prey heavily on nauplii [29–31], was probably responsible for its decline. In the early years of this study, predation pressure on *C. kolensis* used to be minor, because there was a general absence of invertebrate predators during winter. Based on a non-linear relationship (Fig. 12), the mean winter densities of *C. vicinus* explained 73% of variation in the maximum numbers of *C. kolensis*.

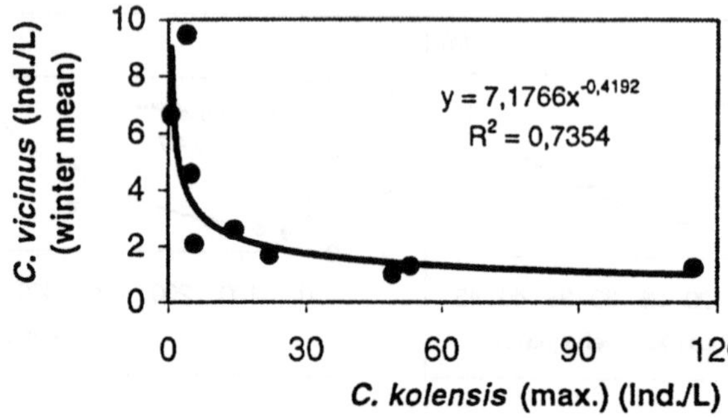

FIGURE 12. Correlation between mean numbers (l^{-1}) of adult *C. vicinus* in the Heiligensee during winter and maximum numbers of *C. kolensis* during winter–spring.

Since *C. kolensis* and *C. vicinus* have high dietory overlaps [27], there seems to be a threshold value for *C. vicinus* densites at ca. 2–4 Ind. l^{-1}, beyond which *C. kolensis* is outcompeted by the larger *C. vicinus* (Fig. 12). Thus *C. vicinus*, which evolved a flexible diapausing strategy, seems to be the superior competitor under more eutrophic conditions. There is no evidence that direct temperature effects contributed to the observed changes in *C. kolensis* and *C. vicinus*.

Observed changes during summer are less clear. Adult cyclopoid copepods are considered to be omnivorous [14, 26, 32], so *T. oithonoides*, which increased in abundance, could have profited from the decreased numbers of *B. c. coregoni* and *B. c. thersites* during summer, thereby reducing interspecific overlaps for the same food resources. While algae were the key dietory factor for reproductive success of the small *Tropocyclops prasinus mexicanus*, the availability of invertebrate prey had a greater influence on reproduction of larger species such as *Mesocyclops edax* and *Diacyclops thomasi* [14, 33]. Thus, since the role of herbivory seems to be dependent on species body size, the small *T. oithonoides* may have profited most, compared to the larger *M. leuckarti* and *D. bicuspidatus*. However, the increase in algal mass during spring may have been favourable for the early population development of all three species (*T. oithonoides*, *M. leuckarti*, and *D. bicuspidatus*). The significant rising trend in algal mass in November (see section 4.3) may have contributed to an extension of the population of *T. oithonoides* into December (Fig. 7).

The role of nauplii and copepodite stages in accounting for the observed changes is unclear. As far as we can tell from monthly sampling intervals their abundances have not changed between 1986 and 1994. Cyclopoid nauplii seem to rely on phytoflagellates in order to successfully develop to the copepodite stage [13, 34]. Early naupliar stages seem to be especially sensitive to low food concentration. Hansen & Santer [33] give threshold concentrations of 0.2 and 0.3 mg C l^{-1} beyond which nauplii of *C. vicinus* and *M. leuckarti* developed to the copepodite stage. Available data for the total mass of phytoflagellates in the Heiligensee ranged between 0.08 and 1.31 mg C l^{-1} in 1986

(March–June), between 0.04 and 0.21 mg C l^{-1} in 1993 (January–September), and between 0.01 and 0.40 mg C l^{-1} in 1994 (March–December). According to this, nauplii in the Heiligensee were probably food-limited during severe periods in 1993 and 1994, while during the winter–spring of 1986 limitation was basically restricted to the clearwater period (mid/end of May). However, since there are both species- and food-specific differences for successful nauplii development [13, 34], the question of whether the above given thresholds translate to field conditions, remains open.

The overall changes in the Heiligensee, which coincided with a period of consecutive mild winters, suggest that plankton communities in shallow, eutrophic and stratified lake ecosystems will respond sensitively towards climatic change (see also Gaedke et al. this volume). To what extent direct versus indirect temperature effects contributed to the observed changes, can not be derived from field data alone. For zooplankters, changes in the food conditions (density, quality) highly influence the competitive potential of coexisting species and may, at times, be even more important than direct temperature effects. This applies especially to effects that influence the life-history traits of species which proceed in time. Alterations in the pool of resting stages will have memory effects for zooplankton communities. Furthermore, effects of climatic changes early in the season, such as the forecasted trend towards mild winter air temperatures in the northern European area, are likely to propagate into summer and autumn. The non-linear character of the observed responses, in conjunction with the species-specific responses, indicate the individual character of lakes and the probabilistic nature of ecological predictions. Moreover, as climate or weather conditions exhibit high variability at small spatial scales, we need a broad range of case studies in order to be able to distinguish more general patterns of changes from individualistic patterns in ecosystems.

Acknowledgements

We wish to thank Rita Stellmacher for performing the time-series analyses and Bärbel Arndt and Kristin Heisterkamp for phytoplankton and zooplankton analyses. N. Walz and Guntram Weithoff provided valuable comments on the manuscript. The study was support by the Freie Universität Berlin (1975–1993) and the Institut für Gewässerkunde und Binnenfischerei (1994).

References

[1]. Edmondson, W. T. & Litt, A. H. (1982). *Daphnia* in Lake Washington. *Limnol. Oceanogr.* **27**, 272-293.

[2]. George, D. G., Hewitt, D. P., Lund, J. W. G. & Smyly, W. J. P. (1990). The relative effects of enrichment and climate change on the long-term dynamics of *Daphnia* in Esthwaite Water, Cumbria. *Freshwater Biology* **23**, 55-70.

[3]. Vollenweider, R. (1982). *Eutrophication of waters*. OECD, Paris. 1-155.

[4]. Adrian, R., Deneke, R., Mischke, U., Stellmacher, R. & Lederer, P. (1995) A long term study of the Heiligensee (1975-1992). Evidence for effects of climatic change on the dynamics of eutrophied lake ecosystems. *Arch. Hydrobiol.* **133**, 315-337.

[5]. Schumpelick, B. (1983). Untersuchungen zur Populationsökologie von *Bosmina longirostris, Eubosmina c. coregoni* und *Eubosmina c. thersites* im Heiligensee. Diplom-Thesis, Freie Universität Berlin.

[6]. Müller, B. (1986). Das Frühjahrsmaximum von *Bosmina longirostris* im Heiligensee: Populationsdynamische Untersuchungen und in-situ Experimente. Diplom-Thesis, Freie Universität Berlin.

[7]. Schubert, M. (1978). Saisonale und vertikale Verteilung der Cyclopiden (Crustacea, Copepoda) im Pelagial des Heiligensees. Diplom-Thesis, Freie Universität Berlin.

[8]. Deutsche Einheitsverfahren (1992). *Deutsche Einheitsverfahren zur Wasser-, Abwasser- und Schlammuntersuchung.* Verlag Chemie GmbH (Weinheim).

[9]. Sneyers, R. (1975). *Sur l'analyse statistique des series d'observation.* WMO TN.

[10]. Berliner Wetterkarten (1975-1994). Meteorological data published daily by the Institut für Meteorologie. Freie Universität Berlin.

[11]. Schubert, M. (1986). Der Einfluß einiger Faktoren auf die Koexistenz von Cyclopiden im Heiligensee (Crustacea, Copepoda) im Heiligensee, Landschaftsentwicklung und Umweltforschung. *Schriftenreihe des Fachbereichs Landschaftsentwicklung der TU Berlin* **40**, 141-148.

[12]. Maier, G. (1989). Variable life cycles in the freshwater copepod *Cyclops vicinus* (Uljanin 1875): support for the predator avoidance hypothesis? *Arch. Hydrobiol.* **115**, 203-219.

[13]. Santer, B. (1990). Lebenszyklusstrategien cyclopoider Copepoden. PhD-Thesis, Christian-Albrecht-Universität Kiel.

[14]. Adrian, R. & Frost, T. M. (1993). Omnivory in cyclopoid copepods: comparison of algae and invertebrates as food for three, differently sized species. *J. Plankton Res.* **15**, 643-658.

[15]. Tilman, D., Mattson, M. & Langer, S. (1981). Competition and nutrient kinetics along a temperature gradient: an experimental test of a mechanistic approach to niche theory. *Limnol. Oceanogr.* **26**, 1020-1033.

[16]. Orcutt, J. D. J. & Porter, K. G. (1983). Diel vertical migration by zooplankton: constant and fluctuating temperature effects on life history parameters of *Daphnia*. *Limnol. Oceanogr.* **28**, 720-730.

[17]. Webster, K. E. & Peters, R. H. (1978). Some size-dependent inhibitions of larger cladoceran filterers in filamentous suspensions. *Limnol. Oceanogr.* **23**, 1238-1245.

[18]. Reynolds C.S. (1993). *The ecology of freshwater phytoplankton.* Cambridge University Press.

[19]. Geller, W. & Müller, H. (1981). The filtering apparatus of cladocera: filter mesh-sizes and their implications on food selectivity. *Oecologia* **49**, 316-321.

[20]. Bogdan, K. G. & Gilbert, J. J. (1987). Quantitative comparison of food niches in some freshwater zooplankton. A multi-tracer-cell approach. *Oecologia* **72**, 331-340.

[21]. DeMott, W. R. (1982). Feeding selectivities and relative ingestion rates of *Daphnia* and *Bosmina*. *Limnol. Oceanogr.* **27**, 518-527.

[22]. Stimpfig, A. (1986). Experimentelle Untersuchungen zur Beendigung der Diapause von *C. kolensis* und *C. vicinus* (Crustacea, Copepoda), Landschaftsentwicklung und Umweltforschung. *Schriftenreihe des Fachbereichs Landschaftsentwicklung der TU Berlin* **40**, 135-139.

[23]. Hansen, A. M. & Jeppesen, E. (1992). Life cycle of *Cyclops vicinus* in relation to food availability, predation, diapause and temperature. *J. Plankton Res.* **14**, 591-605.

[24]. Adrian, R. (1987). Viability of phytoplankton in fecal pellets of two cyclopoid copepods. *Arch. Hydrobiol.* **110**, 321-330.

[25]. Adrian, R. (1991). Filtering and feeding rates of cyclopoid copepods feeding on phytoplankton. *Hydrobiologia* **210**, 217-223.

[26]. Santer, B. (1993) Potential importance of algae in the diet of adult *Cyclops vicinus*. *Freshwater Biology* **30**, 269-278.

[27]. Adrian, R. (1988). Untersuchungen zur herbivoren und carnivoren Ernährungsweise von *Cyclops kolensis* und *C. vicinus* (Crustacea, Copepoda). PhD-Thesis, Freie Universität Berlin.

[28]. George, D. G. (1976). Life cycle and production of *Cyclops vicinus* in a shallow eutrophic reservoir. *Oikos* **27**, 101-110.

[29]. Adrian, R. (1991). The feeding behaviour of *Cyclops kolensis* and *C. vicinus* (Crustacea, Copepoda). *Verh. Internat. Verein. Limnol.* **24**, 2852-2863.

[30]. Brandl, Z. & Fernando, C. H. (1978). Prey selection by the cyclopoid copepods *Mesocyclops edax* and *Cyclops vicinus*. *Verh. Internat. Verein. Limnol.* **20**, 2505-2510.

[31]. Zankai, N. P. (1984). Predation of *Cyclops vicinus* (Copepoda: Cyclopoida) on small zooplankton animals in Lake Balaton (Hungary). *Arch. Hydrobiol.* **99**, 360-378.

[32]. Tóth, L. G. & Zankai, N. P. (1985). Feeding of *Cyclops vicinus* (Uljanin) (Copepoda: Cyclopoida) in Lake Balaton on the basis of gut content analyses. *Hydrobiologia* **122**, 251-260.

[33]. Adrian, R. & Frost, T. M. (1992). Comparative feeding ecology of *Tropocyclops prasinus mexicanus* (Copepoda, Cyclopoida). *J. Plankton Res.* **14**, 1369-1382.

[34]. Hansen, A. M. & Santer, B. (1995). The influence of food on the development, survival and reproduction of the two cyclopoid copepods: *Cyclops vicinus* and *Mesocyclops leuckarti*. *J. Plankton Res.* **17**, 631-646.

THE INFLUENCE OF YEAR-TO-YEAR CHANGES IN POSITION OF THE ATLANTIC GULF STREAM ON THE BIOMASS OF ZOOPLANKTON IN WINDERMERE NORTH BASIN, UK

D. G. GEORGE AND D. P. HEWITT
Institute of Freshwater Ecology
Windermere Laboratory
Far Sawrey, Ambleside,
Cumbria LA22 0LP, UK

Abstract

Zooplankton samples have been collected from Windermere, the largest lake in the English Lake District (Cumbria), at weekly intervals since 1940. In this paper, we explore the physical, chemical and biological factors influencing the year-to-year variations in the summer biomass of zooplankton collected from the North Basin of Windermere between 1966 and 1991. The most important factor influencing the long-term change in the summer biomass of zooplankton was the progressive enrichment of the lake by treated sewage. Nutrient loadings to the lake increased significantly in the 1970s when many properties were connected to the main sewer. Long-term changes in the numbers of perch present in the lake over the same period appeared to have little effect on the zooplankton biomass. There was no significant correlation between the abundance of zooplankton and the numbers of perch caught ($r = -0.04$) but the year-class-strength of perch was positively correlated ($r = 0.44$) with the summer biomass of zooplankton. The most important factor influencing year-to-year variations in the biomass of zooplankton was the early summer weather. Larger biomass was consistently recorded in relatively cool, windy years where the early summer thermocline was relatively unstable and large numbers of edible algae were present in the epilimnion. The summer zooplankton biomass was thus positively correlated with the average wind speed ($r = 0.46$) and the abundance of edible algae ($r = 0.55$). However, an even higher proportion of year-to-year variations in zooplankton biomass could be explained by a general "weather" index based on the position of the Gulf Stream in the North Atlantic Ocean ($r = -0.59$). Latitudinal movements of the Gulf Stream appear to influence the

D.G. George et al.(eds.), Management of Lakes and Reservoirs during Global Climate Change, 223–244.

dynamics of the lake by regulating the weather systems that influence the timing and intensity of mixing. Thus "north" Gulf Stream years were characterised by lower average wind speeds, shallower thermoclines, smaller biomass of edible algae and lower concentrations of herbivorous zooplankters. Whilst comparable physical effects have been recorded recently in the seas around the European continental shelf, lake Windermere appears to be a particularly effective integrator of recent climatic trends.

1. Introduction

The thermal and biological characteristics of deep, temperate lakes are strongly influenced by year-to-year changes in the weather [1]. Changes in the timing and intensity of wind-induced mixing have a profound effect on the seasonal dynamics of plankton [2] but very little is known about the climatic drivers that influence these year-to-year variations. Recent studies in the lakes of the English Lake District suggest that many of these changes are quasi-cyclical in nature and are related to the movement of weather systems from the North Atlantic Ocean. In 1985, George & Harris [3] demonstrated that year-to-year fluctuations in the biomass of crustacean zooplankters in Windermere were correlated with an 8- to 10-year cycle in sea surface temperatures. More recently, George & Taylor [4] have correlated these cycles with changes in the weather that can be related to north–south movements of the Gulf Stream in the eastern Atlantic. In this paper, we examine some of the physical forcing factors responsible for these "bottom–up" effects and assess the extent to which the zooplankton populations are also influenced by the "top–down" effects associated with fish predation.

In recent years, it has become clear that year-to-year changes in the weather have a major effect on the seasonal succession of lake phytoplankton [5, 6]. The phytoplankton populations of Windermere have been the subject of detailed study for a number of years [7–9]. The spring plankton is usually dominated by the diatom *Asterionella formosa* Hass, but a variety of small flagellates appear in early summer and serve as an important source of food for filter-feeding crustaceans.

The fish populations of Windermere have been studied intensively for more than fifty years [10–13]. The most abundant planktivore is the perch *Perca fluviatilis* L. Young perch feed exclusively on zooplankton for a short time in their life but the adults are facultative planktivores that also feed on larger invertebrates. The perch populations of Windermere have been monitored by systematic trapping every year since 1941. Growth of perch is known to be affected by summer water temperatures and population density, and their survival is influenced by both intraspecific and interspecific predation.

2. Description of study site

Windermere is a typical narrow, glacial valley lake situated in the northwest of England (54°22'N, 2°56'W). It is 16.9 km long and is divided into two distinct basins by a large island and a region of shallow water. The water, zooplankton and fish samples analysed here were collected from the North Basin (Fig. 1) which covers an area of 8.1 km^2 and has a mean depth of 25 m and a maximum depth of 64 m. Samples of phytoplankton and zooplankton have been collected from a fixed station in the North Basin at

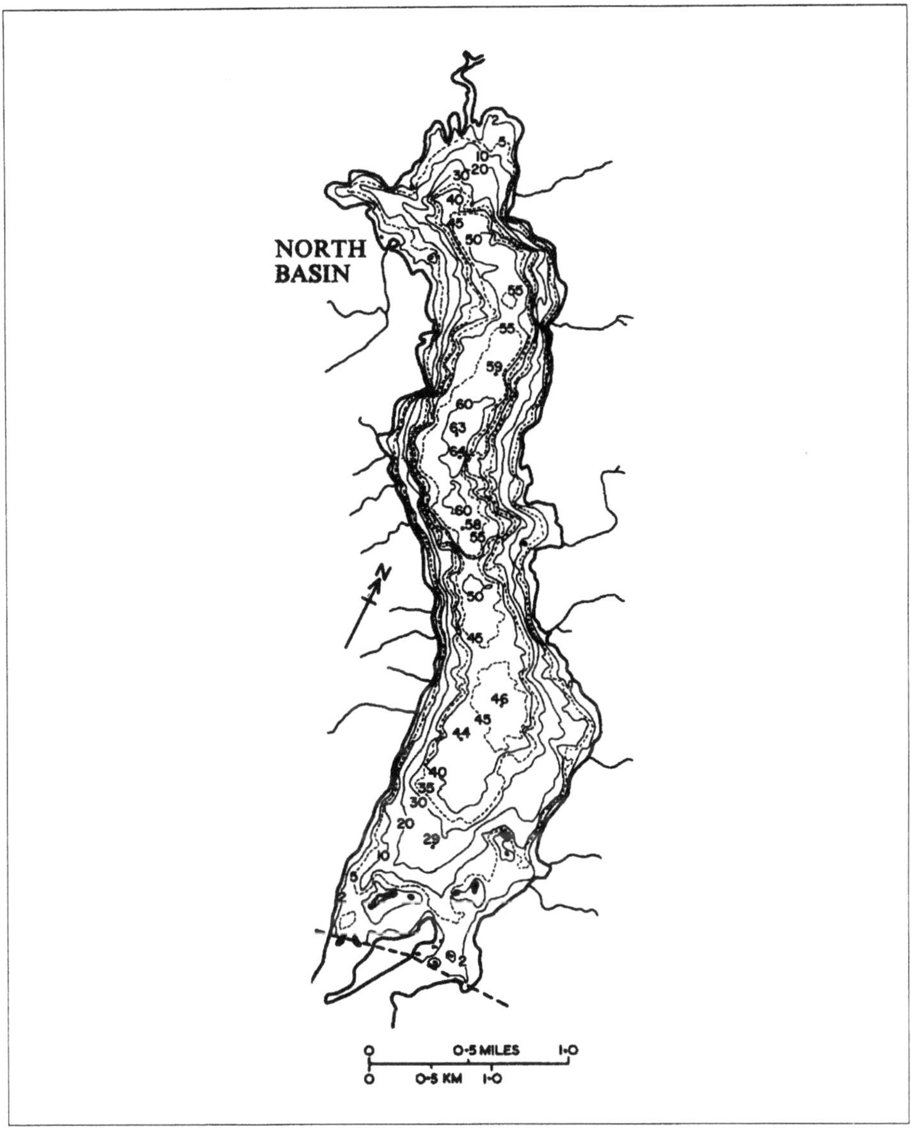

FIGURE 1. A bathymetric map of the North Basin of Windermere, English Lake District.

fortnightly intervals since 1940. The data analysed here cover the period 1966 to 1991, a time-period during which monthly charts of the position of the Atlantic Gulf Stream have been made available from shipboard and satellite observations.

3. Field and laboratory methods

Wind speed, rainfall, air temperature and the number of hours of bright sunshine were measured daily at an open site ca. 1 km from the shores of the lake. Wind speeds were estimated using the Beaufort scale and the mean of two daily observations was converted to metres per second using a standard conversion factor. The mean daily air temperatures are the average of the daily maxima and minima, and the hours of sunshine are the numbers recorded on a standard "burning paper" trace. Seasonal variations in water temperature were recorded at a central station, using a thermistor attached to an oxygen electrode. Early measurements (1966–1984) were taken with a Mackereth Oxygen Electrode and more recent measurements were obtained with a Yellow Springs Instrument oxygen probe and a Windermere Profiler [14]. Water samples for chemical analysis were collected at the same station using a 7-m long weighted tube [15]. The only chemical results reported here are those for dissolved reactive phosphorus, measured by the solvent extraction method first described by Stephens [16]. Chlorophyll-*a* concentrations were estimated by filtering known volumes of water through Whatman GF/C filter papers and extracting the pigment with hot aqueous ethanol [17]. Phytoplankton samples were collected using the same 7-m tube and counted using the inverted microscope technique described by Lund, Kipling & Le Cren [18]. The palatability of the different species was also assessed by referring to previous studies, and estimates of the biomass of edible algae were derived using the individual cell volumes listed by [5].

Samples of crustacean zooplankton were collected by hauling a net vertically through the water column. The net had a mouth area of 700 cm^2 and a mesh-size of 240 nm; it captured most species and life-stages larger than 1st-stage copepodites. The animals captured by the net were preserved in 4% formalin and transferred to neutralised formalin for long-term storage. Zooplankton samples were counted and measured in a circular trough fitted to the stage of a stereo-zoom microscope. Only a small proportion of the archived samples have hitherto been counted but the biomass of every sample has been measured using the simple "settled volume" technique described by George & White [19].

The perch populations of Windermere have been monitored by systematic trapping every year since 1941. The standard "Windermere" traps are constructed of wire-netting on a rigid frame and are deployed at selected sites during the spawning season. The traps are unselective for perch between 9 and 30 cm in length but are highly selective for spawning males which typically account for more than 80% of the catch. Year-to-year variations in the numbers of perch are normally expressed as total catches-per-unit-effort and as year-class strengths estimated from the number of 2-year old fish caught.

4. Data processing

4.1. METEOROLOGICAL AND LIMNOLOGICAL DATA

Meteorological averages were calculated for four defined seasons: winter (December to February), spring (March to May), summer (June to August) and autumn (September to November). This sub-division of the year is limnologically more appropriate since it allows us to contrast spring weather patterns, when the lake is isothermal, with summer weather patterns when the lake is stably stratified. The weekly temperature profiles were reduced to depth–time plots that showed both the absolute temperature and the rate of change in temperature. The depth of the seasonal thermocline was estimated by inspection and average depths were calculated and related to the meteorological driving variables.

4.2. GULF STREAM INDEX

Monthly charts showing the position of the north "wall" of the Atlantic Gulf Stream have been produced from surface, satellite and aircraft measurements since 1966 [20]. These charts were first published by the US Naval and Oceanographic Office in the *Gulf Stream Monthly Summary* (1966–1974) and then by the US National Oceanic and Atmospheric Administration in *Gulf Stream* (1975–1980) and in *Oceanographic Monthly Summary* (1981–1994). Data on the year-to-year variations in the position of the north wall of the Gulf Stream were provided by colleagues at the UK Plymouth Marine Laboratory, using a procedure that is essentially the same as that described by Taylor & Stephens [21]. In this procedure, the latitude of the north wall was read from each chart (Fig. 2) at six longitudes 79° W, 75° W, 72° W, 70° W, 67° W and 65 °W, and an index of position was constructed using principal components analysis. This more complex statistical procedure was chosen in preference to simple averaging which would emphasise the measurements taken at more western longitudes where the north–south movements are larger and more irregular. The broken line drawn in Fig. 2 shows the mean position of the north wall between 1966 and 1992. The associated error bars show the 95% confidence intervals of the north–south movements recorded at each position over the same period. Correlation coefficients were calculated between the time-series at the six longitudes and principal components were calculated from the resulting matrix. The first principal component accounted for 25% of the recorded variation and provided the best estimate of the latitudinal displacement of the whole section of the Gulf Stream.

4.3. ZOOPLANKTON BIOMASS

In this paper, 26 years of data for zooplankton biomass have been reduced to two summary variables: (1), a raw time-series showing the year-to-year variations in the average summer biomass of zooplankton; (2), a de-trended time-series showing the relative year-to-year variation in the average summer biomass of zooplankton. The summer period has been arbitrarily defined as the period between 1 June and 31 August. Appreciable numbers of crustacean zooplankters seldom appear in the lake before the end of May but the period of population decline is more variable and sometimes may be delayed until late September.

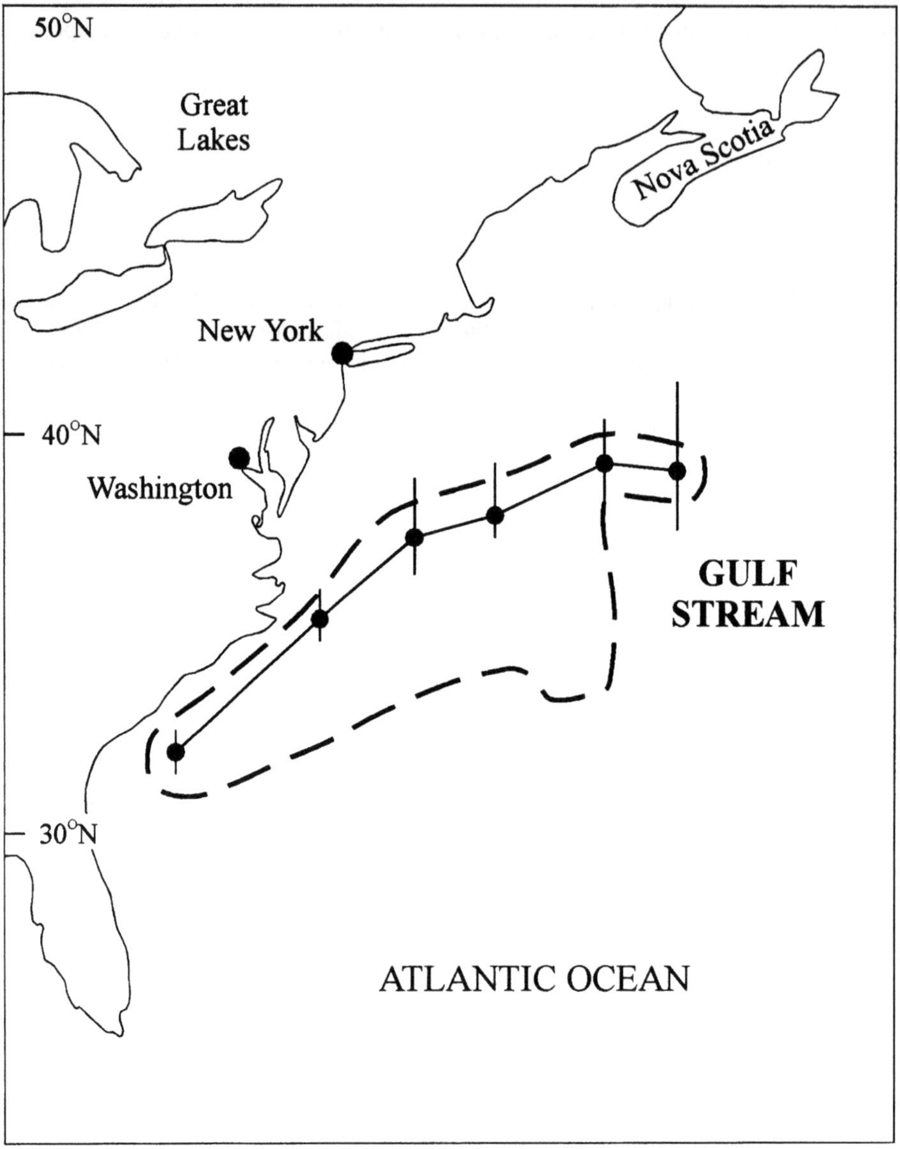

FIGURE 2. A line-drawing of the position of the Gulf Stream in the western Atlantic, taken from a satellite image of sea-surface temperature. The position of the north wall is indicated by the solid line connecting six points, with vertical lines representing the approximate ranges of the north–south movements of the north wall at six standard longitudes. The heavy broken line encompasses an area occupied by the warmest water in the Gulf Stream.

5. Results

5.1. CRUSTACEAN ZOOPLANKTON POPULATIONS IN WINDERMERE

Table 1 lists the most common species of crustacean plankters found in Windermere North Basin and gives a general indication of their relative abundances. The crustacean plankton in Windermere is typical of many large lakes in that it consists of a few species belonging to a small number of genera. The winter and spring plankton is dominated by the calanoid copepods *Eudiaptomus gracilis* and *Eudiaptomus laticeps*. The summer zooplankton is dominated by the cladoceran *Daphnia hyalina* var *galeata* Sars, with smaller numbers of the cyclopoids *Cyclops strenuus abyssorum* Sars and *Mesocyclops leuckarti* (Claus). The most important factor influencing the year-to-year variations in summer biomass is the number of *Daphnia* found in each sample. Counts on a representative series of samples demonstrate that the first summer biomass peak is dominated by *Daphnia*, with small numbers of carnivorous cladocerans like *Leptodora* and *Bythotrephes*. Numbers of *Daphnia* usually decline in mid summer but a second population maximum is frequently recorded in late summer.

TABLE 1. The dominant planktonic microcrustaceans found in the North Basin of Windermere between 1966 and 1991. Values given for their abundances (numbers of individuals per litre) indicate the average population densities between May and September over the 26-year period.

Species	Abundance
Daphnia hyalina var *lacustris* Sars	4.6
Bosmina coregoni var *obtusirostris* (Sars)	0.9
Leptodora kindtii (Focke)	0.1
Bythotrephes longimanus Leydig	0.1
Eudiaptomus gracilis Sars	5.6
Eudiaptomus laticeps Sars	3.8
Cyclops strenuus abyssorum Sars	4.2
Mesocyclops leuckarti Claus	3.0

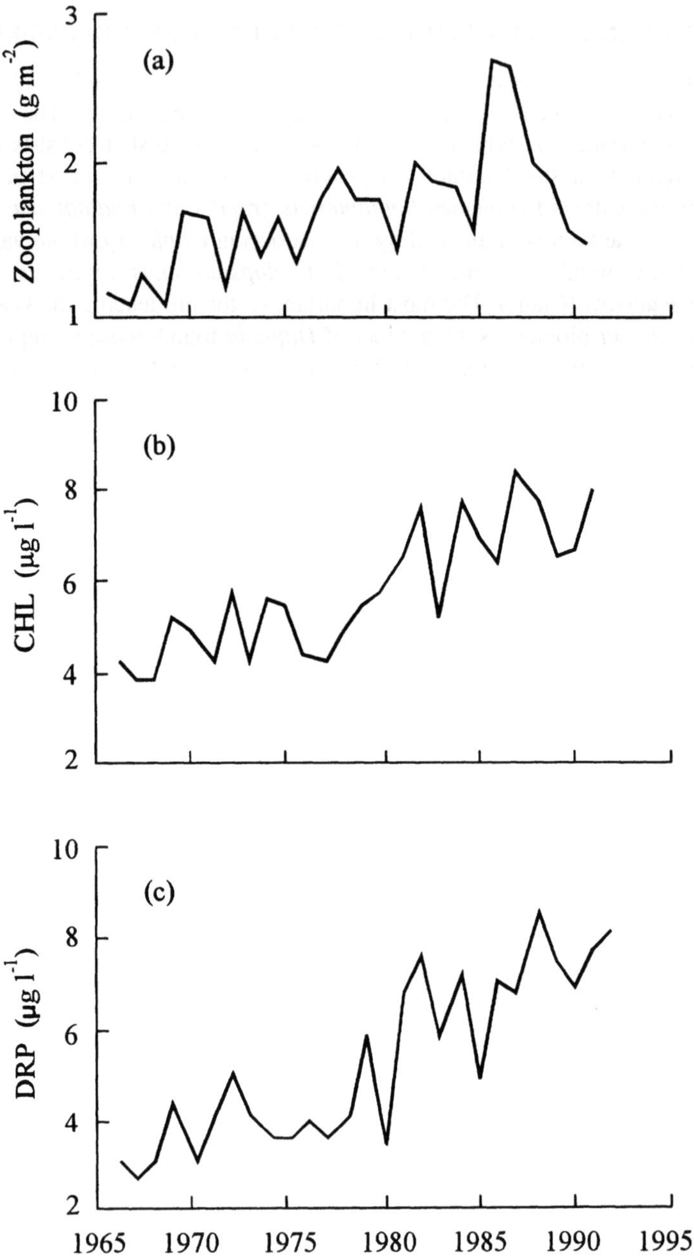

FIGURE 3. Plots of year-to-year variations in: (a) zooplankton biomass (g m^{-2}), (b) phytoplankton chlorophyll (µg l^{-1}) and (c) dissolved reactive phosphorus (µg l^{-1}) in Windermere North Basin from 1966 to 1991/92.

5.2. LONG-TERM TREND IN ZOOPLANKTON BIOMASS

The most important factor influencing a long-term change in the zooplankton biomass of Windermere is a slow but progressive increase in the general productivity of the lake. The North Basin of Windermere receives treated sewage effluent from a small town (Ambleside) and several villages and hamlets. Nutrient loadings to the lake increased substantially in the early 1970s when many rural properties that had previously used septic tanks were connected to main sewers. Fig. 3a–c compares the long-term trend in the annual biomass of the zooplankton with parallel trends in the annual biomass of phytoplankton and the winter concentration of dissolved reactive phosphorus (DRP). For much of the year, the concentration of DRP in the North Basin is very close to the limits of detection, but relatively high values are recorded in winter when the external loading is high and there is little internal assimilation. A statistical analysis of the three time-series showed that there was a significant positive correlation ($r = 0.53$, $p<0.01$) between the annual biomass of zooplankton and the annual biomass of phytoplankton which was, in turn, strongly correlated with the winter concentration of DRP ($r = 0.82$, $p<0.01$).

5.3. YEAR-TO-YEAR VARIATIONS IN ZOOPLANKTON BIOMASS

Fig. 4 shows the average seasonal change in zooplankton biomass recorded in Windermere North Basin between 1 January 1966 and 31 December 1991. The 95% confidence intervals in Fig. 4 show the year-to-year variations in the seasonal cycle and have been calculated by averaging the biomass estimates recorded in the same week for 26 successive years. Most of the confidence intervals are relatively narrow, but the quantity of zooplankton produced in late summer is more variable than that produced in early summer. The horizontal bar in Fig. 4 shows the 3-month "block" of weeks included in our estimates of mean summer biomass. The period of high biomass actually extends beyond the June to August period but our "summer" average captures the most important feature of the annual production cycle. Figs. 5a and 5b show the mean summer biomasses respectively plotted as raw and de-trended time-series for each of the 26 years studied. The raw time-series has been used in all calculations relating the year-to-year variations in biomass to changes in fish and phytoplankton, and the de-trended time-series has been used where correlations are drawn with changes in the weather.

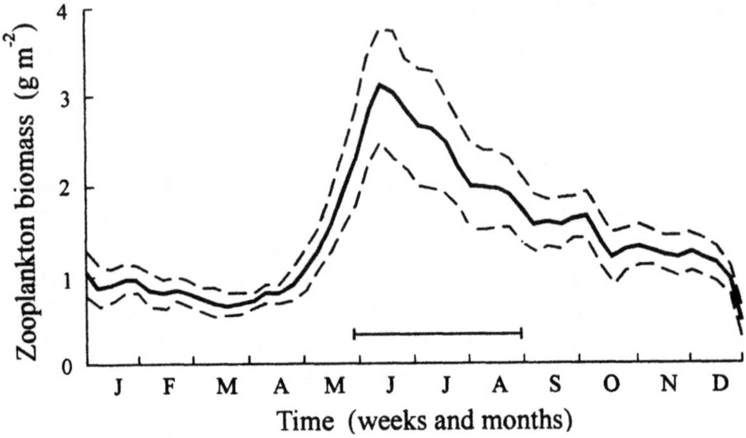

FIGURE 4. Mean zooplankton biomass in Windermere North Basin, based on 52 weekly samples for the period 1 January 1966 to 31 December 1991. Means (solid line) and 95% confidence limits (broken lines) were calculated from biomass in each particular week for the 26-year period. The horizontal bar indicates a 3-month block of weeks included in estimates of mean "summer" biomass (see the text).

5.4. IMPACT OF PERCH PREDATION ON YEAR-TO-YEAR VARIATIONS IN ZOOPLANKTON BIOMASS

The perch populations in the North Basin of Windermere have undergone a number of major fluctuations since recording began in 1941. During the period considered here, the most dramatic changes were recorded in the mid-1970s when an epidemic disease of perch resulted in the death of almost 98% of the adult fish in Windermere [22]. The young perch were less seriously affected and soon increased in numbers once the cannibalistic pressure of the larger fish was removed. Fig. 5c shows year-to-year fluctuations in the total numbers of perch caught in the North Basin between 1966 and 1991. There was relatively little change in total numbers caught between 1970 and 1975, but numbers declined sharply in 1976 and since then have fluctated in an irregular manner from year to year. Fig. 5d shows year-to-year fluctuations recorded in the year-class strength of 2-year old perch in the North Basin between 1966 and 1991. There was a marked improvement in the survival of young perch after the adults were depleted by disease in the late 1970s. The strongest year-classes were usually recorded when summer temperatures were relatively high, but weak year-classes occur under a wide range of temperatures [23].

FIGURE 5 (*on the opposite page*). Mean zooplankton biomass in Windermere North Basin (a, b) during "summer" over a 26-year period (1966–1991), annual estimates for populations of perch (c), and estimated year-class strength of 2+ perch, expressed as a site-specific recruitment index (d).

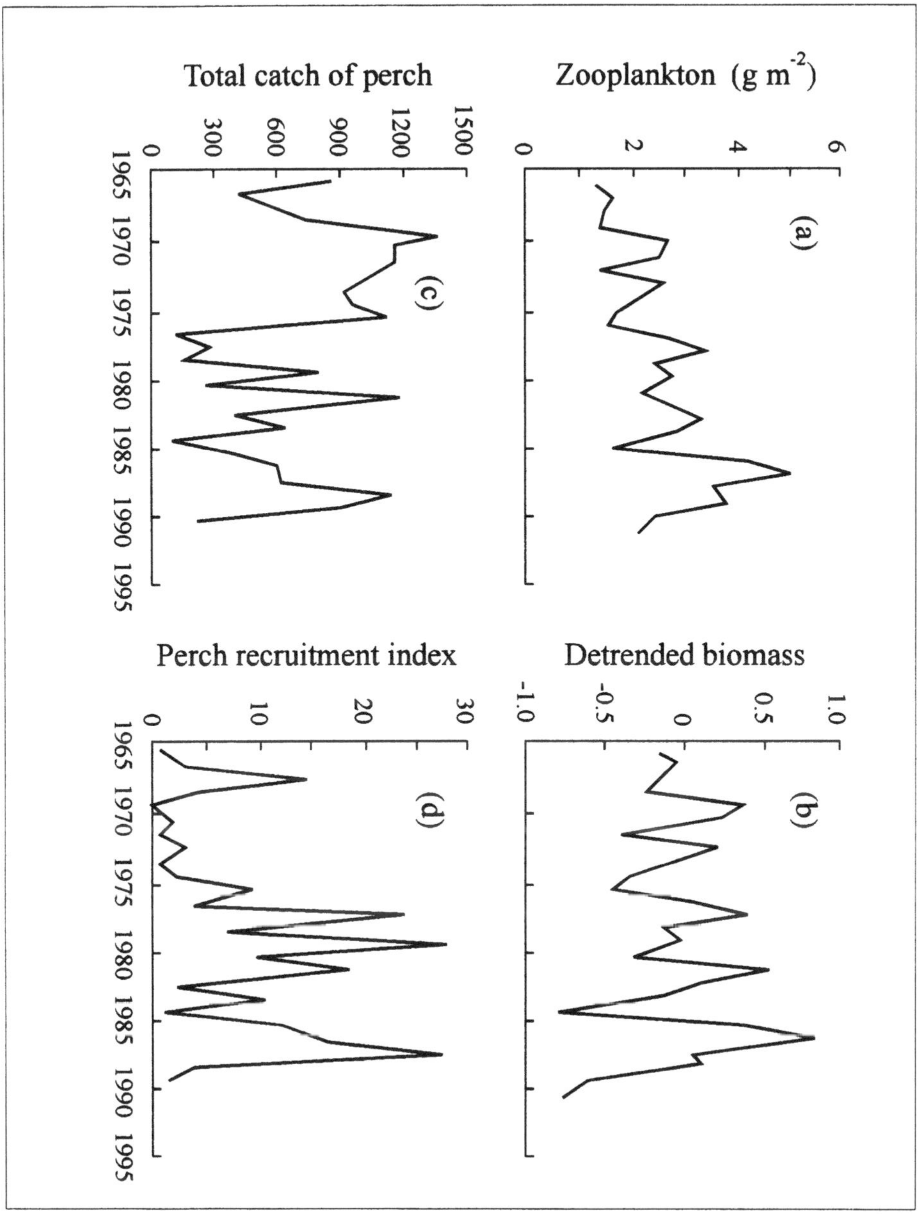
Zooplankton (g m-2)
(a)
Detrended biomass
(b)
Total catch of perch
(c)
Perch recruitment index
(d)
1965 1970 1975 1980 1985 1990 1995

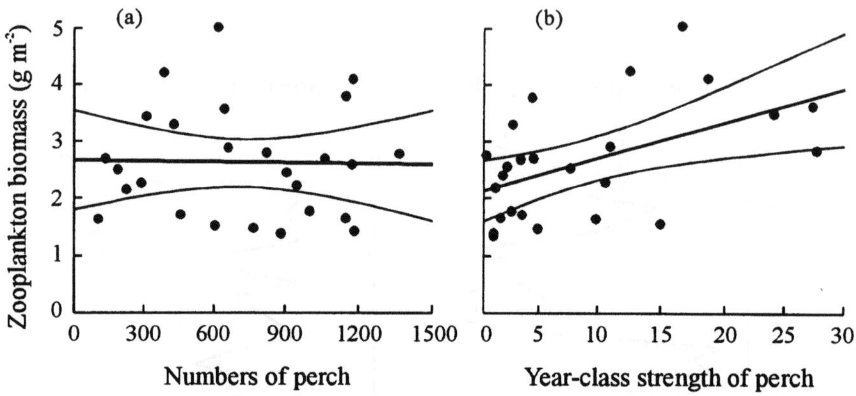

FIGURE 6. Relationships between mean zooplankton biomass (g m^{-2}) during 26 summers (1966–1991) in Windermere North Basin and: (a) total numbers of perch caught in each year, and (b) estimated year-class strength of 2+ perch, expressed as a site-specific recruitment index. Solid lines represent regression means and 95% confidence limits.

Fig. 6a shows the relationship between the average summer biomass of zooplankton in Windermere North Basin and the total numbers of perch caught in any particular year. There is no systematic relationship between the abundance of zooplankton and the numbers of perch caught ($r = -0.04$) and no evidence that changes in the standing crop of fish were having any effect on year-to-year changes in the zooplankton. Fig. 6b shows the relationship between the average summer biomass of zooplankton and the estimated year-class strength of 2+ perch expressed as a site-specific recruitment index (see Winfield et al. this volume, pp. 245–261). A number of investigators have shown that the young perch in Windermere feed on zooplankton in early summer [24, 25] but there is no evidence of any "top–down" effect in our 26-year time-series. The significant positive correlation in the fitted regression (Fig. 6b; $r = 0.44$, $p<0.01$) even suggests that more perch survive in years when the zooplankton is relatively abundant.

5.5. IMPACT OF "BOTTOM–UP" EFFECTS ON YEAR-TO-YEAR VARIATIONS IN ZOOPLANKTON BIOMASS

The cladocerans that dominate the summer zooplankton community feed selectively on flagellates which typically appear when the stratified water column is periodically mixed by wind. Year-to-year variations in the frequency and timing of wind-induced mixing thus have an indirect effect on the growth and survival of the zooplankters, by regulating the seasonal succession of these important food organisms. Fig. 7a shows the relationship between the de-trended summer biomass of zooplankton in Windermere North Basin and the average wind speed in summer. The plot demonstrates that

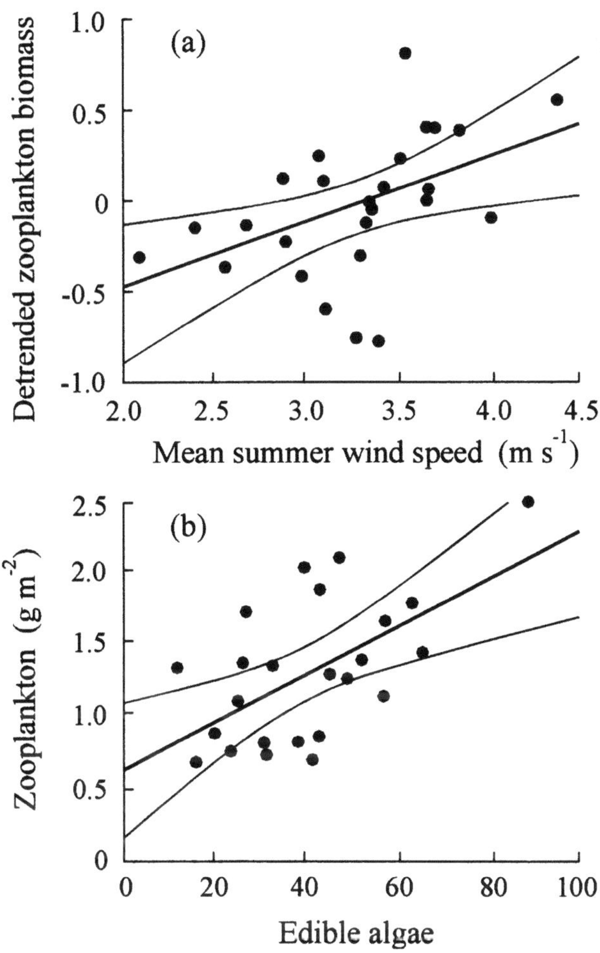

FIGURE 7. Mean biomass of zooplankton in Windermere North Basin during summer over a 26-year period (1966–1991). (a) De-trended summer biomass plotted against mean wind speeds (m s^{-1}) in summer. (b) Summer biomass (g m^{-2}) plotted against mean concentrations of edible algae. Solid lines represent a regression mean and 95% confidence limits.

there is a significant positive correlation ($r = 0.46$, $p<0.05$) between the two variables but there is a wide scatter of points around the fitted regression. Fig. 7b shows the relationship between the summer biomass of zooplankton and the average concentration of edible algae during the same period. There are large year-to-year variations in the quantity of edible algae present in early summer, but the fitted regresssion explains a significant proportion of measured variation in zooplankton biomass ($r = 0.55$, $p<0.01$).

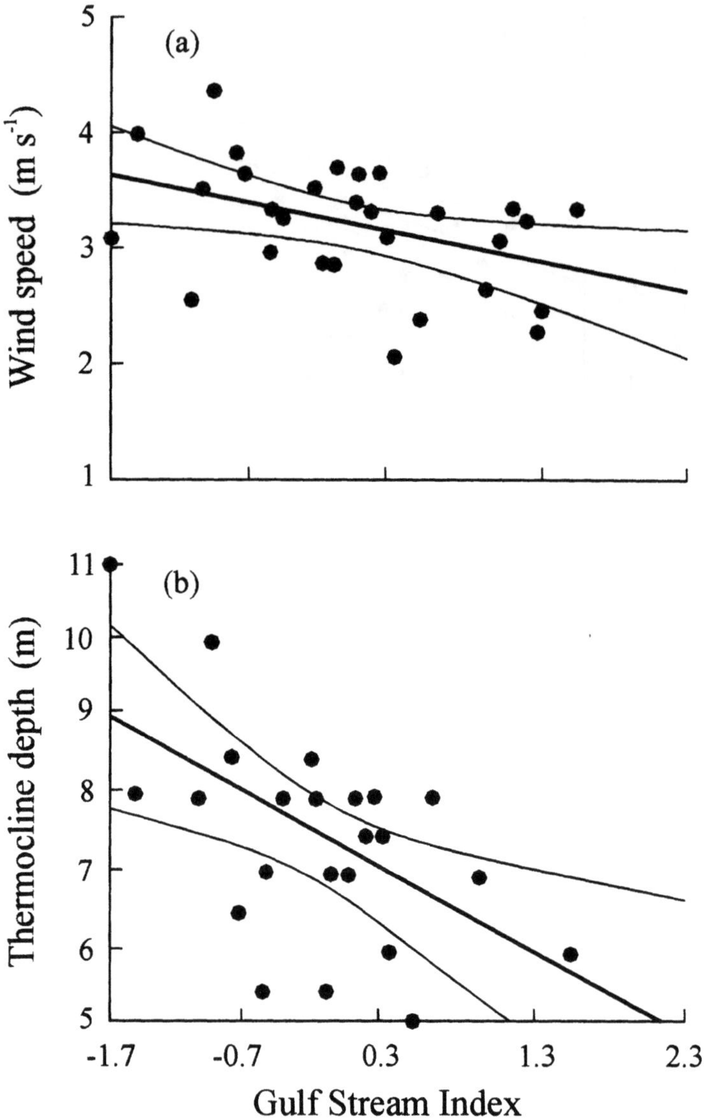

FIGURE 8. (a) Relationship between average summer wind speed near Windermere (m s^{-1}) and the position of the north wall of the Atlantic Gulf Stream. (b) Relationship between the depth of the early June thermocline in Windermere North Basin and the position of the north wall of the Gulf Stream. Solid lines represent regression means and 95% confidence limits.

5.6. INFLUENCE OF NORTH–SOUTH MOVEMENTS OF THE GULF STREAM ON MIXING CHARACTERISTICS OF WINDERMERE

Changes in the position of the Gulf Stream are now known to influence a number of climatic variables and most of these effects can be related to the movement of storm tracks across the Atlantic Ocean [26]. Fig. 8a shows the extent to which north–south movements of the Gulf Stream influence wind speed measured in the Windermere area. There is a weak negative correlation ($r = -0.39$, $p<0.05$) between the average summer wind speed and position of the Gulf Stream, which appears to have a particularly pronounced effect on the thermal characteristics of the lake in early summer. Fig. 8b shows the relationship between the depth of the early June thermocline in Windermere North Basin and position of the Gulf Stream. There is a much stronger negative correlation ($r = -0.54$, $p<0.01$) between the depth of the thermocline and the Gulf Stream index. When the Gulf Stream moves north the seasonal thermocline is shallow and relatively well defined. When the Gulf Stream moves south the thermocline is very much deeper and sometimes difficult to identify. The strength of this correlation implies that the lake is integrating (and perhaps amplifying) the relatively weak "signals" found in the raw weather variables. This is not particulary surprising since the depth of the early summer thermocline is determined by the complex interaction of a range of driving variables. For example, the quantity of heat transferred across the air–water interface depends on the vertical penetration of solar radiation as well as the ambient air temperature. The downward transport of heat in the water column is largely controlled by wind-mixing but substantial quantities of heat also can be lost during the night if the lake surface is in contact with very cold air.

5.7. INFLUENCE OF NORTH–SOUTH MOVEMENTS OF THE GULF STREAM ON YEAR-TO-YEAR VARIATIONS IN ZOOPLANKTON BIOMASS

Fig. 9 shows the relationship between the de-trended summer biomass of zooplankton in Windermere North Basin and year-to-year variations in the position of the Atlantic Gulf Stream. When the Gulf Stream is located well to the south (negative index values) the biomass of zooplankton is relatively high, but when it moves north (positive index values) there is a marked reduction in the summer biomass ($r = -0.59$, $p<0.01$). The time-series plot presented in Fig. 10 shows the extent to which these fluctuations follow a quasi-cyclical pattern of variation. In this plot, the de-trended time-series for zooplankton biomass has been used again to remove the effects of lake enrichment, and the Gulf Stream index has been inverted to produce a positive relationship between the two time-series. In the 1960s, 1970s and 1980s, the north–south movements of the Gulf Stream and variations in zooplankton biomass appeared to follow a quasi-cyclical pattern of variation with a "return time" of 8 to 9 years. In recent years, however, this predictable pattern of behaviour has broken down and the Gulf Stream has now stayed well to the north for five successive years.

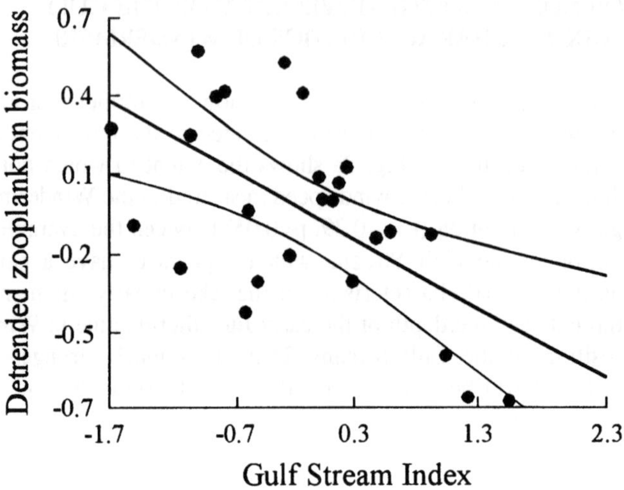

FIGURE 9. Relationship between the de-trended summer biomass of zooplankton in Windermere North Basin over a 26-year period and the position of the north wall of the Atlantic Gulf Stream. Solid lines represent a regression mean and 95% confidence limits.

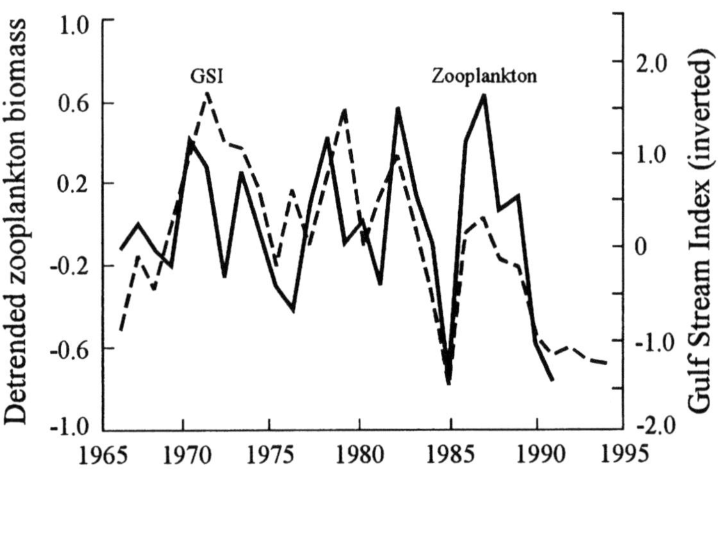

FIGURE 10. Year-to-year variations in the de-trended summer biomass of zooplankton in Windermere North Basin over a 26-year period (solid line) and the position of the north wall of the Atlantic Gulf Stream, represented by an inverted version of the Gulf Stream index (broken line).

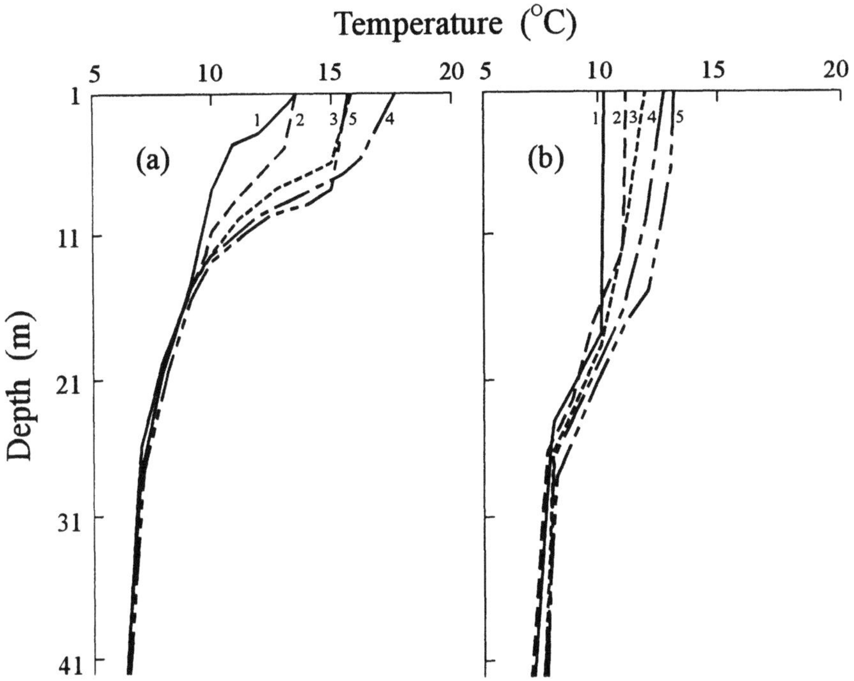

FIGURE 11. Examples of temperature profiles recorded in Windermere North Basin when the Gulf Stream was located (a) well to the north, and (b) well to the south. Each profile represents a single set of measurements recorded during five successive weeks (labelled 1–5) at the end of May and during June in 1972 (a) and 1966 (b).

5.8. PHYSICAL AND BIOLOGICAL DIFFERENCES BETWEEN "NORTH" AND "SOUTH" GULF STREAM YEARS

In the above sections, we have highlighted the general effect of the Gulf Stream on the physical and biological characteristics of Windermere North Basin. In this section, we contrast the behaviour of the lake in the more extreme years and show how these conditions are related to the recorded position of the Gulf Stream. Fig. 11a shows variations in the depth of the early summer thermocline in the lake, in a year when the Gulf Stream was located well to the north. Each profile represents a single set of measurements recorded during a particular week in June and shows that a well defined thermocline was present at depths of ca. 5 to 10 m. Fig. 11b shows similar week-to-week variations in the depth of the early summer thermocline in the lake, in a year when the Gulf Stream was located well to the south. The seasonal thermocline is now very poorly defined and appears as a density gradient that covers the top 20 m or more of the water column.

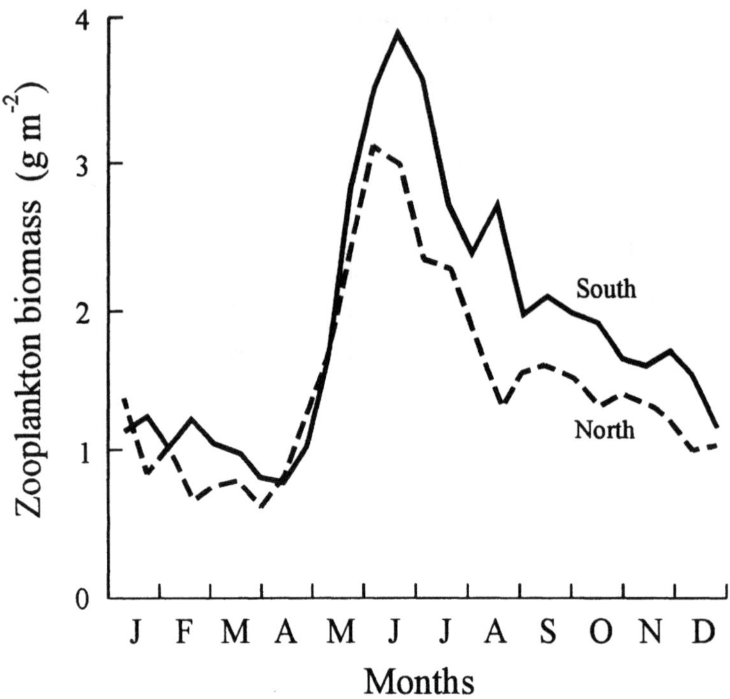

FIGURE 12. The "average" effect that physical variations in development of the thermocline (see Fig. 11) have on the dynamics of zooplankton in Windermere North Basin. The plot compares seasonal variations in zooplankton biomass recorded in an average "north" and an average "south" Gulf Stream year.

Fig. 12 shows the "average" effect that these physical variations have on the seasonal dynamics of the zooplankton in Windermere. The plot contrasts the variations in biomass that characterise "north" and "south" Gulf Stream years and has been produced by averaging the biomass for particular weeks from different years, using the Gulf Stream index as a weighting factor. When the Gulf Stream is located well to the south, the summer zooplankton maximum and the average biomass are distinctly higher. Preliminary observations on the seasonal succession of phytoplankton in these extreme years suggest that the year-to-year variations reflect "matches" and "mismatches" between the reproductive state of the zooplankters and the growth of their phytoplankton food. When the Gulf Stream is located well to the north, the growth of edible algae occurs before the animals are able to lay their full complement of eggs. When the Gulf Stream moves towards the south, the growth of edible algae is delayed and there is a better match between the zooplankters and their food supply.

6. Discussion

Seasonal and interannual changes in the zooplankton populations of lakes are typically regulated by three distinct types of driving variables: (1), those that are related to changes in the supply of nutrients; (2), those related to changes in the structure of the pelagic food web; (3), those related to short- and long-term changes in the weather.

In this paper, we have shown that changes in the catchment and short-term predation effects have relatively little effect on the year-to-year changes recorded in Windermere North Basin. Increased nutrient loadings were clearly responsible for a progressive increase in biomass recorded over the 26-year period considered here but not the large year-to-year variations that were superimposed on this long-term trend. A catastrophic collapse in the numbers of perch in the mid-1970s similarly appears to have had relatively little effect on year-to-year changes recorded in the zooplankton. Population dynamics of zooplankters are not presented here, but studies to be reported elsewhere show that the only significant changes recorded in the late 1970s were those associated with body size at first reproduction in the zooplankters.

Throughout the period of study, the most important factor influencing the seasonal dynamics of the plankton was the timing and intensity of thermal stratification. Thermally stratified lakes are particularly sensitive to year-to-year changes in the weather and frequently amplify the effects of short-lived mixing events. Simple averages of the meteorological variables often fail to capture the impact of these episodic events, but indices that characterise the general weather pattern usually are more effective. In recent years, there has been a renewed interest in examining weather patterns and analysing time-series of empirical indices that appear to vary in a quasi-cyclical way. The most celebrated of these quasi-cyclical phenomena are the El Niño mixing events recorded in the Pacific Ocean. This phenomenon, which involves major shifts in sea-surface temperature as well as atmospheric pressure, influences the periodicity of droughts in several countries and has been shown to have an effect on the dynamics of lakes in Australia [27] and North America [28]. Rather less attention has been paid to the long-range impact of weather patterns over the Atlantic Ocean, but the evidence currently being assembled by Taylor and his co-workers [26, 29] suggests that quasi-cyclical oceanic processes in the Atlantic have a major effect on weather patterns in western Europe.

In this paper, we have used an index that records the north–south movements of the Gulf Stream in the western Atlantic as a general measure of summer conditions in the English Lake District. Weather conditions in this mountainous area of northwest England are notoriously variable and local, but the results presented here suggest that Windermere is a particularly effective integrator of the regional climate. The year-to-year variations highlighted by the Gulf Stream index have also served to demonstrate a hitherto unsuspected link between biological variations in the UK and oceanic events at the other side of the Atlantic. The strength of this "teleconnection" is quite surprising but comparable effects have been identified recently in other UK ecosystems. The most comprehensive analyses published to date are those of Taylor and his co-workers [29] on the abundance of plankton in the seas around the UK. These long-term variations were also related to changes in the intensity of mixing and were most pronounced in areas like the northern North Sea which remains thermally stratified for much of the summer. However, similar quasi-cyclical patterns of variation have also been recorded in the annual biomass of several plant species collected from a roadside verge in Bibury, Gloucestershire [30]. In this unique long-term dataset, total above-ground biomass was positively correlated with Gulf Stream northerliness, which appeared to favour the growth of robust perennial plants. The processes responsible for these effects are still being investigated but the critical factor appears to be the vertical variation in soil moisture where dry "north" Gulf Stream years limit the growth of annual species that have relatively shallow root systems.

Acknowledgement

We thank Trevor Furnass for drawing the text-figures.

7. References

[1]. George, D. G. (1989). The thermal characteristics of lakes as a measure of climate change. In: *Conference on climate and water, Helsinki.* Publications of the Academy of Finland, Volume 1, pp. 402-412.

[2]. Reynolds, C. S. (1993). Scales of disturbance and their role in plankton ecology. *Hydrobiologia* **249**, 157-171.

[3]. George, D. G. & Harris, G. P. (1985). The effect of climate on long-term changes in the crustacean zooplankton biomass of Lake Windermere, U.K. *Nature* **316**, 536-539.

[4]. George, D. G. & Taylor, A.H. (1995). UK lake plankton and the Gulf Stream. *Nature* **378**, 139.

[5]. Reynolds, C. S. (1984). *The ecology of freshwater phytoplankton.* Cambridge University Press.

[6]. Reynolds, C. S. (1989). Physical determinants of phytoplankton succession. *Plankton ecology* (ed. U. Sommer), pp. 9-56. Springer-Verlag, Berlin.

[7]. Lund, J. W. G. (1950). Studies on *Asterionella formosa* Hass. II. Nutrient depletion and the spring maximum. *Journal of Ecology* **38**, 1-35.

[8]. Lund, J. W. G. (1961). The periodicity of algae in three English lakes. *Verh. Internat. Verein. Limnol.* **14**, 147-154.

[9]. Maberly, S. C., Hurley, M. A. et al. (1994). The rise and fall of *Asterionella formosa* in the South Basin of Windermere: analysis of a 45-year series of data. *Freshwater Biology* **31**, 19-34.

[10]. Worthington, E. B. (1950). An experiment with populations of fish in Windermere, 1939-48. *Proceedings of the Zoological Society of London* **120**, 113-149.

[11]. Craig, J. F., Kipling, C., Le Cren, E. D. and McCormack, J. C. (1979). Estimates of the numbers, biomass and year-class-strength of perch (*Perca fluviatilis* L.) in Windermere from 1967 to 1977 and some comparisons with earlier years. *Journal of Animal Ecology* **48**, 315-325.

[12]. Le Cren, E. D. (1987). Perch (*Perca fluviatilis*) and pike (*Esox lucius*) in Windermere from 1940 to 1985; studies in population dynamics. *Canadian Journal of Aquatic Sciences* **44**, 216-228.

[13]. Mills, C. A. & Hurley, M.A. (1990). Long-term studies on the Windermere populations of perch (*Perca fluviatilis*), pike (*Esox lucius*) and Arctic charr (*Salvelinus alpinus*). *Freshwater Biology* **23**, 119-136.

[14]. Rouen, M.A. (1989). The design and development of the Windermere Profiler. *Annual Report of the Freshwater Biological Association* **57**, 93-106.

[15]. Lund, J. W. G. & Talling, J. F. (1957). Botanical limnological methods with special reference to the algae. *Botanical Review* **23**, 489-583.

[16]. Stephens, K. (1963). Determination of low phosphate concentrations in lake and marine waters. *Limnology and Oceanography* **8**, 361-362.

[17]. Talling, J. F. (1974). Photosynthetic pigments. General outline of spectrophotometric methods: specific procedures. In: *A manual on measuring primary production in aquatic environments*. 2nd ed. (ed. R. A. Vollenweider), pp. 22-26. Blackwell Scientific Publications, Oxford.

[18]. Lund, J. W. G., Kipling, C. & Le Cren, E. D. (1958). The inverted microscope method of estimating algal numbers and the statistical basis of estimations by counting. *Hydrobiologia* **11**, 143-170.

[19]. George, D. G. & White, N. J. (1985). The relationship between settled volume and displacement volume in samples of freshwater zooplankton. *Journal of Plankton Research* **7**, 411-414.

[20]. Miller, J. L. (1994). Fluctuations of Gulf Stream frontal position between Cape Hatteras and the Straits of Florida. *Journal of Geophysical Research* **99**, 5057-5064.

[21]. Taylor, A. H. & Stephens, J. A. (1980). Latitudinal displacements of the Gulf Stream (1966 to 1977) and their relation to changes in temperature and zooplankton abundance in the NE Atlantic. *Oceanol. Acta* **3**, 145-149.

[22]. Buck, D., Cawley, G. D., Craig, J. F., Pickering, A. D. & Willoughby, L. G. (1979). Further studies on an epizootic of perch, *Perca fluviatilis* L., of uncertain aetiology. *Journal of Fish Diseases* **2**, 297-311.

[23]. Kipling, C. (1976). Year-class strengths of perch and pike in Windermere. *Annual Report of the Freshwater Biological Association* **44**, 68-75.

[24]. McCormack, J. C. (1970). Observations on the food of perch (*Perca fluviatilis* L.) in Windermere. *Journal of Animal Ecology* **39**, 255-267.

[25]. Craig, J. F. (1978). A study of the food and feeding of perch, *Perca fluviatilis* L. in Windermere. *Freshwater Biology* **8**, 59-68.

[26]. Taylor, A. H. (1996). North-south shifts of the Gulf Stream: Ocean-atmosphere interactions in the North Atlantic. *International Journal of Climatology* **16**, 559-583.

[27]. Harris, G .P., Davies, P., Nunez, M. & Meyers, G. (1988). Interannual variability in climate and fisheries in Tasmania. *Nature* **333**, 754-757.

[28]. Anderson, W. L., Robertson, D. M. & Magnuson, J. J. (1996). Evidence of recent warming and El Nino-related variations in the ice breakup dates of Wisconsin lakes. *Limnology and Oceanography* **41**, 815-821.

[29]. Taylor, A. H. (1995). North-South shifts of the Gulf Stream and their climatic connection with the abundance of zooplankton in the U.K. and its surrounding seas. *ICES Journal of Marine Science* **52**, 711-721.

[30]. Willis, A. J., Dunnett, N. P., Hunt, R. & Grime, J. P. (1995). Does Gulf Stream position affect vegetation dynamics in Western Europe? *Oikos* **73**, 408-410.

ENVIRONMENTAL FACTORS INFLUENCING THE RECRUITMENT AND GROWTH OF UNDERYEARLING PERCH (*PERCA FLUVIATILIS*) IN WINDERMERE NORTH BASIN, UK, FROM 1966 TO 1990

IAN J. WINFIELD, D. GLEN GEORGE,
JANICE M. FLETCHER AND DIANE P. HEWITT
NERC Institute of Freshwater Ecology
Windermere Laboratory
Far Sawrey, Ambleside
Cumbria LA22 0LP, UK

Abstract

In this study, we examine the influences of year-to-year variations in water temperature, changes in the weather (expressed by an index that measures the position of the North Atlantic Gulf Stream), and zooplankton biomass, on the recruitment and growth of underyearling perch (*Perca fluviatilis*) in the North Basin of Windermere, a relatively large lake in northwest England, UK. Between 1966 and 1990, there were considerable year-to-year variations in perch recruitment, superimposed on a more general increase following the outbreak of a disease that killed many adults in 1976. Underyearling perch growth showed less variation, no significant overall trend, and no significant relationship with recruitment. There were no significant relationships between recruitment and water temperature (expressed as the number of degree-days above 14°C) or Gulf Stream position, but a significant positive relationship with zooplankton biomass explained 27% of the observed variation in recruitment. There was no significant relationship between underyearling perch growth and zooplankton biomass, but there were significant positive and negative relationships with water temperature and the index of Gulf Stream position. Together, these two variables explained 40% of the observed annual variations in underyearling growth. The weather patterns associated with the north–south movements of the Gulf Stream have yet to be analysed in detail, but these year-to-year variations are known to have a particularly pronounced effect on the timing and intensity of thermal stratification in lakes.

D.G. George et al.(eds.), Management of Lakes and Reservoirs during Global Climate Change, 245–261.

1. Introduction

Studies on the population dynamics of fish species require long-term datasets spanning several decades, because they have relatively long life-spans in comparison with most other aquatic organisms. Unfortunately, few such studies have been undertaken in lakes that do not support commercial fisheries and even less attention has been paid to the impact of climate on the dynamics of lacustrine fish. However, a long-term study of the perch (*Perca fluviatilis*) and pike (*Esox lucius*) populations in Windermere, a lake in northwest England (Cumbria, UK), is a notable exception [1] and offers an opportunity to determine the influence of environmental factors on the population biology of these two important fish species. The present paper is concerned with the perch, for which Le Cren [1] emphasised that lakes in northwest Britain may provide examples of the increasing importance of abiotic factors as species reach the edge of their geographical ranges. The Windermere study thus offers the possibility of investigating the effects of variations in climate for this and perhaps related percids, with implications for issues of fisheries management during global change.

Sampling of the Windermere perch population, or more properly populations because the fish of the two basins of this lake are discrete [2], began in the 1940s as an exploratory fishery. After a period of intensive fish removal in the 1940s and 1950s, the fishery closed in the mid 1960s and evolved into a scientific monitoring programme which has since been the only significant agency of fish removal. During the 1960s, perch abundance was relatively stable and remained so until the outbreak of a disease (probably a form of furunculosis [3]) in 1976 that killed 98% of the population. Subsequent analyses of the perch populations (e.g. [4, 5]) have been concerned primarily with the pre-disease era, although it is clear that even in the 1990s the perch populations have not fully recovered to their former stability ([6] and unpublished data).

The numerous analyses summarised in [1], [5] and [7] have shown that, in addition to the major piscine factors of adult perch acting as spawners and cannibals, and pike acting as predators, the dynamics of young perch in Windermere are also influenced by water temperature, with the number of degree-days above 14°C being particularly important. While strong perch year-classes have always been associated with relatively warm years, poor year-classes have appeared under a range of temperature conditions. Moreover, Le Cren [1] concluded that the influence of temperature may be indirect by affecting the timing of events in the food web and the availability of food for the young perch at critical periods. Le Cren also noted that both Smyly [8] and Craig [7] had observed what appeared to be shortages of such food in July, and more recent studies by George & Hewitt (pp. 223–244 in this volume) have confirmed that numbers of zooplankters are often very low in late summer. The preliminary analyses of zooplankton and perch data reported by George & Harris [4] found no significant relationship between zooplankton abundance and perch year-class strength, but there was some indication that poor perch recruitment was associated with low zooplankton abundance. Moreover, the analysis of George & Harris indicated that in addition to algal food resources for zooplankters, their development in Windermere is also influenced by variations in water temperature that are related to year-to-year variations in the timing and intensity of thermal stratification. More recently, George & Hewitt (in this volume) have shown that the timing of stratification, and thus the timing of other food-web events, is also influenced by wind, and they have demonstrated a link between these local conditions and the position of the Gulf Stream in the western North Atlantic Ocean. Climate thus appears to play a pivotal role in the chronology of at least the lower levels of the Windermere food web.

The aims of the present study were twofold. Firstly, to describe the trends recorded in the recruitment and growth of underyearling perch in Windermere over a 25-year period from 1966 to 1990. Secondly, to relate the observed variations to a series of driving variables such as zooplankton abundance, water temperature and Gulf Stream position, in order to determine if the climatic effects detected in the lower trophic levels also influence the long-term dynamics of perch.

2. Methods

2.1. STUDY SITE

Windermere is situated at 54°22'N, 2°56'W, at an altitude 39 m in the English Lake District, northwest England. The lake is 14.8 km^2 in surface area, elongate in shape and divided into north and south basins by an area of islands and shallows less than 5 m deep. The North Basin has an area of 8.1 km^2 and a maximum depth of 64 m, while the South Basin has an area of 6.7 km^2 and a maximum depth of 44 m. Following several decades of increasing nutrient levels, the North Basin may now be classed as mesotrophic and the South Basin as eutrophic.

The fish community of Windermere is relatively simple and contains only a few species of numerical importance. In addition to perch and pike, both basins contain significant populations of Arctic charr (*Salvelinus alpinus*) and brown trout (*Salmo trutta*). Eel (*Anguilla anguilla*), minnow (*Phoxinus phoxinus*) and three-spined stickleback (*Gasterosteus aculeatus*) are found in the lake, while Atlantic salmon (*Salmo salar*) pass through on migration. Roach (*Rutilus rutilus*) are also present, but only in small numbers and with a localised distribution (during an extensive inshore gill-net survey during September 1995, roach accounted for only 2.5% of 1696 fish caught, 96.8% of which were perch: I.J.W. unpublished data). A few other fish species are also present in very small numbers.

Although Windermere has a long history of fisheries, the only extant commercial operation is a small plumb-line fishery for Arctic charr. Salmon, trout, perch and pike are all fished to varying degrees by recreational anglers, but negligible numbers of the last two species are removed from the lake.

2.2. PERCH RECRUITMENT AND GROWTH

Data on perch recruitment and growth were taken from a long-term monitoring programme undertaken by the Freshwater Biological Association from the 1940s to 1988, and by the Institute of Freshwater Ecology from 1989 to the present. As the full suite of environmental factors considered in the present analyses was only available for the North Basin of Windermere between 1966 and 1990, only corresponding perch data were considered. Four decades of tagging studies have shown that the perch populations of the two basins are discrete [2]. A detailed account of the full history of the perch study is given by Le Cren [1], but the main methodological features for the period between 1966 and 1990 were as follows.

Perch were sampled by unbaited, wire-netting traps each 740, 590 and 640 mm in length, width and height, respectively, with an opening on one end narrowing to 8 cm. The body of the trap consisted of 12-mm hexagonal mesh wire-netting, resulting in a size-unbiased sampling of perch between 90 and 300 mm in total body length [1]. Traps were deployed in gangs of five between late April and early June of each year on four known perch spawning grounds at depths of 2 to 7 m, although the present analysis considers only data from the single North Basin site of Green Tuft (National Grid Reference NY 374 019). Traps were set during daylight and subsequently lifted during daylight, usually 7 days later, the precise time of lifting being occasionally influenced by bad weather. All perch taken in the traps were pooled for each site and taken to the laboratory for processing. The emptied gang of traps was then redeployed at the same site for the next set of samples. Over the course of a single spawning season, a total of six sets was thus usually obtained for each site.

In the laboratory, the entire catch was enumerated and then processed as follows in its entirety, except when large numbers of fish were subsampled – stratified on the basis of length-classes (see [9]). Each perch was measured (total length to 1 mm), weighed (wet weight to 1 g) and dissected to determine sex, and the left opercular bone was removed and aged as described by Le Cren [10], with a nominal birthday taken as the beginning of the sampling period. Annuli spacings were projected and measured at x10 magnification to allow back-calculations of length-at-age using the formula described by [10] and the constants given by Craig [11].

Over the last ca. 20 years, catches have been dominated by male perch aged 2+ years and so the present analysis has considered only this age and sex-class in order to keep potentially confounding variables to a minimum. A simple recruitment "index" for each year was obtained from the numbers of 2+ males that were caught two years after their actual recruitment into the population as underyearling fish (aged 0+ years), and the mean body length attained at the end of their first summer was derived from back-calculations.

2.3. ENVIRONMENTAL FACTORS

Full methodological details for the environmental factors of zooplankton and Atlantic Gulf Stream position are given by George & Hewitt (pp. 223–244 in this volume), but essential details were as follows.

Crustacean zooplankton has been collected by the FBA and IFE at intervals of 2 weeks or more frequently from 1935 to the present, by vertical net hauls made at the deepest point of the North Basin of Windermere. The present analysis considered only data for total crustacean zooplankton biomass, measured as settled volume, from 1966 to 1990. For each year, mean zooplankton biomass was calculated for each month from May to September, inclusive, and an overall summer mean was calculated for the same 5-month period.

The position of the Gulf Stream between 1966 and 1990 was taken from an index of the latitude of its north wall (A. H. Taylor, Plymouth Marine Laboratory, UK, unpublished data), from which a mean summer position was calculated for the period May–September, for each year (Fig. 2 in George & Hewitt, this volume p. 228). Positive values of this index indicate a displacement to the north over its long-term mean location, while negative values indicate a movement to the south.

Since previous analyses of the recruitment and growth of perch in Windermere [1] have used the number of degree-days above 14°C as a convenient climatic index, we have used the same method to summarise summer conditions between 1966 and 1990. Early values of this index were obtained from the listings given in [12] and more recent estimates were calculated from near-daily readings of inshore surface temperature recorded between 1981 and 1990. A temperature of 14°C was only reached during the months of May to October, inclusive, for which the accumulations of degree-days were calculated on a monthly basis, in addition to an overall annual value.

3. Results

3.1. TRENDS IN PERCH RECRUITMENT AND GROWTH

Fig. 1 shows year-to-year variations in underyearling perch recruitment and mean body length in the period 1966 to 1990. The recruitment index varied from 5 to 791, and has shown an increasing trend over the whole period ($F_{1,23}$ = 4.075, p = 0.055, r^2 = 0.150). However, this change has not been gradual and the periods before (1966 to 1975) and after (1976 to 1990) the outbreak of perch disease showed significantly different mean indices, with the higher value in the latter period (mean ± 1 S.E.: pre-disease 87.900 ± 36.132, post-disease 291.533 ± 59.091; t = 2.594, df = 23, p = 0.016). Although mean length varied considerably between 69 and 97 mm over the same period, there was no significant trend ($F_{1,23}$ = 0.051, p = 0.823, r^2 = 0.002) and mean values before and after the disease outbreak showed no significant difference (pre-disease 81 ± 2 mm, post-disease 85 ± 2 mm; t = 1.142, df = 23, p = 0.265). There was also no significant relationship between recruitment and mean length ($F_{1,23}$ = 0.129, p = 0.722, r^2 = 0.006).

3.2. TRENDS IN ENVIRONMENTAL FACTORS

Fig. 2 shows year-to-year variations in the mean summer biomass of zooplankton, the annual numbers of degree-days above 14°C, and Gulf Stream position, in the period 1966 to 1990. There was a significant increase in zooplankton abundance, from values of ca. 0.7 g m^{-3} in the 1960s up to a peak of 2.2 g m^{-3} in the 1980s ($F_{1,23}$ = 18.608, p = 0.001, r^2 = 0.447), but no corresponding shift in annual degree-days above 14°C ($F_{1,23}$ = 0.004, p = 0.950, r^2 = 0.001) or Gulf Stream position ($F_{1,23}$ = 0.571, p = 0.458, r^2 = 0.024). Both, however, showed marked variations over this period, with minima and maxima of 165 and 511 for annual numbers of degree-days above 14°C, and –2.041 and +2.046 for Gulf Stream position. There were no significant relationships between the mean summer zooplankton biomass and annual numbers of degree-days above 14°C ($F_{1,23}$ = 0.443, p = 0.512, r^2 = 0.019) or the Gulf Stream position ($F_{1,23}$ = 0.788, p = 0.384, r^2 = 0.033), and none between the annual numbers of degree-days above 14°C and the Gulf Stream position ($F_{1,23}$ = 0.003, p = 0.955, r^2 = 0.001).

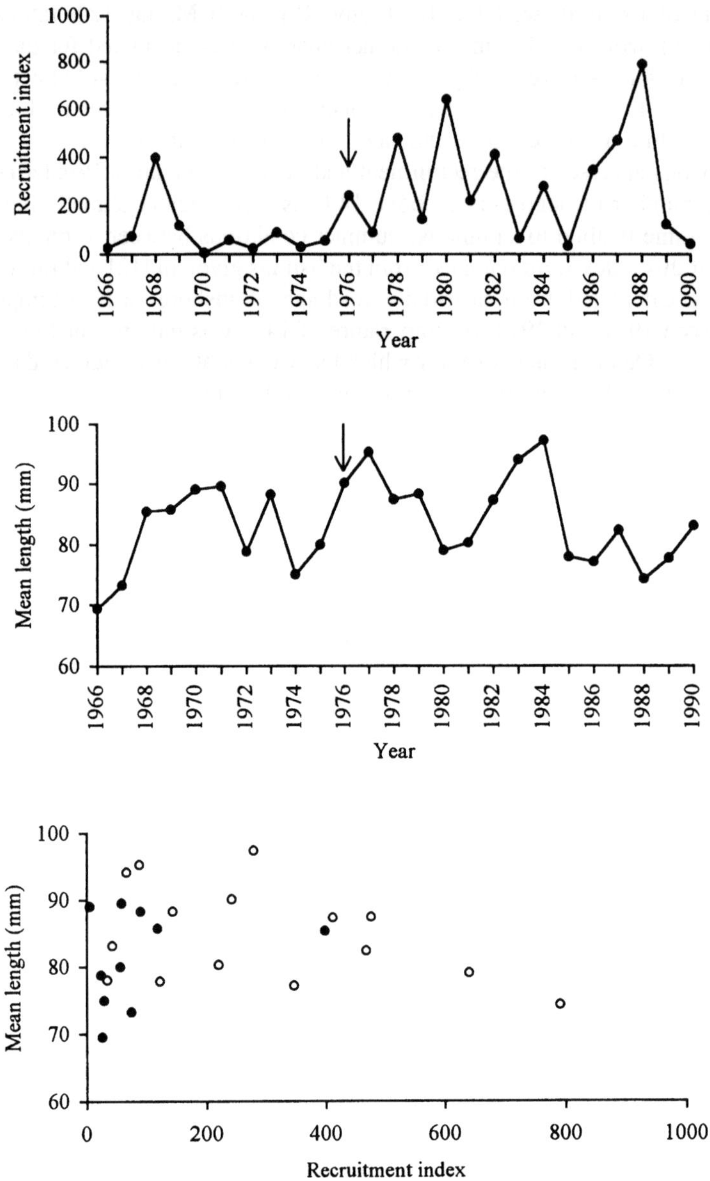

FIGURE 1. Trends in the recruitment index (see Section 2.2) (upper figure) and the mean body length (mm) of underyearling perch in Windermere North Basin (centre figure) for the period 1966 to 1990, and a plot of the relationship between mean length and the recruitment index (bottom figure). In the time-series, arrows indicate the outbreak of perch disease in early 1976. The bottom figure shows data from before (1966–1975, closed circles) and after (1976–1990, open circles) the outbreak of perch disease.

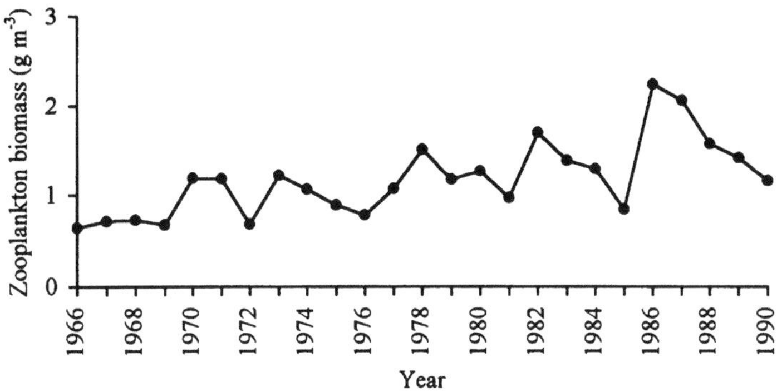

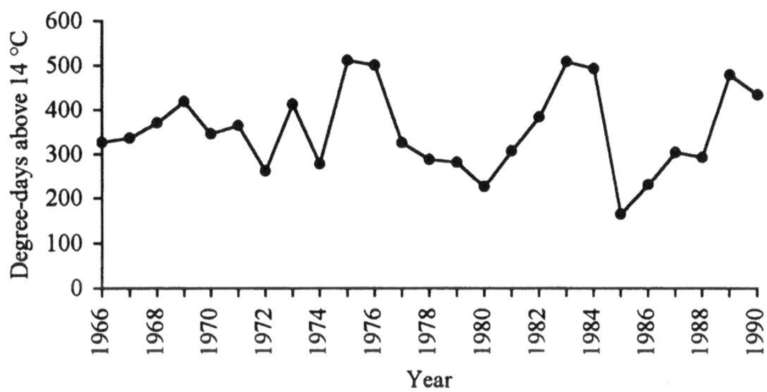

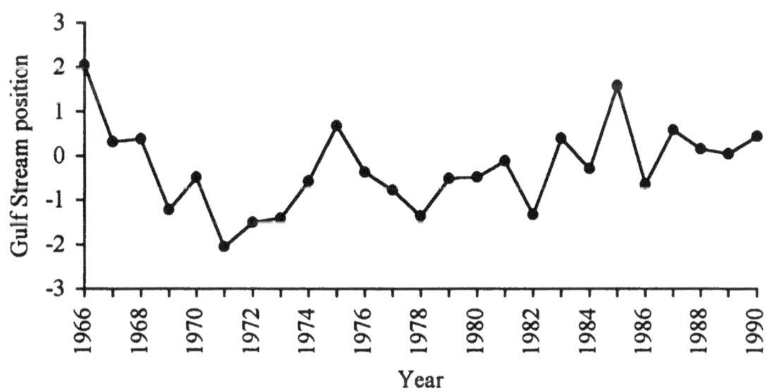

FIGURE 2. Trends in the mean summer zooplankton biomass (g m^{-3}) (upper figure) and annual numbers of degree-days above 14°C in Windermere North Basin (centre figure), and the position of the Atlanic Gulf Stream (see text for details) (lower figure), for the period 1966 to 1990.

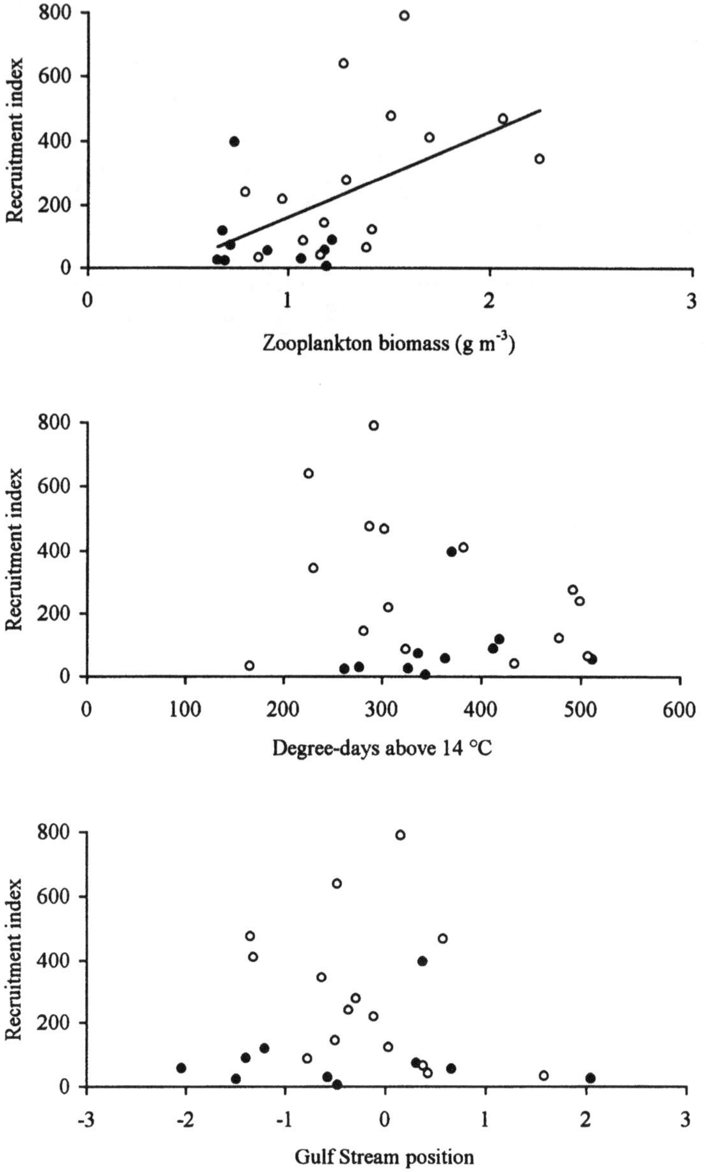

FIGURE 3. Relationships between the recruitment index for underyearling perch in Windermere North Basin and mean summer zooplankton biomass (g m^{-3}) (upper figure), the annual numbers of degree-days above 14°C (centre figure), and the position of the Atlantic Gulf Stream (bottom figure), for the period 1966 to 1990. The regression line indicates a statistically significant relationship between recruitment and mean summer zooplankton biomass (see text for details). Data from before (1966 to 1975) and after the outbreak of perch disease (1976 to 1990) are shown as closed and open circles, respectively.

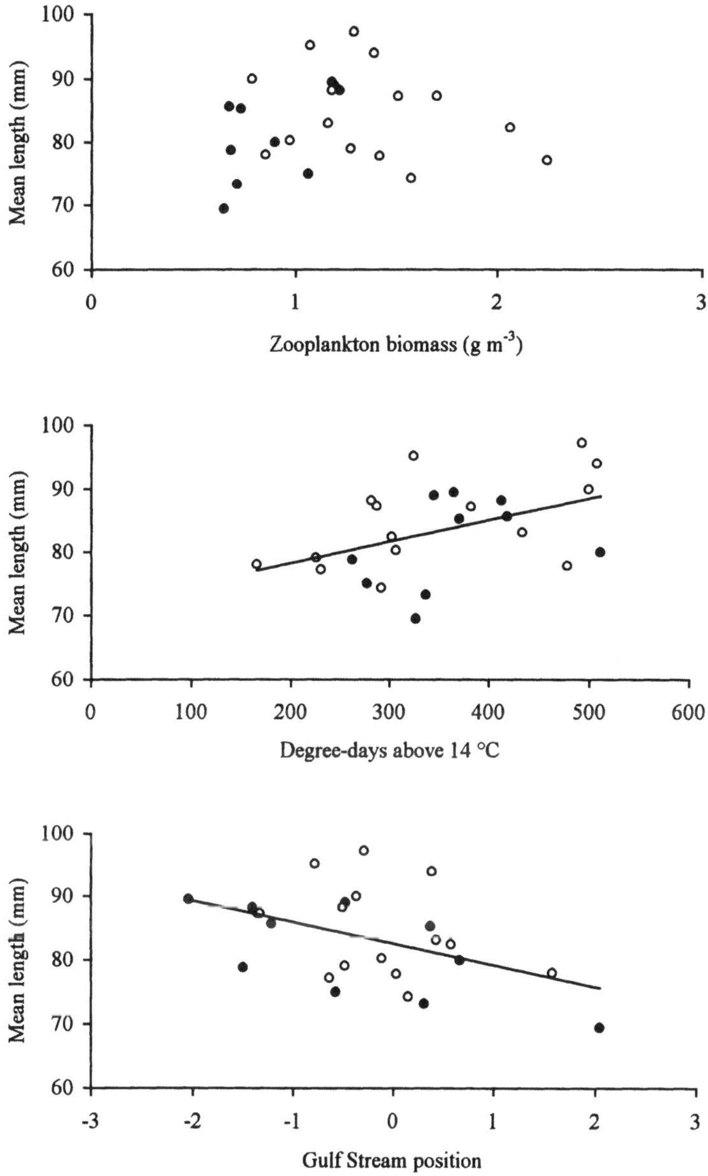

FIGURE 4. Relationships between mean body length (mm) of underyearling perch in Windermere North Basin and mean summer zooplankton biomass (g m^{-3}) (upper figure), the annual numbers of degree-days above 14°C (centre figure), and the position of the Atlantic Gulf Stream (bottom figure), in the period 1966 to 1990. The regression lines indicate statistically significant relationships between mean body length and annual numbers of degree-days above 14°C (centre) and Gulf Stream position (bottom). Data from before (1966 to 1975) and after the outbreak of perch disease (1976 to 1990) are shown as closed and open circles, respectively.

3.3. PERCH RECRUITMENT AND ENVIRONMENTAL FACTORS

TABLE 1. Values of the variance ratio (F), probability (p) and regression coefficient (r^2) from regressions of (Y) the annual recruitment index for underyearling perch in Windermere North Basin versus (X_1) monthly values for mean zooplankton biomass in May to September and (Y) versus (X_2) the number of degree-days above 14°C in May to October, for the 25-year period 1966 to 1990.

Variables X_1 and X_2	Month	$F_{1,23}$	p	r^2
Zooplankton biomass	May	11.477	0.003	0.333
	June	3.260	0.084	0.124
	July	9.350	0.006	0.289
	August	6.081	0.022	0.209
	September	3.915	0.060	0.145
Degree-days above 14°C	May	0.432	0.517	0.018
	June	0.076	0.786	0.003
	July	2.063	0.164	0.082
	August	1.445	0.242	0.059
	September	0.370	0.549	0.016
	October	1.277	0.270	0.053

Fig. 3 shows the relationships between underyearling perch recruitment (Y) and mean summer zooplankton biomass (X_1), the annual numbers of degree-days above 14°C (X_2), and the Gulf Stream position (X_3) in the 25-year period 1966 to 1990. There was no significant relationship with annual degree-days above 14°C ($F_{1,23}$ = 1.455, p = 0.240, r^2 = 0.060) or Gulf Stream position ($F_{1,23}$ = 0.137, p = 0.714, r^2 = 0.006), but there was a significant positive relationship with mean summer zooplankton biomass ($F_{1,23}$ = 8.717, p = 0.007, r^2 = 0.275).

When perch recruitment (Y) was regressed on all three of the above variables (X_1–X_3), using a stepwise multiple regression procedure [13] with a minimum tolerance of 0.01, the model retained the only biological variable, mean summer zooplankton biomass, and reproduced the statisical results given in the previous paragraph.

Finally, recruitment was regressed individually on monthly mean values for zooplankton biomass in May to September, and on monthly accumulations of the number of degree-days above 14°C from May to October (Table 1). Significant relationships were found with May, July and August mean zooplankton biomass (all $p<0.05$), while those for June and September approached significance ($p<0.10$). The regression explaining the greatest amount of variation (33.3%) was that for May (r^2 = 0.333, Table 1). All regressions on the numbers of degree-days above 14°C in each summer month were not significant ($p>0.10$).

3.4. PERCH GROWTH AND ENVIRONMENTAL FACTORS

TABLE 2. Values of the variance ratio (F), probability (p) and regression coefficient (r^2) from regressions of (Y) mean body length of underyearling perch in Windermere North Basin versus (X_1) monthly values for mean zooplankton biomass in May to September and (Y) versus (X_2) the number of degree-days above 14°C in May to October, for the 25-year period 1966 to 1990.

Variables X_1 and X_2	Month	$F_{1,23}$	p	r^2
Zooplankton biomass	May	0.502	0.486	0.021
	June	0.902	0.352	0.038
	July	0.001	0.998	0.000
	August	0.002	0.962	0.001
	September	3.637	0.069	0.137
Degree-days above 14°C	May	0.182	0.674	0.008
	June	0.005	0.946	0.001
	July	4.847	0.038	0.174
	August	7.790	0.010	0.253
	September	1.168	0.291	0.048
	October	0.142	0.710	0.006

Fig. 4 shows the relationships between the mean body length of underyearling perch (Y) and mean summer zooplankton biomass (X_1), the annual numbers of degree-days above 14°C (X_2), and the Gulf Stream position (X_3), in the 25-year period 1966 to 1990. There was no significant relationship with mean summer zooplankton biomass ($F_{1,23} = 0.247$, $p = 0.624$, $r^2 = 0.011$), but there was a significant positive relationship with annual degree-days above 14°C ($F_{1,23} = 6.086$, $p = 0.022$, $r^2 = 0.209$) and a significant negative relationship with Gulf Stream position ($F_{1,23} = 5.785$, $p = 0.025$, $r^2 = 0.201$); i.e. northerly displacements of the Atlantic Gulf Stream are associated with smaller mean body lengths of perch in Windermere.

When mean body length (Y) was regressed on all three of the above variables (X_1–X_3) using a stepwise multiple regression procedure [13] with a minimum tolerance of 0.01, the model retained the two abiotic variables: annual numbers of degree-days above 14°C and Gulf Stream position ($F_{2,22} = 7.499$, $p = 0.003$, $r^2 = 0.405$). Thus 40.5% of variation in the annual growth of underyearling perch was explained by multiple regression on water temperature in summer and the Gulf Stream position.

Finally, mean fish length was regressed individually on monthly mean values for zooplankton biomass in May to September, and on monthly accumulations of the number of degree-days above 14°C from May to October (Table 2). All regressions on mean zooplankton biomass were not significant ($p>0.05$), although the regression for September approached significance ($0.10>p>0.05$). Significant relationships were also found with July and August degree-days above 14°C (both $p<0.05$), but those of all other months were not significant ($p>0.10$). The regression explaining the greatest amount of variation (25.3%) was that for August ($r^2 = 0.253$, Table 2).

4. Discussion

The population dynamics of perch in Windermere have been the subject of numerous analyses over the last three decades. Most of this work has already been analysed and discussed at length [1, 5, 14] and will not be repeated here. In this discussion we will summarise the reported trends, examine some critical interactions within the fish community and consider some of the more general factors that influence the recruitment and growth of perch.

The reported year-to-year variations in the recruitment of perch in the North Basin of Windermere are comparable to those observed in other large lakes, e.g. in Lake Ijssel in the Netherlands [15]. Most of these variations are almost certainly due to external factors, but the significant increase in recruitment recorded since the outbreak of perch disease in 1976 was clearly related to the reduction in abundance of adults and thus the risk of cannibalism. Growth of underyearling perch over the same period has shown less variation and no overall trend, although it must be acknowledged that its assessment was based on back-calculations from fish that had survived two winters. Size-selective mortality during these periods [16, 17], however, may have had a disproportionate effect on individuals that grew unusually slowly or quickly as underyearlings, and may thus have masked some of the environmental effects on first-summer growth. Overwintering mortality of underyearling perch in Windermere is at present poorly understood and requires further study before the back-calculation of individual growth histories can be interpreted in full.

Some of the factors reponsible for the year-to-year changes in the biomass of zooplankton are discussed elsewhere in this volume (George & Hewitt, pp. 223–244). Here it is sufficient to note that the real impact of these climatic factors can only be identified after the zooplankton biomass time-series has been de-trended to remove the long-term effects of progressive enrichment. Although many studies on the likely effects of climate change on fish populations have taken thermal effects as their main theme (see below), the water temperature of Windermere North Basin, as expressed in the number of degree-days above 14°C, showed no upward trend from 1966 to 1990. The absence of significant two-factor interactions between zooplankton biomass, water temperature and Atlantic Gulf Stream position simplifies the examination of the effects of these environmental variables on the ecology of young perch.

The present analysis has been conducted with little attention to intra-community effects on the population biology of young perch in Windermere. However, it is not suggested that such influences are of no importance and their incorporation is a goal of future analysis. Spawning stock and predation pressures from both adult perch and pike are no doubt of considerable importance, but the current analyses lend further weight to the conclusion of Craig [14] that events during the first two years of life are particularly critical. For some other perch populations, such events have been shown by field observations and experiments (reviewed by Persson [18]) to include both intra- and inter-specific competition for zooplanktonic prey. Although young roach commonly out-compete young perch, particularly in the open-water habitat of eutrophicated lakes, such interactions in Windermere are unlikely because of the persistently low densities of roach in both basins. Furthermore, the positive relationship observed between zooplankton biomass and perch recruitment suggests that the latter is controlled by the former, i.e. the zooplankters are being controlled "from the bottom–up" by the availability of food, and not "from the top–down" by predatory perch. The lack of a negative relationship between recruitment and underyearling growth also indicates an absence of competition among the young perch. Elsewhere, Willemsen [15] and Mooij et al. [19] have also found perch populations with no indication of competition involving young stages, suggesting that the importance of competition as shown by Persson [18] is not universal for this species.

It is also of interest to note that zooplankton biomass was the only environmental factor that had a significant effect on the recruitment of perch during the period of study. A regression on the mean summer biomass of zooplankton explained 27.5% of the year-to-year variation in perch recruitment, whilst the average biomass in May accounted for 33.3% of the recorded variation. Earlier studies (e.g [1] and [9]) have always emphasised the importance of temperature effects but the results were strongly influenced by the very high recruitment rates recorded in one extreme year (1959). This particular year was not included in the present analysis because data for the Gulf Stream position was not available until 1966. In contrast, the growth of underyearling perch had no significant relationship with any measure of zooplankton biomass, but growth was positively related to water temperature and negatively related to the northerly movement of the Gulf Stream. Remarkably, when these two environmental variables were considered together, a multiple regression explained 40.5% of the variation observed in the mean body length of perch at the end of their first summer.

It is not yet clear how changes in the position of the Gulf Stream influence the growth but not the recruitment of perch in Windermere. The analyses presented by George & Hewitt (in this volume) imply that year-to-year variations in the availability of edible algae have a major effect on the seasonal dynamics of the zooplankton. Similar match–mismatch effects could well be responsible for the variations recorded in the growth of perch, which feed on different components of the plankton at different times of the year. Detailed accounts of the food of young perch in Windermere have been given by Smyly [8] and Guma'a [20], while Furnass [21] examined some behavioural aspects of this interaction. These studies confirm the importance of gape-limitation in the feeding of perch in early summer and demonstrate that small cladocerans and copepods were preferred prey at certain times. This is in striking contrast to the typical situation with zooplanktivorous fish, where large cladocerans are the preferred prey [22]. This dynamic situation makes the interaction between young perch and zooplankton in Windermere very complex and requires further integrated studies of both trophic levels to identify the key processes. An analysis of the growth of underyearling perch in the eutrophic Tjeukemeer in the Netherlands [19] found a much stronger influence of water temperature (about 94% of variance in body size explained) than that found in the present analysis. However, in contrast to Windermere, the extremely shallow Tjeukemeer does not stratify and so complicating mechanisms equivalent to those suggested for Windermere will not operate.

The climatic effects identified in the present study are certainly more subtle than those reported elsewhere. Previous analyses of the possible mechanisms by which global change may have an impact on fish populations in lakes, have been restricted largely to the direct effects of increased temperature and its interaction with levels of dissolved oxygen [23–26], or the more indirect effects of temperature on prewinter size of underyearlings and thus overwinter survival [17, 27]. Some investigators (e.g. 28]) have already incorporated timing effects into predictive models of direct thermal influences, but more detailed studies are required to explore the wider implications of the match–mismatch effects suggested here.

All of the evidence accumulated so far suggests that the most important effects of the Atlantic Gulf Stream are those related to the local wind speed, and the timing and intensity of thermal stratification. Similar effects have been reported in the seas around the UK [29] and are most pronounced in areas that are stably stratified. Comparable quasi-cyclical patterns of variation linked to the Gulf Stream have also been recorded in other systems where strong correlations have been reported with other climatic variables. For example, in a correlation analysis which parallels that used in the present study, there is a significant positive relationship between Gulf Stream northerliness and the total productivity of above-ground terrestrial vegetation at Bibury, England [30]. Several other species-specific effects were also detected and Willis et al. [30] concluded that the critical factor was the impact of dry "north Gulf Stream" summers on the performance of shallow-rooted plants. We do not know whether perch populations elsewhere have been affected by these climatic variations but there is an urgent need to examine long-term data from other locations in western Europe, in the context of variations in the Gulf Stream position.

In addition to the above wider issue of future research, several areas specific to Windermere are worthy of future attention. Firstly, much of the present analysis could and should be repeated for data obtained with perch traps set at three sites in the South Basin of the lake, which is shallower and usually stratifies rather earlier in the year. Secondly, the scope of the analyses for both basins should also be extended within the fish community to take into account the influences of adult perch and pike [5]. Finally, a much better understanding is required of the spatial and temporal aspects of the interaction between underyearling perch and zooplankton. Recent developments in quantitative echo-sounding technology, applicable to both small fish and zooplankton [31], hold much promise for the study of such interactions.

Acknowledgements

The data on which the present paper is based were gathered over many years by many people and we are grateful to all of them. We thank in particular our current colleagues Peter Cubby and Ben James, and we are indebted to Arnold Taylor of Plymouth Marine Laboratory for making available unpublished data on the Gulf Stream. We also appreciate work undertaken on the management of these data by Sophie des Clers and others at the Renewable Resources Assessment Group of Imperial College, London, during periods of this project partially funded by the Ministry of Agriculture, Fisheries and Food.

References

[1]. Le Cren, E. D. (1987). Perch (*Perca fluviatilis*) and pike (*Esox lucius*) in Windermere from 1940 to 1985; studies in population dynamics. *Canadian Journal of Fisheries and Aquatic Sciences* **44** (Supplement 2), 216-228.

[2]. Kipling, C. & Le Cren, E. D. (1984). Mark-recapture experiments in Windermere, 1943-1982. *Journal of Fish Biology* **24**, 395-414.

[3]. Bucke, D., Cawley, G. D., Craig, J. F., Pickering, A. D. & Willoughby, L. G. (1979). Further studies of an epizootic of perch, *Perca fluviatilis* L., of uncertain aetiology. *Journal of Fish Diseases* **2**, 297-311.

[4]. George, D. G. & Harris, G. P. (1985). The effect of climate on long-term changes in crustacean zooplankton biomass of Lake Windermere, U.K. *Nature* **316**, 536-539.

[5]. Mills, C. M. & Hurley, M. A. (1990). Long-term studies on the Windermere populations of perch (*Perca fluviatilis*), pike (*Esox lucius*) and Arctic charr (*Salvelinus alpinus*). *Freshwater Biology* **23**, 119-136.

[6]. des Clers, S., Winfield, I. J., Fletcher, J. M., Kirkwood, G. P., Cubby, P. R. & Beddington, J. R. (1993). Further analysis of the long-term Windermere perch and pike data. Unpublished commissioned report from Institute of Freshwater Ecology to Ministry of Agriculture, Fisheries and Food.

[7]. Craig, J. F. (1978). A study of the food and feeding of perch, *Perca fluviatilis* L., in Windermere. *Freshwater Biology* **8**, 59-68.

[8]. Smyly, W. J. P. (1952). Observations on the food of the fry of perch (*Perca fluviatilis* L.) in Windermere. *Proceedings of the Zoological Society of London* **122**, 407-416.

[9]. Le Cren, E. D., Kipling, C. & McCormack, J. C. (1977). A study of the numbers, biomass and year-class strengths of perch (*Perca fluviatilis* L.) in Windermere from 1941 to 1966. *Journal of Animal Ecology* **46**, 281-307.

[10]. Le Cren, E. D. (1947). The determination of the age and growth of the perch (*Perca fluviatilis*) from the opercular bone. *Journal of Animal Ecology* **16**, 188-204.

[11]. Craig, J. F. (1980). Growth and production of the 1955 to 1972 cohorts of perch, *Perca fluviatilis* L., in Windermere. *Journal of Animal Ecology* **49**, 291-235.

[12]. Kipling, C. & Roscoe, M. E. (1977). Surface water temperature of Windermere. *Freshwater Biological Association Occasional Publications* No. 2.

[13]. Wilkinson, L., Hill, M., Welna, J. P. & Birkenbeuel, G. K. (1992). *SYSTAT for Windows: statistics, version 5 edition.* SYSTAT Inc., Evanston.

[14]. Craig, J. F. (1987). *The biology of perch and related fish.* Croom Helm, London.

[15]. Willemsen, J. (1977). Population dynamics of percids in Lake Ijssel and some smaller lakes in The Netherlands. *Journal of the Fisheries Research Board of Canada* **34**, 1710-1719.

[16]. McCauley, R. W. & Kilgour, D. M. (1990). Effect of air temperature on growth of largemouth bass in North America. *Transactions of the American Fisheries Society* **119**, 276-281.

[17]. Shuter, B. J. & Post, J. R. (1990). Climate, population viability, and the zoogeography of temperate fishes. *Transactions of the American Fisheries Society* **119**, 314-336.

[18]. Persson, L. (1991). Interspecific interactions. In: *Cyprinid fishes: systematics, biology and exploitation* (eds. I. J. Winfield & J. S. Nelson). Chapman & Hall, London, pp. 530-551.

[19]. Mooij, W. M., Lammens, E. H. R. R. & Van Densen, W. L. T. (1994). Growth rate of 0+ fish in relation to temperature, body size, and food in shallow eutrophic Lake Tjeukemeer. *Canadian Journal of Fisheries and Aquatic Sciences* **51**, 516-526.

[20]. Guma'a, S. A. (1978). The food and feeding habits of young perch, *Perca fluviatilis*, in Windermere. *Freshwater Biology* **8**, 177-187.

[21]. Furnass, T. I. (1979). Laboratory experiments on prey selection by perch fry, *Perca fluviatilis*. *Freshwater Biology* **19**, 33-43.

[22]. Lazzaro, X. (1987). A review of planktivorous fishes: their evolution, feeding behaviours, selectivities and impacts. *Hydrobiologia* **146**, 97-167.

[23]. Schertzer, W. M. & Sawchuk, A. M. (1990). Thermal structure of the lower Great Lakes in a warm year: implications for the occurrence of hypolimnion anoxia. *Transactions of the American Fisheries Society* **119**, 195-209.

[24]. Northcote, T. G. (1991). Prediction and assessment of potential effects of global environmental change on freshwater sport fish habitat in British Columbia. *Geojournal* **28**, 39-49.

[25]. Chang, L. H., Railsback, S. F. & Brown, R. T. (1992). Use of a reservoir water quality model to simulate global climate change effects on fish habitat. *Climate Change* **20**, 277-296.

[26]. Shuter, B. J. & Meisner, J. D. (1992). Tools for assessing the impact of climate change on freshwater fish populations. *Geojournal* **28**, 7-20.

[27]. Johnson, T. B. & Evans, D. O. (1990). Size-dependent winter mortality of young-of-the-year white perch: climate warming and invasion of the Laurentian Great Lakes. *Transactions of the American Fisheries Society* **119**, 301-313.

[28]. Minns, C. K. & Moore, J. E. (1992). Predicting the impact of climate change on the spatial pattern of freshwater fish yield capability in eastern Canadian lakes. *Climate Change* **22**, 327-346.

[29]. Taylor, A. H., Prestidge, M. C & Allen, J. I. (1996). Modelling seasonal and year-to-year changes in the ecosystems of the NE Atlantic Ocean and the European Shelf Seas. *Journal of the Advances in Marine Science and Technology Society* **2**, 133-150.

[30]. Willis, A. J., Dunnett, N. P., Hunt, R. & Grime, J. P. (1995). Does Gulf Stream position affect vegetation dynamics in Western Europe? *Oikos* **73**, 408-410.

[31]. Winfield, I. J. & Bean, C. W. (1995). The applications and limitations of new developments in echo-sounding technology in studies of lake fish populations. *Proceedings of the Institute of Fisheries Management 1994 Annual Study Course*, 227-242.

FLUCTUATIONS IN LAKE FISHERIES AND GLOBAL WARMING

DANIEL GERDEAUX
I.N.R.A.
Station d'Hydrobiologie Lacustre
75 Avenue de Corzent BP 511
F74203 Thonon Cedex, France

1. Introduction

Data collected in the great European lakes during recent decades indicate a warming of the waters [1, 2]. Since 1960, the heat content of the lake waters has increased the same as the mean air temperature. In general, simulations of climate suggest that winter and summer heat contents of the lakes would be higher than at present. The higher winter heat content could cause an earlier thermal stratification and the period of stratification would increase. The models suggest that the turnover in winter may be incomplete and oxygen in the deeper layers of the lake would decrease [3].

Fish communities in the lakes are composed of warm-, cool- and cold-water fishes. The influence of global climate warming would have different effects on the autecology of each species. The potential changes in thermal habitat, growth, fecundity and survival rate would change the population dynamics of each species. Extrapolation of these changes to multispecies assemblages (fish and zooplankton) is not straightforward, given the complexity of community-level phenomena [4].

Changes in the lakes during the last three decades have depended on the management of the watershed or catchment (affecting the inflow of nutrients), on bottom–up effects, and on fisheries management through top–down effects. A decrease of the phosphorus load, natural interannual fluctuations and increased fish stocking, are perhaps predominant causes of the observed changes, which make the perception of global climatic warming difficult.

In Lake Geneva, the last recorded overturn occurred in 1986. Since then the winters have not been cold and windy enough to mix the water in the 300-m column, and the mean weighted temperature of the whole lake has gradually increased (Fig. 1). The absence of total overturn induces a decrease in the oxygen content of the bottom layers of water (Fig. 2). This modification has no direct effect on fish that do not live in this

D.G. George et al.(eds.), Management of Lakes and Reservoirs during Global Climate Change, 263–272.

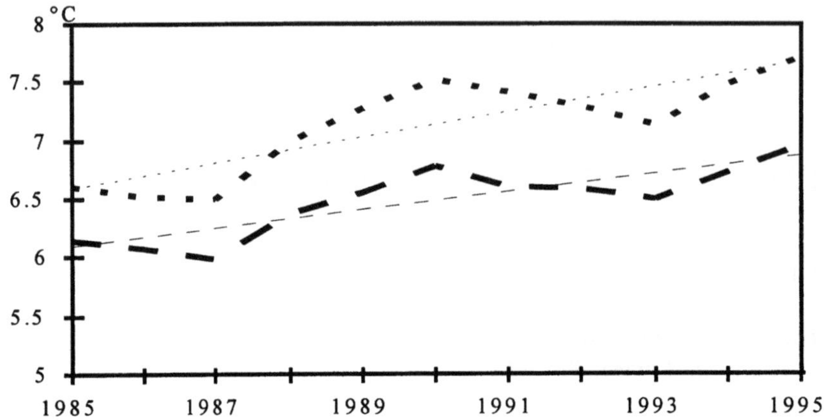

FIGURE 1. Changes in weighted temperatures (°C) in water layers at 0–200 m (upper points) and 0–309 m (lower dashes) in Lake Geneva during the period 1985–1995 [2].

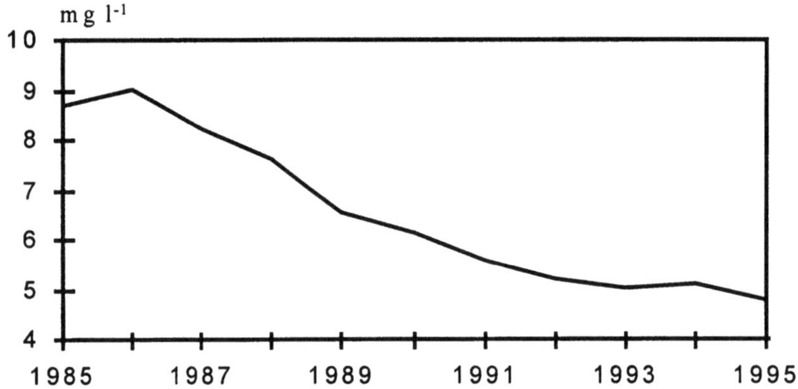

FIGURE 2. Changes in oxygen concentrations (mg l^{-1}) in the deep water layer at 200–309 m in Lake Geneva during the period 1985–1995 [2].

part of the lake. Nevertheless, the decrease of oxygen could affect the benthic fauna, such as chironomids, which are briefly important as a food item during their ascent through the water column for emergence at the surface.

This climate trend is not the only one that has occurred in Lake Geneva during recent decades. The restoration of water quality, managed by the Commission Internationale pour la Protection des Eaux du Léman, is quite efficient. The total

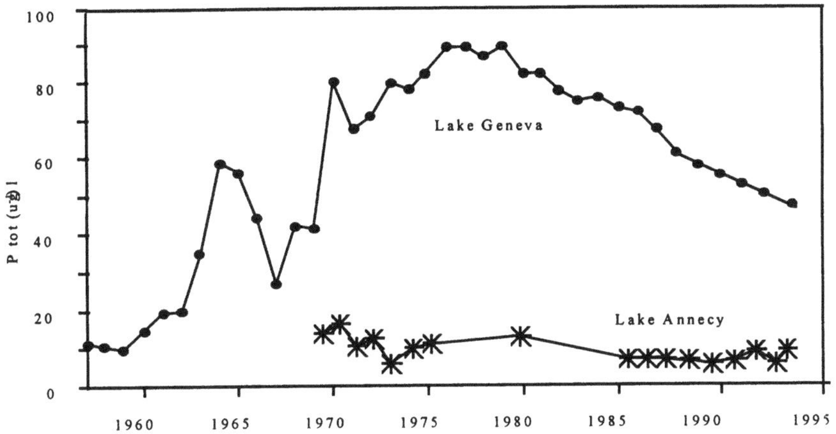

FIGURE 3. Changes in total phosphorus concentrations (μg l^{-1}) in lakes Geneva and Annecy, from 1957 to 1993.

phosphorus concentration had increased from 10 μg l^{-1} to 90 μg l^{-1} during the period 1950 to 1975, but after building a sewage treatment plant to remove phosphorus the concentration decreased to 40 μg l^{-1} (Fig. 3).

There are few lakes where eutrophication is not a problem nowadays; Lake Annecy is probably the only one of its kind in France. The building of a collector around the lake, with a sewage plant downstream, protects the lake from eutrophication (Fig. 3). This kind of lake could be a good choice for studying the influence of climate changes.

2. Effects of fisheries management

Fisheries management can modify the natural trends in fish communities, especially in lakes such as Annecy where the oligotrophic waters are very favourable to salmonids. This lake has a very important fishery, and professional fishermen share the resource with numerous sport fishermen. The fishery modifies changes in the fish community through selectivity towards the particular species and sizes of the fish that are caught. But fish-stocking should be the most important factor in modifying the trends. The stocking policies are in fact not continuous and the effectiveness of stocking is not the same for different species. Catches of Arctic char in lakes Geneva and Annecy are strongly correlated to the stocking levels (Fig. 4).

Natural reproduction is not very efficient for the Arctic char, doubtless because of the degradation of spawning grounds caused by eutrophication. Fluctuations in catches of trout in Lake Geneva since 1950 (Fig. 5) are not so easy to explain. Stocking is not the principal cause of variation in numbers. Reproduction takes place in rivers, and in order to explain the observed trends the most important fact to take into account is probably a change in quality of riverwater.

Numbers of pike caught in Lake Geneva decreased in the 1970s and 1980s (Fig. 5),

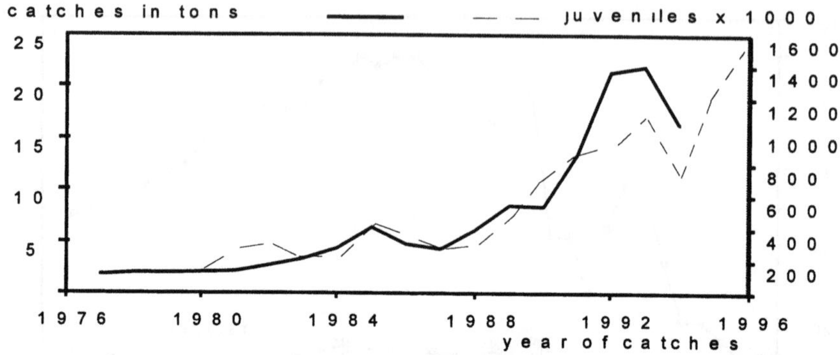

FIGURE 4. Changes in Arctic char catches (tonnes; solid line) and stocking rates of juveniles (numbers x 10^3; dashed line) (3 years before catches) in the French part of Lake Geneva since 1977.

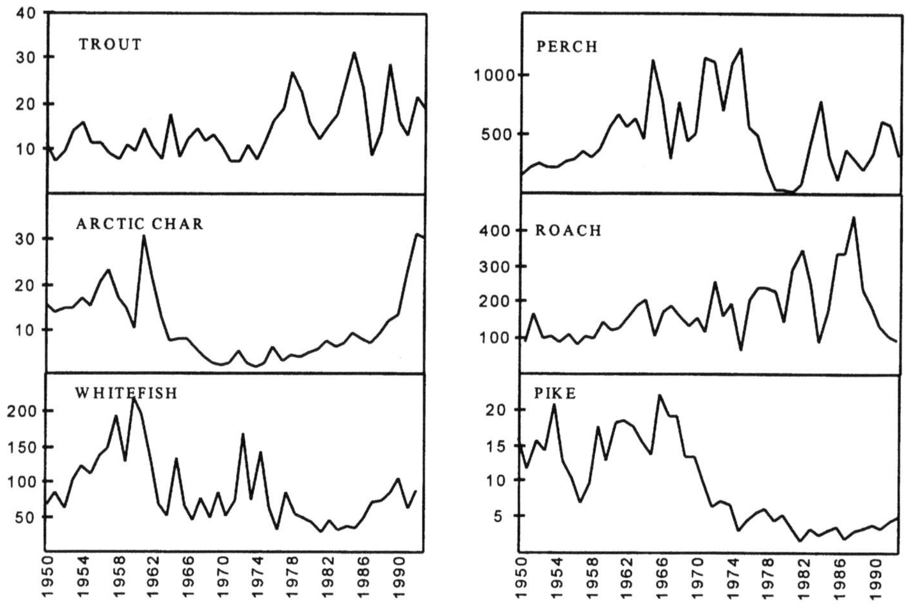

FIGURE 5. Changes in the catches (tonnes) by professional fishermen in Lake Geneva since 1950.

like those of Arctic char; these changes may be related to degradation of the littoral zone, where macrophytes have disappeared. For coregonids, perch and roach in Lake Geneva (Fig. 5), recruitment of young fish depends primarily on the natural stock and on climate.

3. Potential effects of global warming on whitefish, perch and roach populations in lakes

3.1. WHITEFISH

In Europeans lakes, whitefish occur in the southern part of the present distribution area. Whitefish are considered to be cold-water planktivorous fish inhabiting waters below the thermocline in summer, at temperatures of ca. 10–12°C. They spawn during winter, in December, when the water temperature is below 7°C. In Lake Geneva, spawning takes place in shallow waters, on gravel bottom, and eggs hatch by late February. The larvae stay in the surface waters and feed on zooplankton. During the first months of life, whitefish prefer the warm water near the surface and they often stay along the shoreline. This period is the critical phase of the life cycle. During spring, storms can cause heavy mortality of larvae and severely reduce recruitment. Daily growth is strongly correlated with the water temperature [5] and survival rate is also dependent on water temperature [6, 7]. In June, water at the surface becomes too warm and larvae move to deeper layers, where the temperature is around 10 to 12°C. Mortality is then not so dependent on climate.

Currently, the stock-recruitment relationship is the principal parameter explaining fluctuations in the whitefish populations in Lake Geneva [8]. Then, climate affects the strength of the cohorts. A cold winter and a warm spring are favourable for good recruitment. Ekmann et al. [6] found that a warm spring is the most important factor in explaining recruitment. Trippel [9] developed a model for the whitefish in Lake Constance, using the predicted mean surface water temperature during April for three warming scenarios. The model shows that the fish stock increases and interannual fluctuations are greater, but in this model there is no possibility for density-dependent regulation. No doubt the density level of the stock could not increase without limits. If the stock is very big, cannibalism by adults on eggs and larvae could be widespread as competition for food develops between adults. The authors also consider that increased warming could cause fish to abstain from spawning if the winter temperature is too high. This occurs only when the increase is 4.5°C in 50 years. Other fish populations could also interfere with the population dynamics of the whitefish and the fishery must also to be taken into account. However, if the stock is very substantial, but fluctuating, it is difficult to predict the effects of fishery management. In conclusion, it seems very difficult to develop a good model of changes in whitefish populations in lakes affected by global warming. In deep peri-alpine lakes, global warming should be favourable for populations of whitefish, but in shallow lakes it may not be so favourable.

3.2. PERCH

Perch is a ubiquitous species, widely distributed in Europe. Perch spawn in spring, when the water temperature is near 10°C, thermal stratification is not installed in lakes, and the warm surface layer is very unstable. Models of global warming predict a greater instability of weather in spring, with occasional strong storms. This may cause high mortality of perch eggs. Then, water temperature is very important for the absorption

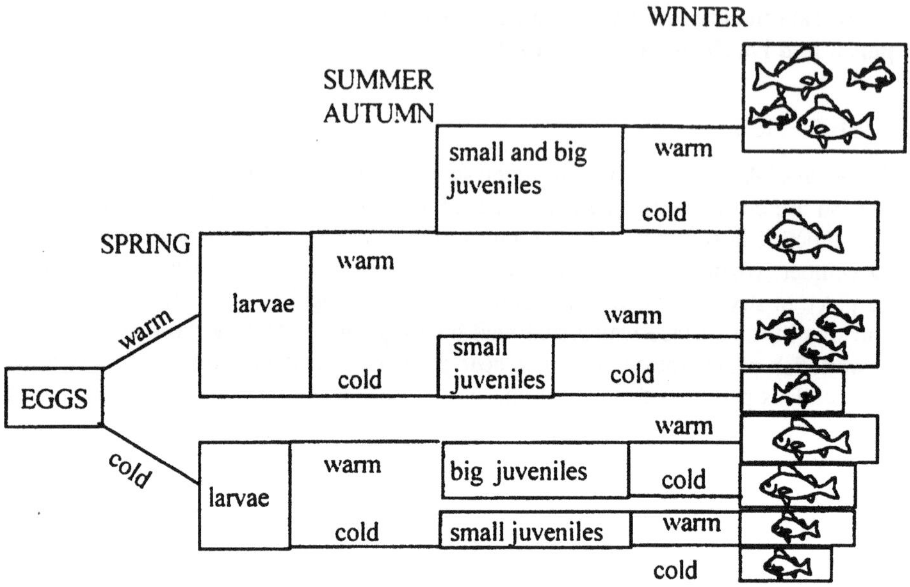

FIGURE 6. The recruitment of perch depends on the level of the spawning stock and climate. For the same quantity of eggs, recruitment differs according to the temperature regime during the first year of life. The drawings of perch at the right of the figure indicate relative sizes of 0+ fish, and the areas of the rectangles indicate relative strengths of the cohorts.

of air in the swim bladder (Egloff, personal communication). If the water is warmer, the survival rate of the larvae may be better. Also, a warm summer is good for the growth of perch. The instantaneous natural mortality rate is negatively correlated with the temperature in Lake Windermere [10], and Shuter & Post [11] found the same relationship for yellow perch. We also know that growth is lower when the size of the cohort is too big. Finally, it is difficult to predict the annual growth of young perch. If the survival rate of eggs is good, the number of juveniles is very high and growth may be slow because of density-dependent variations. The overwinter mortality depends on temperature and the duration of the winter [12]; overwinter mortality is higher for the smaller perch. This mortality is negligible for fish which are longer than 80 mm (fork length) held at 2.5°C, while the mortality is very high for small fishes (50 mm long).

The stock–recruitment relationship seems to be very important for perch. The observed fluctuations of perch populations in Lake Geneva (Fig. 5) are mainly explained by the population density and then by climate [13]. In fact, the same recruitment could be obtained through different scenarios (Fig. 6).

As is shown in Fig. 6, climate modulates the effects of population density. An average stocking level should produce different levels of recruitment in accordance with the climate. A high stocking level produces a low recruitment, because of the density-dependent relationship. In this case, the influence of climate is not preponderant.

Population dynamics during the first year could be different according to the prevailing climate but the result is always the same after the first winter. For a small spawning stock, climate is the principal factor affecting recruitment and population size.

3.3. ROACH

The distribution area of roach is a little smaller than that of perch, and roach often inhabit the same waters. However, water temperature during the spawning time of roach is higher than for perch. The consequences of global warming could be good for the roach, which is more of a warm-water species. There is no cannibalism in this species and the consequences of global warming should be a better survival rate and a lower growth rate if the population is too big.

The factors having an important effect on roach populations are predators such as perch, and climate, rather than intraspecific factors. The spring water temperature scenario is important in this case. If the warming of the lake is regular and slow, spawning of perch will occur some time before the roach spawn, so the young perch can eat the roach larvae. In this case the cohort of roach will be very small and the survival rate of perch is good. If the spring warming is regular and rapid, the interval between the reproduction of perch and roach will be short, the young 0+ perch can not eat the roach, and the survival rate of roach larvae will be good. If the temperature of the water drops suddenly after a storm, the larvae of both perch and roach may die. However, if water temperature drops before the roach have spawned, the perch could disappear and the roach larvae could then survive well. On the other hand, if water temperature drops just after the roach have spawned, their larvae will die but not those of the perch which are big enough to survive.

Thus the scenario for increasing water temperatures during spring is very significant for the population dynamics of both perch and roach, but for roach the survival rate at the end of the summer depends on the amount of predation by perch. So it is impossible to predict the future population dynamics of such a fish assemblage.

4. Discussion

The influence of global warming in European lakes is easy to forecast if the temperature conditions become incompatible with a particular biological point in the life cycle of the fish. Otherwise it is very difficult to estimate the potential effects. If the temperature of the lakes is above 7°C in winter, reproduction becomes impossible for Arctic char and probably for coregonids. One of the results of the global climate models is the predicted increase in variation. If we model, as Ricker [14], the changes in a population, the result could be very different according to the intrinsic regulatory factors in each population (Fig. 7). The modelled population has a 5-year life span, the same as the exploited populations in Lake Geneva. Reproduction takes place in years 4 and 5. We use two Ricker models for the stock–recruitment relationship. The first, with parameters $a = 0.005$ and $b = 10$, is more-or-less a Beverton model, and the second is a typical Ricker model with a density-dependent depletion for high stock levels ($a = 0.03$ and $b = 80$). In each model, we introduce a random variation from a Gaussian curve where

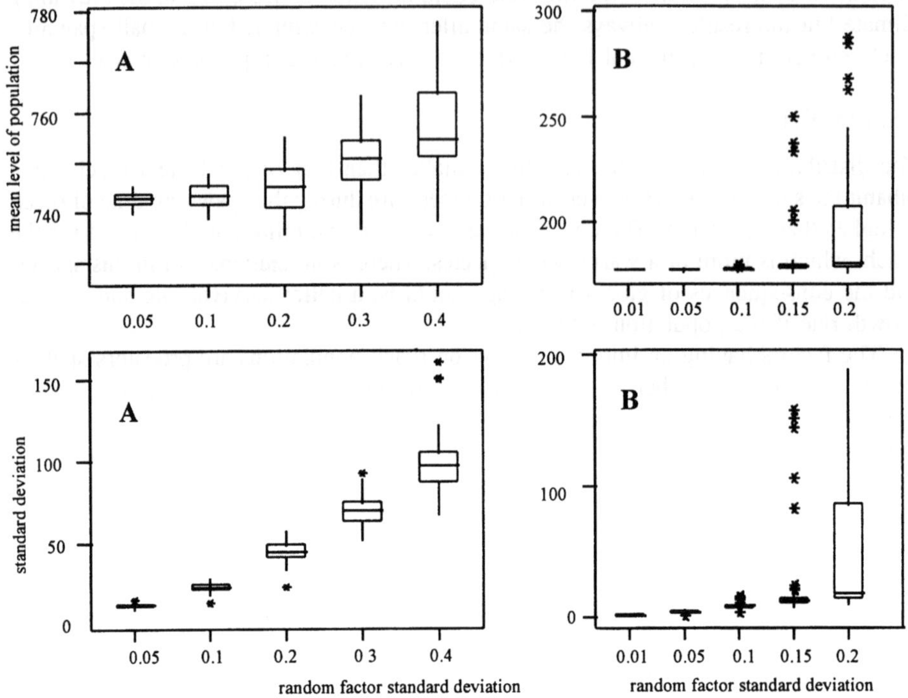

FIGURE 7. Boxplots showing mean population levels and standard deviations obtained from simulations of two fish populations based on (A) a Beverton-type model and (B) a Ricker model. Further explanation in the text.

the mean is 1. The simulation is made for 100 years with different variances of the random alea. The mean results of 50 simulations in each model are summarised in a boxplot graph (Fig. 7).

With the Beverton-type model, the variation of the mean increases slowly with the alea variance and the mean of the deviation also increases slowly (Fig. 7 A). With the Ricker-type model, the mean and variance of the population density do not increase, but its variation is substantial for a small variance of the alea (Fig. 7 B). If the alea is too high, the population could collapse.

So, we can predict that global warming, through its effects on water temperature, will influence the population dynamics of fish communities. But the future depends on the relative importance of the regulatory factors for each species-population, and on the relationships between the populations in a community. Fluctuations in numbers could be much higher than they are at present, and some populations of fish could disappear if natural fecundity is too low to restore population numbers when the reproductive stock falls to very low levels.

References

[1]. Ambrosetti, W. & Barbanti, L. (1989). The physical limnology of Lago Maggiore: a review. *Memorie dell' Istituto Italiano di Idrobiologia* **46**, 41-68.

[2]. CIPEL (1996). Rapports sur les études et recherches entreprises dans le bassin Lémanique. Programme Quinquennial 1991-1995. Campagne 1995.

[3]. McCormick, M. J. (1990). Potential changes in thermal structure and cycle of Lake Michigan due to global warming. *Transactions of the American Fisheries Society* **119**, 183-194.

[4]. Tonn, W. M. (1990). Climate change and fish communities: a conceptual framework. *Transactions of the American Fisheries Society* **119**, 337-352.

[5]. Eckmann, R. & Pusch, M. (1989). The influence of temperatures on growth of young coregonids (*Coregonus lavaretus*) in a large prealpine lake. *Rapp. P. v. Réun. Cons. Perm., Int. Explor. Mer* **191**, 201-208.

[6]. Eckmann, R., Gaedke, U. & Wetzlar, H. J. (1988). Effects of climatic and density-dependent factors on year-class-strength of *Coregonus lavaretus* in Lake Constance. *Canadian Journal of Fisheries and Aquatic Sciences* **45**, 1088-1093.

[7]. Shepherd, J. G. , Pope, J. G. & Cousens, R. D. (1985). Variations in fish stocks and hypotheses concerning their links with climate. *Rapp. P. v. Réun. Cons. Perm., Int. Explor. Mer* **185**, 255-267.

[8]. Caranhac, F. & Gerdeaux, D. (1996). Factors affecting the recruitment of lake whitefish (*Coregonus lavaretus*) in Lake Geneva and Lake Annecy. In: *6th International Symposium on Biology and Management of Coregonid Fishes*. Constance.

[9]. Trippel, E. A. (1991). Potential effects of global warming on whitefish in Lake Constance, Germany. *Ambio* **20**, 226-231.

[10]. Craig, J. F., Kipling, C., Le Cren, E. D. & McCormack, J. C. (1979). Estimates of the numbers, biomass and year-class strengths of perch (*Perca fluviatilis*) in Windermere from 1967 to 1977 and some comparisons with earlier years. *Journal of Animal Ecology* **48**, 315-325.

[11]. Shuter, B. J. & Post, J. R. (1990). Climate, population variability, and the zoogeography of temperate fishes. *Transactions of the American Fisheries Society* **119**, 314-336.

[12]. Johnson, T. B. & Evans, D. O. (1990). Size-dependent winter mortality of young-of-the-year White Perch: climate warming and invasion of the Laurentian Great Lakes. *Transactions of the American Fisheries Society* **119**, 301-313.

[13]. Gillet, C., Dubois, J. P. & Bonnet, S. (1995). Influence of temperature and size of females on the timing of spawning of perch, *Perca fluviatilis*, in Lake Geneva from 1984 to 1993. *Environmental Biology of Fishes* **42**, 355-363.

[14]. Ricker, W. E. (1954). Stock and recruitment. *Journal of the Fisheries Board of Canada* **11**, 559-623.

INTERANNUAL SYNCHRONOUS DYNAMICS IN NORTH TEMPERATE LAKES IN WISCONSIN, USA

TIMOTHY K. KRATZ[1], PATRICIA A. SORANNO[1],
STEPHEN B. BAINES[2], BARBARA J. BENSON[2],
JOHN J. MAGNUSON[2], THOMAS M. FROST[1]
AND RICHARD C. LATHROP[2,3]

[1]*Trout Lake Station, Center for Limnology*
University of Wisconsin-Madison
10810 County Highway N
Boulder Junction
Wisconsin 54512, USA

[2]*Center for Limnology, University of Wisconsin-Madison*
680 N. Park Street, Madison
Wisconsin 53716, USA

[3]*Wisconsin Department of Natural Resources*
1350 Femrite Drive, Monona
Wisconsin 53716, USA

Abstract

One way of assessing the potential influence of climate change on lakes is to take advantage of the fact that neighbouring lakes are exposed to nearly identical climatic variations. Limnological features most directly influenced by climatic processes would be expected to vary synchronously across neighbouring lakes. By examining patterns of synchronous behaviour, i.e. temporal coherence, across lakes, we can identify the extent to which we would expect various features of lakes to shift regionally in response to climatic change. We assessed the degree to which 61 limnological parameters varied synchronously across eleven north temperate lakes, seven in northern Wisconsin USA and four in southern Wisconsin, over a 13-year period. Physical variables such as dates of ice-on and ice-off, water level, and water temperature, varied synchronously across

D.G. George et al.(eds.), Management of Lakes and Reservoirs during Global Climate Change, 273–287.

years. Chemical variables were in general less coherent than physical variables; however, selected chemical variables, such as major ions, were highly coherent. The coherence of major ion data is likely due to similar responses to a several-year drought that occurred midway through the period. Biological data exhibited little or no coherence. Lake pairs containing at least one dystrophic lake did not vary synchronously, whereas lake pairs containing two clearwater lakes showed moderate levels of coherence. Our results suggest that regional generalizations for effects of climate change on lakes should be possible for many physical and some chemical features of clearwater lakes, but regional generalizations for dystrophic lakes or biological features will be difficult.

1. Introduction

Some understanding of climate change effects on lakes can be derived from climate change's regional nature. Lake districts, i.e. relatively small geographic regions containing tens, hundreds or thousands of lakes, occur throughout the globe. Within some of these lake districts, especially those without a large elevational gradient, the lakes all experience nearly identical weather. The range of responses by adjacent lakes to similar variations in weather can define the nature of climatically induced shifts in lake conditions.

We might expect all lakes in a district to respond similarly to climatic events such as a multi-year drought, a series of cool, wet years, or systematic year-to-year variations. We call a pattern of synchronous variation across lakes in a region, temporal coherence [1]. However, neighbouring lakes often differ in morphometry, hydrology, nutrient status, community composition, and other limnological characteristics, and these differences can cause lakes exposed to the same climatic signal to respond in different ways. For example, in seven northern Wisconsin USA lakes [1], temporal coherence decreased as the considered variables progressed from physical factors such as water level and water temperature, to chemical factors whose main source is atmospheric, to chemical features whose main source is the landscape, and then to biotic responses. The more removed a variable was from direct climatic influence, the less likely it would be coherent among neighbouring lakes. The more similar two lakes were in their ratio of surface area to mean depth, an index of a lake's exposure to the atmosphere, the more likely they were to respond similarly to each other [1].

Here we extend the analysis of Magnuson et al. [1] using more extensive data. The original analysis used data for 15 limnological variables measured in the lakes over eight years. Here we consider 61 variables measured in the same lakes over thirteen years. This 13-year span, from 1982 to 1994, included a 3-year drought (1988–1990) that was not covered by the earlier analysis. We also analyze a smaller dataset from four southern Wisconsin lakes for the same 1982–1994 period. We ask three basic questions. First, which variables are temporally coherent? Second, what attributes make lakes more or less likely to be coherent with their neighbours? Third, what effect does the inclusion of additional lakes, years and variables have on our perception of temporal coherence?

2. Methods

2.1. STUDY AREAS

The seven northern Wisconsin lakes used in our analyses are in the forested Northern Highland Lake District (46'01°N, 89'40°W). Data from these seven lakes serve as the primary focus of our analyses. The other four lakes are in agricultural and urban watersheds in southern Wisconsin near Madison (43'06°N, 89'25°W). Nutrient and major ion data from the southern lakes are compared with the northern lakes. These two lake districts are about 350 km apart.

TABLE 1. Characteristics of the study lakes.

Lake	Area (ha)	Mean depth (m)	Trophic status	Total P ($\mu g\ l^{-1}$)	Specific conductance ($\mu S\ cm^{-1}$)
Northern Wisconsin					
Allequash	168.4	2.9	Mesotrophic	29	88
Big Muskellunge	396.3	7.5	Oligotrophic	22	49
Crystal Bog	0.5	1.7	Dystrophic	19	11
Crystal Lake	36.7	10.4	Oligotrophic	9	14
Sparkling	64.0	10.9	Oligotrophic	15	80
Trout Bog	1.1	5.6	Dystrophic	40	23
Trout Lake	1607.9	14.6	Oligotrophic	17	93
Southern Wisconsin					
Kegonsa	1299	5.1	Eutrophic	86	442
Mendota	3985	12.7	Eutrophic	51	412
Monona	1326	8.3	Eutrophic	46	434
Waubesa	843	4.7	Eutrophic	70	439

The seven lakes in northern Wisconsin lie within 10 km of each other and the elevation difference between the highest and lowest lake is ca. 10 m. The lakes are exposed to essentially identical weather, but differ in other attributes including size (0.5 to 1608 ha), mean depth (1.7–14.6 m), ion chemistry (11–93 $\mu S\ cm^{-1}$ for specific conductance), and trophic status (dystrophic, oligotrophic, and mildly mesotrophic) (Table 1). Many of these differences can be attributed to the positions of lakes in the regional hydrologic flow system, in which groundwater flow plays a major role [2]. Lakes high in the flow system tend to be smaller, more dilute chemically, and have lower species richness than those lower in the flow system [2, 3]. Only two of the lakes, Allequash and Trout, are directly connected by streams; the others are seepage lakes which have water residence times of 5 to 10 years. Two of the lakes, Crystal Bog and Trout Bog, are dystrophic and are enclosed by a surrounding peat mat. All data for

TABLE 2. Coherence values for the 61 limnological variables. Where appropriate, coherence values for both the epilimnion (numbers before the slash) and hypolimnion (numbers after the slash) are given. Single asterisk indicaties hypsometrically weighted average; double asterisk indicates one sample taken per year either in winter or open water season.

Variable Type	Variable Name	Spring Epi/Hypo	Summer Epi/Hypo	Fall Epi/Hypo	Open Water Epi/Hypo	Average
Physical						
Ice	Ice on**	-	-	-	0.63	0.63
	Ice off**	-	-	-	0.94	0.94
	Ice duration**	-	-	-	0.74	0.74
Water Level	Water level	0.58	0.50	0.54	0.56	0.55
Temperature	Water temperature	-/-	0.91/0.34	-/-	-/-	0.63
	Water column temperature*	0.56	0.73	0.53	0.47	0.57
Thermocline	Thermocline depth	0.22	0.25	-0.08	0.06	0.11
	Epilimnion depth	-	0.42	-	-	0.42
	Maximum temperature change per m.	0.17	0.25	0.15	0 21	0.19
Chemical						
Ions	Calcium	0.33/0.31	0.48/0.40	0.68/0.63	0.57/0.31	0.46
	Magnesium	0.19/0.41	0.25/0.26	0.50/0.46	0.33/0.41	0.35
	Sodium	0.76/0.79	0.57/0.58	0.73/0.73	0.74/0.81	0.71
	Potassium	0.57/0.40	0.47/0.44	0.46/0.41	0.55/0.41	0.46
	Iron	0.00/0.10	0.35/0.07	0.09/0.42	0.11/0.33	0.18
	Manganese	0.06/0.23	0.10/0.27	0.02/0.27	0.03/0.35	0.17
	pH - air equilibrated	0.22/0.18	0.57/0.56	0.49/0.65	0.59/0.60	0.48
	pH - closed cell	0.52/0.47	0.54/0.64	0.67/0.68	0.74/0.75	0.63
	Total cations	0.40/0.37	0.50/0.39	0.68/0.64	0.63/0.57	0.52
	Sulfate	0.16/0.04	0.19/0.12	0.26/0.32	0.19/0.02	0.16
	Chloride	0.42/0.36	0.08/0.25	0.27/0.12	0.31/0.35	0.27
	Acid Neutralizing Capacity	0.31/0.19	0.47/0.02	0.64/0.61	0.57/0.19	0.37
	Conductivity	0.33/0.30	0.41/0.22	0.33/0.36	0.21/0.28	0.30
Nutrients	Total P	0.12/0.18	0.06/0.12	0.06/0.13	-0.05/0.15	0.10
	Total Dissolved P	0.32/0.12	0.08/0.01	0.18/0.00	0.17/-0.02	0.11
	Total N	0.21/-0.06	0.37/0.06	0.05/-0.04	0.17/-0.01	0.09
	Total Dissolved N	0.24/-0.01	0.40/0.00	-0.04/0.02	0.20/0.02	0.10
	Nitrate	0.21/0.09	0.33/0.01	0.05/0.12	0.11/0.02	0.12
	Ammonia	0.39/0.05	0.63/0.06	0.35/0.04	0.54/-0.02	0.26
	Dissolved reactive silica	-0.04/0.03	-0.05/0.22	0.10/0.13	0.04/0.14	0.07
	Bicarbonate reactive silica (total)	-0.03/0.08	-0.06/0.21	-0.03/0.10	-0.04/0.20	0.05
	Bicarbonate reactive silica (dissolved)	-0.05/0.00	0.00/0.11	-0.05/0.16	-0.04/0.07	0.03

Carbon	Total Organic Carbon	0.12/0.14	0.22/0.15	0.23/0.26	0.26/0.24	0.20
	Dissolved Organic Carbon	0.07/0.17	0.29/0.03	0.14/0.10	0.23/0.08	0.14
	Total Inorganic Carbon	0.42/0.50	0.41/0.17	0.24/0.05	0.45/0.19	0.30
	Dissolved Inorganic Carbon	0.41/0.37	0.38/0.36	0.22/0.09	0 41/0.17	0.30
Biological						
Clarity	Secchi	0.15	0.12	0.24	0.23	0.18
	Extinction coefficient	0.03	-0.01	0.21	-0.11	0.03
Crayfish	Orconectes virilis density	-	0.04	-	-	0.04
	Orconectes propinquus density	-	-0.01	-	-	-0.01
	Total crayfish density	-	0.05	-	-	0.05
Fish	Fish species richness	-	0.12	-	-	0.12
	Fish density per fyke net set	-	0.11	-	-	0.11
Oxygen	Min. depth of D.O. < 1 mg/L, winter**	-	-	-	0.10	0.10
	Min. depth of D.O. < 1 mg/L, open	-	-	-	0.00	0.00
Particulates	Total Particulated Matter	0.24/0.22	0.48/0.07	0.47/0.18	0.49/0.08	0.28
Algal Biomass	Total pigment concentration	0.07/-	-0.02/-	0.23/-	0.00/-	0.07
Primary Production	Volumetric primary productivity	-0.04	0.63	0.16	0.59	0.34
	Areal primary productivity	0.28	0.40	0.23	0.32	0.31
Sedimentation	Sedimentation rate - P	-0.33	0.23	-0.09	-0.11	-0.07
	Sedimentation rate - N	-0.03	0.38	0.47	0.34	0.29
	Sedimentation rate - total mass	-0.21	0.44	-0.05	0.01	0.05
	Sedimentation rate - C	-0.29	0.65	0.12	0.56	0.26
Zooplankton	Total zooplankton biomass	-0.01	0.05	0.13	0.06	0.06
	Rotifer biomass	-0.03	0.14	0.22	0.22	0.14
	Copepod biomass	-0.05	-0.04	0.05	0.00	-0.01
	Cladoceran biomass	-0.14	0.16	-0.05	0.07	0.01
	Cladoceran/Copepod biomass ratio	0.16	-0.14	-0.12	-0.15	-0.06
	Leptodora sp. density (/m3)	-	0.19	-	-	0.19
	Leptodora sp. density (/m2)	-	0.20	-	-	0.20
	Chaoborus density (/m3)	-	-0.02	-	-	-0.02
	Chaoborus density (/m2)	-	-0.04	-	-	-0.04

the northern lakes were from the North Temperate Lakes Long-Term Ecological Research project [4]; [http://limnosun.limnology.wisc.edu/].

The four Madison-area lakes (Mendota, Monona, Waubesa and Kegonsa) are connected by the Yahara River and lie within 15 km of each other, differ by less than 10 m in elevation, and because of their proximity experience nearly identical weather. These lakes range in area from 843 to 3985 ha, in mean depth from 4.7 to 12.7 m, and are highly eutrophic [5] (Table 1). Lakes Kegonsa, Monona and Waubesa have short residence times (0.45, 1.1 and 0.36 years, respectively). Lake Mendota, the lake farthest upstream in the drainage system, has a longer water residence time of 6.5 years.

2.2. THE VARIABLES

We analyzed 61 limnological variables for temporal coherence in the northern lakes (Table 2). Nine variables described physical attributes of the lakes, including water level, ice-cover duration, water temperature and thermocline characteristics. Twenty-six of the variables described chemical characteristics. These included aspects of major ion, nutrient and carbon chemistry of the lakes. The remaining 26 variables were biological, including abundances of individual organisms (*Chaoborus*, *Leptodora*, and two species of crayfish, *Orconectes*), abundances of groups of organisms (rotifers, cladocerans, copepods, crayfish and fish), primary production, chlorophyll, water clarity, sedimentation of various nutrients, and hypolimnetic oxygen depletion. With the exception of information on ice duration, fish and crayfish, all data were collected from a sampling station at the deepest part of each lake. Ice-on and ice-off were estimated from visits to each lake every other day during critical periods. Fish and crayfish data were collected from littoral-zone sampling locations. Water level, water temperature, dissolved oxygen, primary production, extinction coefficient, Secchi-disc depth and chlorophyll, were measured every two weeks during the ice-free season. Nutrients, pH, carbon chemistry and sedimentation were measured every four weeks. Major ions were measured immediately after ice-out, during August stratification, and in November. Fish, crayfish, *Chaoborus* and *Leptodora* were measured annually in August. Where data were available we calculated spring (ice-out to 31 May), summer (1 June to 31 August), and fall (1 September to ice-on) time-weighted averages for each parameter for each lake and year. Also, for stratified periods, we calculated epilimnetic and hypolimnetic averages. We defined the epilimnion and hypolimnion as zones where temperature changed less than 1°C per metre change in depth. Thermocline depth was the depth of maximum temperature change. Including all seasonal and thermal strata, we calculated coherence for 305 distinct variables for the northern lakes (Table 2).

We analyzed 14 variables from the southern lakes. Except for organic N, each of these 14 were also analyzed in the northern lakes. The variables included spring, summer and fall data for major ions (Ca^{2+}, Mg^{2+}, Na^{+}, K^{+}, Cl^{-}, SO_4^{2-}, acid neutralizing capacity, and pH), and nutrients (dissolved reactive silica, nitrate, ammonia, organic nitrogen, total phosphorus and dissolved reactive phosphorus). Samples were taken from a depth of 2 m. Detailed descriptions of the sampling methods are given in [6]. Data for the southern Wisconsin lakes were provided by the Wisconsin Department of Natural Resources.

2.3. MEASURES OF TEMPORAL COHERENCE

For each of the 305 variables we produced a *year* × *lake* matrix, with each cell in the matrix containing our best estimate of the value of the variable for a particular year and lake. From each of these *year* × *lake* matrices we calculated both the Pearson product-moment correlation coefficient and the Spearman rank correlation coefficient for each lake pair. These two measures of correlation gave nearly identical results and we report only the Pearson product-moment correlation. The average of the correlation coefficients across all lake-pairs was the measure of temporal coherence [1]. For example, in a dataset containing seven lakes, there are 21 lake pairs and the coherence measure is the average of the 21 correlation coefficients, one correlation coefficient for each lake pair. One advantage of using correlation coefficients as a measure of coherence is that it is possible to identify individual pairs of lakes that are consistently more or less coherent than other lake pairs. To assess the coherence of a particular lake pair, we computed the average of the correlation coefficients across all variables for that lake pair.

We used a randomization technique to compare the actual distribution of coherence values against one that might be expected if values were assigned to years randomly. In this test, for each variable's *year* × *lake* matrix we took the values for each lake and randomly assigned them to years. This resulted in 305 randomized *year* × *lake* matrices, one for each variable. We then calculated the coherence for each of the 305 newly-constructed randomized matrices.

3. Results

In the northern lakes, coherence ranged from 0.94 for off-ice date to –0.33 for sedimentaion rate of P in spring. The average for all 305 variables was 0.24. Overall, the variables were more coherent than would be expected from a random model. The frequency distribution of coherence using randomly assigned values had a mean centered around zero with a non-skewed distribution of positive and negative values. In contrast, the distribution of actual coherence values had a positive mean and was highly skewed to the right, with few negative numbers and many values well above the maximum values observed using the randomized data (Fig. 1).

3.1. DIFFERENCES AMONG LAKE PAIRS

Average coherence of individual lake pairs (measured as the average coherence of the 305 variables) varied substantially. Sparkling and Trout Lake were the most coherent lake pair (0.41) and Big Muskellunge and Crystal Bog were the least coherent (0.09) (Table 3). Following [1] we plotted the average coherence of each lake pair against the rank difference in the lake pair's surface area to mean depth ratio, a measure of a lake's exposure to climatic factors. We found that coherence decreased as morphometric differences increased (Fig. 2). Much of this pattern can be attributed to the low coherence of lake pairs containing at least one of the dystrophic lakes. Of the 21 possible lake pairs, 11 have a dystrophic lake as one or both members of the pair, while 10 lake pairs contain only clearwater lakes. Lake pairs with at least one dystrophic lake are significantly less coherent than the clearwater lake pairs ($p<0.01$, Fig. 3). Average coherence between clearwater pairs ranged from 0.28 to 0.41, whereas average coherence for the pairs containing at least one dystrophic lake ranged from 0.09 to 0.28. Interestingly, the two dystrophic lakes had low coherence with each other (0.22). No relation between coherence and the morphometric similarity in pairs was apparent for clearwater lakes (Fig. 2).

TABLE 3. Lake pairs listed in order (1–21) of decreasing average coherence (AvCo).

No.	Lake pair	AvCo	No.	Lake pair	AvCo
1.	Sparkling × Trout Lake	0.41	12.	Sparkling × Trout Bog	0.24
2.	Big Muskellunge × Sparkling	0.37	13.	Crystal Lake × Trout Bog	0.23
3.	Big Muskellunge × Trout Lake	0.35	14.	Crystal Bog × Trout Bog	0.22
4.	Allequash × Big Muskellung	0.35	15.	Allequash × Trout Bog	0.20
5.	Crystal Lake × Sparkling	0.33	16.	Allequash × Crystal Bog	0.19
6.	Allequash × Trout Lake	0.33	17.	Trout Bog × Trout Lake	0.18
7.	Allequash × Sparkling	0.31	18.	Crystal Bog × Sparkling	0.15
8.	Allequash × Crystal Lake	0.29	19.	Big Muskellung × Trout Bog	0.15
9.	Crystal Bog × Crystal Lake	0.28	20.	Crystal Bog × Trout Lake	0.14
10.	Crystal Lake × Trout Lake	0.28	21.	Big Muskellunge × Crystal Bog	0.09
11.	Big Muskellunge × Crystal lake	0.28			

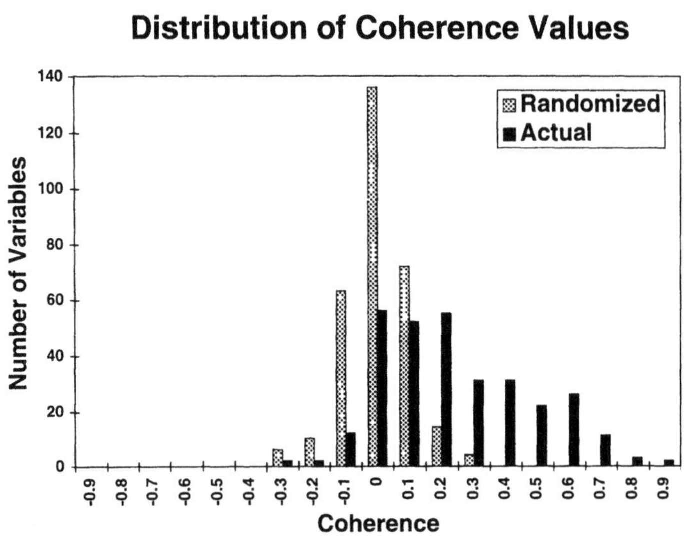

FIGURE 1. Distribution of coherence values for the northern Wisconsin lakes compared with randomized data (see explanation in the text).

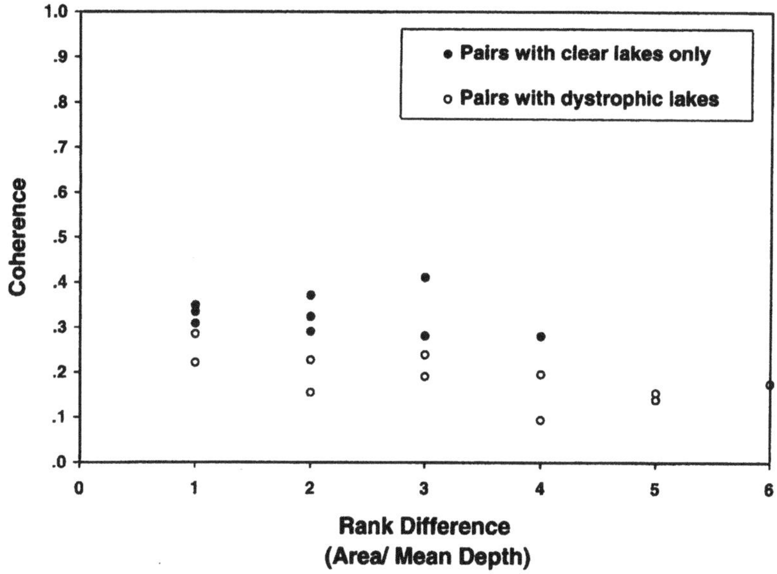

FIGURE 2. Coherence of lake pairs as a function of their rank difference in the ratio of area to mean depth.

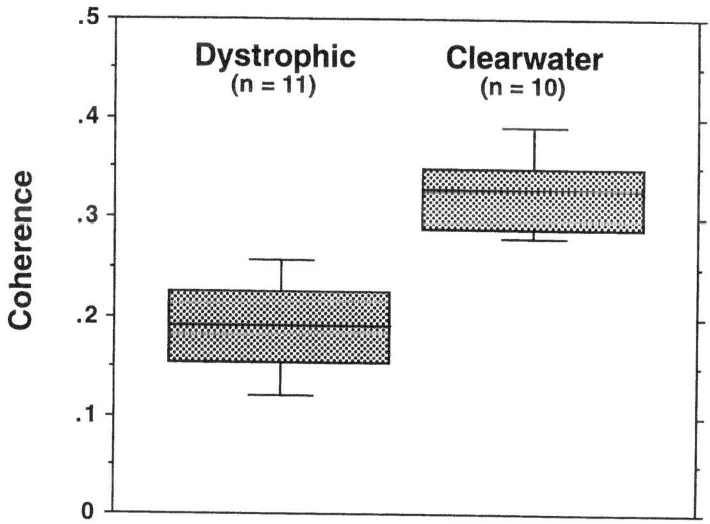

FIGURE 3. Distribution of coherence values for lake pairs containing at least one dystrophic lake versus pairs containing two clearwater lakes.

3.2. DIFFERENCES AMONG VARIABLES

Variables differed markedly in coherence, measured as the average coherence across all lake pairs. Physical variables tended to be the most coherent, chemical variables were intermediate, and biological variables were the least coherent (Fig. 4).

Among the physical features, ice cover, water level and temperature variables were highly coherent, whereas variables describing features of the thermocline such as its depth and temperature gradient were the least coherent (Fig 5). Among water temperature variables, epilimnetic temperature during summer was the most coherent (0.91) and hypolimnetic temperature was the least coherent (0.34).

Although chemical variables were, in general, less coherent than the physical variables, some chemical variables were highly coherent (Fig. 6). Some of the major ion variables, especially sodium, calcium, potassium, pH, alkalinity and total cations, were highly coherent. However, other major ions, especially iron, manganese, sulfate, chloride and conductivity, had low to moderate coherence. With few exceptions, nutrient- and carbon-related variables exhibited low coherence. Among the exceptions were ammonia and inorganic carbon, which had moderately high coherence (Table 2).

These results for chemical variables in the northern lakes contrast with those for the southern lakes. In the latter, not only were the major ion variables coherent but, compared with the northern lakes, the data for nutrients – especially silica and total phosphorus – were also coherent (Fig. 7).

As a group, biological data had low coherence (Fig. 8). The only biological variables with even moderate coherence were variables related to particulate matter in the water column, primary production and sedimentation. Many variables, including algal biomass (measured as chlorophyll), water clarity, and zooplankton biomass, had coherence values close to zero.

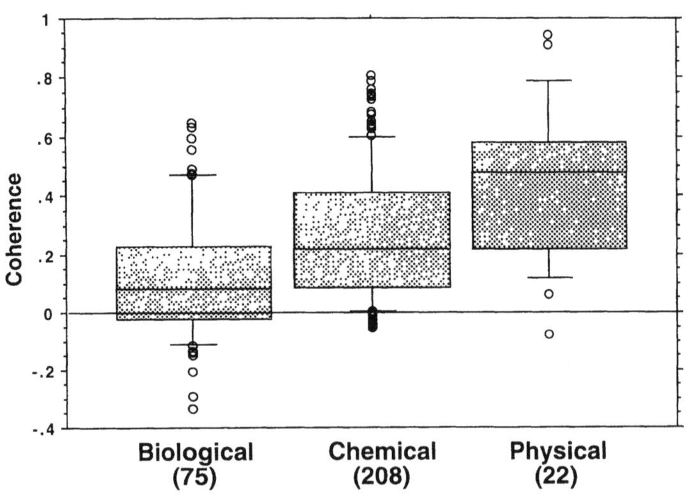

FIGURE 4. Distribution of coherence values for physical, chemical and biological variables.

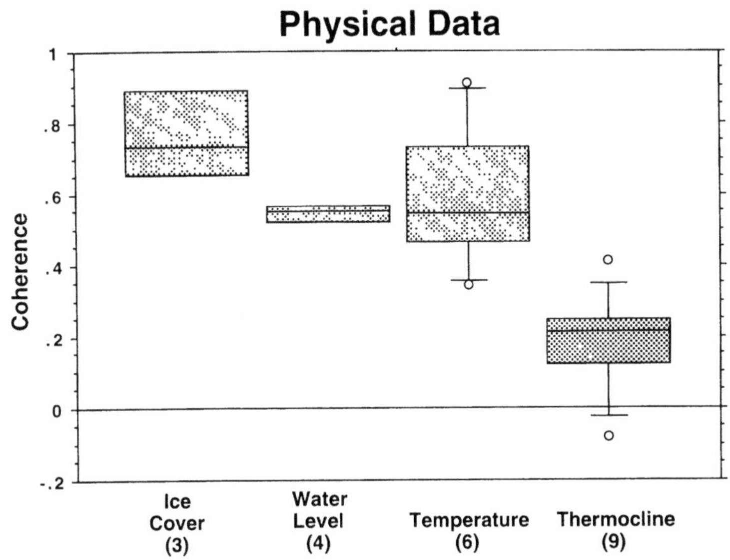

FIGURE 5. Distribution of coherence values for groups of physical variables.

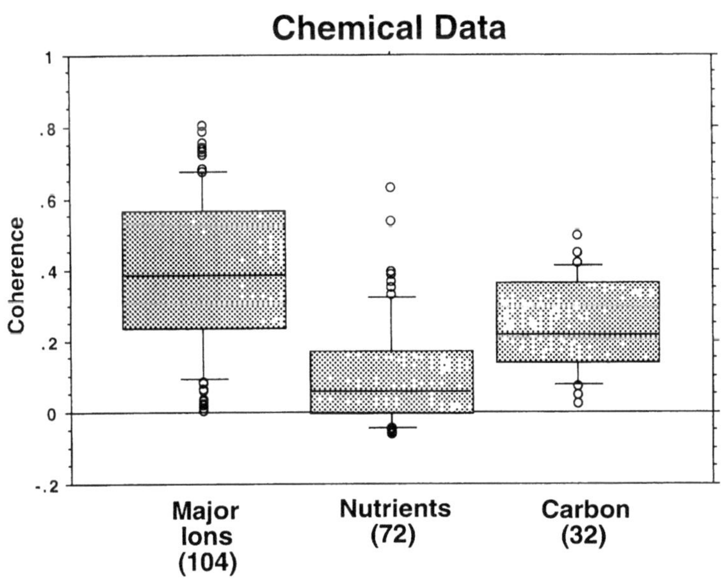

FIGURE 6. Distribution of coherence values for chemical variables.

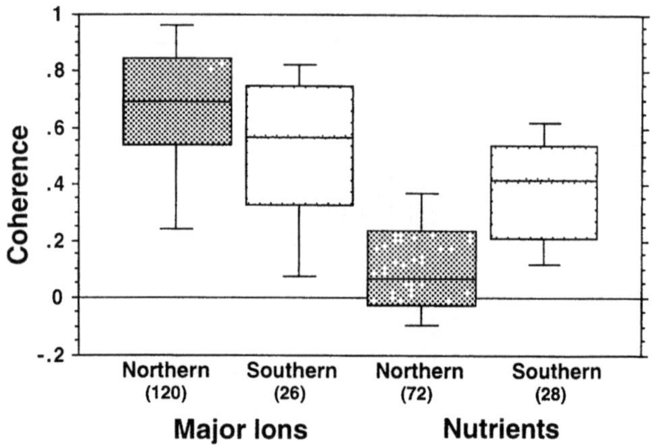

FIGURE 7. Comparison of the coherence of major ion and nutrient variables between the northern and southern Wisconsin lakes.

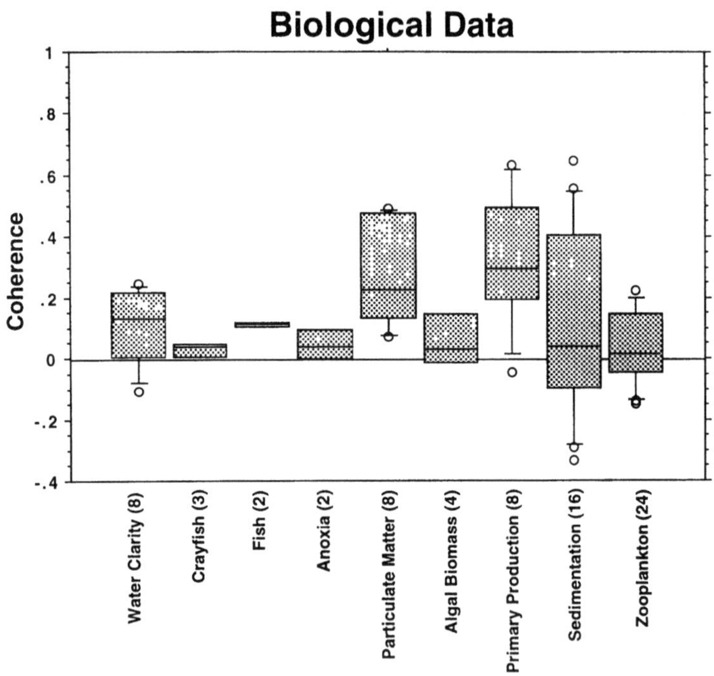

FIGURE 8. Distribution of coherence values for groups of biological variables.

4. Discussion

Our results support the contention that coherence can be used to identify lake features that are most likely to be affected by climatic change. The most coherent variables were those with a direct, mechanistic link to climatic forcing. This pattern can be seen in the physical variables where the most coherent variables included the timing of lake freezing and thawing, epilimnetic temperatures, and water levels. Each of these variables is directly influenced by climate. The physical variables that were least coherent were less directly related to climatic forcing. These less coherent variables included hypolimnetic temperature, thermocline depths, and temperature differential across the thermocline. Although these variables are clearly affected by climate, other factors, including lake morphometry, also influence them strongly, leading to lower coherence.

The same pattern occurred for chemical variables. Major ion data tended to be moderately to highly coherent. Concentrations of major ions respond directly to drought conditions, owing to a combination of evaporative concentration and hydrologically-mediated changes in ion loading to lakes [2]. Nutrients are less tightly coupled to hydrologic inputs in the northern lakes, where they exhibited low coherence. However, in the southern lakes, where overland runoff is an important source of nutrient loading to the lakes [7, 8], the nutrients were more coherent. Another factor leading to higher coherence of nutrients in the southern lakes is that they are connected by streams, and the lower lakes have relatively low residence times, so water from upstream lakes replaces water in the lower lakes relatively rapidly.

Biological variables tend to be the furthest removed from direct climatic forcing, and these variables showed the lowest coherence. Although biological variables are influenced by climate, many other processes including trophic interaction are also important and it is possible that the climatic signal is overwhelmed by these within-lake processes.

Results for major ion data illustrate the scale dependence of coherence. Major ions tended to be highly to moderately coherent in our 13-year dataset, but only moderately to weakly coherent in the 8-year dataset. A major drought occurred at the end of the shorter dataset, but was fully included in the longer one. The coupling between climatic events and major ion concentration became clear only in the longer dataset, while the signal for among-year variation in the shorter dataset was weak. Systematic changes in coherence will be expected as the measure incorporates among-year variation, short-term droughts, El Nino events, interdecadal climate changes and long-term climate change.

Our finding that lake pairs containing at least one dystrophic lake were less coherent than lake pairs containing clearwater lakes suggests that small dystrophic lakes respond to climatic forcing differently than the larger clearwater lakes. This differential response could be for at least two reasons. First, because of the high levels of dissolved humic materials in the dystrophic lakes, light penetration is much reduced and the distribution of heat in the water column is substantially different than in clearwater lakes. Low penetration of light, coupled with the small fetches that are typical in dystrophic lakes, causes higher surface temperatures and shallower thermoclines than in clearwater lakes.

Therefore, the same climatic signals relating to light and heat transfer will be manifested in different ways in dystrophic lakes and clearwater lakes. Second, the peat mats surrounding the dystrophic lakes influence the loading of major ions, nutrients and dissolved organic carbon to the lakes [9]. It is possible that the peat mats cause the dystrophic lakes to respond to wet and dry years differently from clearwater lakes, leading to low coherence of chemical parameters in lake pairs containing a dystrophic lake. The fact that the two dystrophic lakes as a pair were not coherent suggests that individual small lakes may be more independent of variations in annual climate than larger lakes. Because dystrophic lakes may have fundamentally different patterns of temporal variation than clearwater lakes within a lake district, generalizations regarding the influence of climatic fluctuations on dystrophic lakes within a region may be problematic.

With the exception of dystrophic lakes, the coherent response of variables directly influenced by climatic forcing suggests that it should be possible to extrapolate general trends from a small subset of lakes to entire lake districts. However, for effects moderately independent of climate or more complexly related to climate, e.g. biological factors and nutrients, such extrapolation will be difficult. In our dataset, changes and variation in climate appears to be the primary driver leading to coherence. However, simultaneous changes in land-use around a set of lakes also could lead to coherent behaviour among lakes. Regardless of the driving forces, analysis of coherence is an important tool in attempts to regionalize our understanding of long-term lake dynamics.

Acknowledgements

This work was funded by National Science Foundation through grants for the North Temperate Lakes Long-Term Ecological Research project. We thank the myriad field and laboratory personnel who have helped to collect and manage this large dataset over the 13-year period.

References

[1]. Magnuson J. J., Benson B. J. & Kratz T. K. (1990). Temporal coherence in the limnology of a suite of lakes in Wisconsin, U.S.A. *Freshwater Biology* **23**, 145-159.

[2]. Webster K. E., Kratz T. K., Bowser C. J., Magnuson J. J. & Rose, W. J. (1997). The influence of landscape position on lake chemical responses to drought in northern Wisconsin, USA. *Limnology and Oceanography* (in press).

[3]. Kratz, T. K., Benson, B. J., Bowser, C. J., Magnuson J. J. & Webster K. E. (1997). The influence of landscape position on northern Wisconsin lakes. *Freshwater Biology* (in press).

[4]. Magnuson J. J. & Bowser C. J. (1990). A network for long-term ecological research in the United States. *Freshwater Biology* **23**, 137-143.

[5]. Kitchell J. F. (Ed.) (1992). *Food web management: a case study of Lake Mendota.* Springer-Verlag, New York.

[6]. Lathrop R. C., Nehls S. B., Brynildson C. L. & Plass K. R. (1992). The fishery of the Yahara lakes. Technical Bulletin No. 181. Wisconsin Department of Natural Resources, Madison Wisconsin.

[7]. Lathrop R. C. (1992). Nutrient loadings, lake nutrients, and water clarity. In: *Food web management: a case study of Lake Mendota* (ed. J. F. Kitchell), pp. 69-96. Springer-Verlag, New York.

[8]. Soranno P. A., Hubler S. L., Carpenter S. R. & Lathrop R. C. (1997). Phosphorus loads to surface waters: A simple model to account for the spatial pattern of land use. *Ecological Applications* (in press).

[9]. Marin L. E., Kratz T. K. & Bowser C. J.. (1990). Spatial and temporal patterns in the hydrogeochemistry of a poor fen in northern Wisconsin. *Biogeochemistry* **11**, 63-76.

WORKING GROUP TOPIC 1: MODELLING THE RESPONSES OF LAKES TO DAY-TO-DAY CHANGES IN WEATHER

C. S. REYNOLDS, B. DESORTOVÁ AND P. ROSENDORF

Abstract

The paper considers the issues for those attempting to model the impact of alterations in the general atmospheric circulation upon particular lakes and reservoirs ("downscaling"). It shows that models of plankton dynamics can be hardly less stochastic than the weather but a proven capability exists to predict day-to-day changes in the underwater environment from given day-to-day changes in the weather. Approaches to modelling the consequences of hydrographical change upon the composition and periodicity of dominant respondents are also substantially developed but there is still a deficiency in predicting the appropriate responses of relevant species to given variables. Proposals for a simple sensitivity-iteration are advanced to address the particular needs of water managers in determining the scale of future problems at particular sites and, where relevant, the more appropriate options for securing an improvement. The approach is amenable to modelling the biotic responses to global-change scenarios.

1. Introduction

This paper presents the discussion of a small Working Group at the Advanced Research Workshop and deals directly with several issues concerning model development as they relate to management options for lakes and reservoirs during climatic changes. The report reflects the group's subdivision of the topic to address particular questions. Based on the premise that global changes will result in perceptible shifts in the general air circulation patterns and so alter the frequency with which different types of weather prevail at given locations on the Earth's surface, how might it be possible to "downscale" fluctuations to the impact on the attributes of waterbodies and their biota? What modelling approaches are available for predicting biotic responses? And how may managers of lakes and reservoirs expect to determine either how sensitive a given

D.G. George et al.(eds.), Management of Lakes and Reservoirs during Global Climate Change, 289–295.

waterbody is to change, or how successful a particular therapy might be in bringing about a sustainable solution.

2. Climate, weather events and hydrographic responses

Managers responsible for the quality of water in lakes and reservoirs, and conservators of aquatic systems and biota, are entitled to be concerned about the possible impacts of global climate change. The main shortcoming in meeting their concerns is no longer whether such change is about to occur or has already begun to do so; rather, it is in anticipating how change will be manifest at a given location. The problem is succinctly addressed elsewhere in this volume (see Davies et al. pp. 1–13): this century's rise in the Earth's mean temperature is a symptom of "global-warming" but it has not meant a universal small increase in the temperature at all locations. Thus, there can be more warm days and fewer cold ones but the range is unaltered. There can be local cooling, with fewer warm days. Similarly, there may be more or fewer sunny days and fewer or more with cloud. There may be more or fewer rain days, or there may be more or less rain in total but on a similar number of days. There may be stronger or weaker average windspeeds though these may reflect more or fewer storms.

General circulation models help to define the weather patterns established at given locations and permit some alteration in the frequencies of particular "types" (*sensu* Davies et al.) but they are not predictive in the sense that we can say where or when or how often particular types of weather will occur in the future. Contrasting with this vagueness is the recognition that there are robust and well-validated mathematical formulations and models which connect fluctuations in the atmospheric forcing above the lake or reservoir (the weather) to the momentum and penetration of water movements beneath the surface, which are manifest in the density or velocity structure (the hydrography) of the water column. Perhaps the best known of these is the Monin-Obukhov Length formulation which predicts, other factors being equal, the balance between the normally buoyancy-generating heat input and the rate of its dissipation by wind mixing and, in consequence, a predicted depth of mixing.

Where a vertical gradient of density already exists and wind mixing is resisted, a further mathematical description is required. Physical limnologists have used such derivations as the Brunt-Väsälä frequency, Richardson's Number and, most recently, the Wedderburn Number of Imberger & Hamblin [1] as a (quite sensitive) measure of the point where the forcing and the buoyancy forces are approximately balanced, this being the base of the mixed layer.

The application of Wedderburn-tested Monin-Obukhov equations to the simulation of water-column density structure is at the core of the now-widely available programme, DYRESM [2]. Its utility has been demonstrated at latitudes other than the subtropical regions for which it was written. A similar formulation, though related to latitude, Julian day and cloud interference or to real solar irradiance data, drives a Wedderburn calculation which is the basis of the PROTECH2 phytoplankton-growth model of the British National Rivers Authority (Reynolds & Irish [3] and unpublished). The date and duration of stratification and the depths of density gradients can be simulated with reasonable precision and accuracy. Both models also allow entry of inflows, adjustments in volume and outflow at capacity but PROTECH has a subroutine which specifically disperses inflow, dilutes its nutrient load and flushes exponentially, again, with acceptable realism.

We may deduce that the modelling capability to simulate day-to-day limnetic responses of changes in water temperature, stability of stratification and the depth of the mixed layer, as well as the effects of high rain-flushing events, already exists. Moreover, the link to day-to-day variability is substantially proven. It should be perfectly possible to deal with altered climate scenarios operating through proximal weather events (sun, wind, rain) set to occur with increasing relative frequency. Running on consecutive days of variability cumulates some sort of averaged integral structure or period which reflects a greater preponderance of one weather type over another (for instance, the simulated thermal stratification may stabilise at a greater or lesser depth and over a shorter or longer period, with an earlier or later start date). Thus, although the problem of downscaling from climate models persists, the appropriate hydrographic responses can be generated by a quite realistic alteration in the bias to one weather type over another, or just an altered frequency of alternation between them.

In this way, the modeller may attempt to reproduce the likely effects of an altered dominant wind speed or direction, by using randomly or systematically altered atmospheric drivers. The outstanding problem is to select the appropriate weather variable, at least until there is a more certain prediction of which one(s) will alter at any particular location.

3. Biotic responses – the sensitivity problem

Supposing that good, weather-sensitive models of hydrography are available, to what extent may we predict that particular organisms or the assemblage of species will react to the predicted changes? This should not prove to be a great philosophical problem: biological research is generally aimed at determining the sensitivities of species to environmental fluctuations. We accept that the physiological responses and tolerances of individual species, say to temperature or shear, have not been fully investigated; neither have they always been verified by controlled experiments or tested against observed distributions in space and time. So far as aquatic biota are concerned, there is probably still a pressing need to determine species-specific responses and sensitivities. On the other hand, good data are available on the environmental requirements and tolerance-limits of growth rates, related to temperature, among planktonic algae, some common species of zooplankton and several widely-distributed fish species. The sensitivity of fecundity and breeding success to characters of the physical, chemical and biological environment are being elucidated (see Winfield et al. in this volume, pp. 245–261) while the relevance of the interaction of vertical mixing with the gradient of underwater light extinction to the selection and onset of net growth of primary producers is emphasised once more by Gaedke et al. (see pp. 71–84) and Reynolds (see pp. 15–38). We recognise that the physiological understanding and available database require strengthening. For the moment, we can use regression equations relating activity to morphological characters (e.g. [4, 5]) as surrogates for hard factual data.

4. Management deductions – a sensitivity approach

Managers often need advice on strategy options and on the likely success of expensive investment in engineering to overcome a problem associated with the behaviour of aquatic biota, yet experiences on both sides indicate that there is little appreciation of each other's difficulties. This occurs not only in the field of global change but also in other dimensions of current change: is too much phosphorus the problem?; will the quality be better and soon if we invest in removing it from wastewater or will non-point sources destroy my prospects of success? Analogous questions can be put in the context of control of algae by aeration or destratification or biomanipulation. The engineer and the politician want quick, accurate answers – it is little wonder that they ignore the biologist who vaguely requests more time or more investigations or pleads variable responses of biological systems. Yet there is a simple way of determining whether a change in factor application will have much effect or not. We are aware that other topic discussions address the similar issue and argue different approaches from our check-list idea; this particular method could be simply adapted to co-ordinates of multi-dimensional space. On the other hand, this one is amenable to a simple iterative programme or "expert system" which may compare its own entries into the check-list to determine the principal sensitivities of the system in question.

The principle we wish to introduce is that in spite of the complexity of factor interactions that make up the environmental conditions, it is rare that all will be controlling simultaneously or even at all. Thus it is easy to see, for instance, that if the algal biomass of a lake is directly controlled by the meagre supply of phosphorus, halving the nitrogen supply, or indeed, risking its being doubled, will make small impact upon the biomass of algae. A small shift in the phosphorus load, on the other hand, could be crucial. Similarly, acid precipitation has much more impact on a base-poor water than on a base-rich one, while using destratification equipment in a 5-m pond will not have the same effect as extending a 5-m epilimnion in a lake 50-m deep. In this way, we could build a check-list of relevant environmental characteristics of the site to be managed against values used to define that it would be very, slightly or not at all sensitive to an alteration in management. The check-list might then look like the generalised version shown in Table 1.

Table 1. Generalised layout for a check-list of relevant environmental characteristics for managing waterbodies of differing sensitivities.

Environmental characteristics	Sensitivity of waterbody to change in relevant factors:- Very sensitive	Slightly sensitive	Not sensitive
Factor 1	Critical range	Critical range	Critical range
Factor 2	Critical range	Critical range	Critical range
Factor 3	Critical range	Critical range	Critical range

It remains to nominate the factors and their critical ranges. We considered that the list of factors could be extensive and the ranges could vary with the focus of interest. In the context of the workshop, however, we proposed a short list consisting of factors proximating for others in many instances; we suggest values to be sensitive so far as phytoplankton biomass might be affected. We proposed ranges of nutrients consistent with oligotrophic and eutrophic conditions, acidity and base concentrations that would resist (or otherwise) the influx of strong-acid ions or the withdrawal of dissolved carbon dioxide by active aquatic photosynthesis, and depth as an attribute which would allow stratification for short or long periods or one which permitted rapid recycling from shallow water (see [6, 7]). We attach great importance to flushing rate as a counter to algal growth, so flushing terms were also considered.

Suggested critical values are shown in Table 2, for boundary conditions that are briefly considered in the text.

TABLE 2. Suggested ranges for critical values of environmental factors controlling the sensitivity of waterbodies to change in algal biomass. Abbreviations and symbols are explained in the text.

Environmental Factor	Sensitivity of waterbody to change in relevant factors:- Very sensitive	Slightly sensitive	Not sensitive
TP concentration	< 20 µg P l^{-1}	20 – 100 µg P l^{-1}	> 100 µg P l^{-1}
SRP < 5 µg P l^{-1}	For > 4 months	For < 4 months	Never
TN concentration	< 0.3 mg N l^{-1}	0.3 – 1.5 mg N l^{-1}	> 1.5 mg N l^{-1}
DCN < 100 µg N l^{-1}	For > 4 months	For < 4 months	Never
Acidification	TA < 200 µeq l^{-1}	TA 0.2 – 1.0 meq l^{-1}	TA > 1.0 meq l^{-1}
pH drift	CA < 400 µeq l^{-1}	CA 0.4 – 2.0 meq l^{-1}	CA > 2.0 meq l^{-1}
h_m / H	< 0.30	0.3 – 0.8	> 0.8
$(A_{0-5m})/(A_{tot})$	< 0.05	0.05 – 0.25	> 0.25
Continuous flushing, τ	< 2 generation times	2 – 5 generation times	> 5 generation times
Irregular flushing, frequency ϕ	< 10 generation times	10 – 50 generation times	> 50 generation times

Phosphorus. Two approaches were considered. In the first, the load of total phosphorus (TP) was corrected for mean water depth and mean hydraulic retention time as proposed in Vollenweider's later models (e.g. [8]). Lakes with TP > 100 µg P l^{-1} on average throughout the year are not sensitive to a change in loading to the extent that one of < 20 µg P l^{-1} will be. Another view is that algae are not much limited by available dissolved phosphorus concentrations of > 5 µg P l^{-1} as soluble reactive phosphorus (SRP), especially if this condition is rarely breached; we propose separations at more than four months of the year, and never.

Nitrogen. The corresponding critical values for total nitrogen (TN) and dissolved combined nitrogen (DCN) are suggested in Table 2.

Acidity/Alkalinity. The criteria in Table 2 are proposed to represent concentrations of total alkalinity (TA) sensitive to acidification and bicarbonate alkalinity (CA) resistant to pH-drift at rapid rates of CO_2-withdrawal by aquatic photosynthesis.

Depth, recycling, stratification. Critcal values relate to the natural summer thermocline depth (h_m) as a factor of water depth (H) and the areas of shallow sediment ($A_{0\text{-}5m}$) as a proportion of total area (A_{tot}). Sensitivity is judged according to whether a change in mixing intensity would substantially alter the growth conditions in the waterbody.

Retention time. This character relates to generation times that can be sustained against removal by flushing (τ is the mean time to displace one volume). Because algal growth is not always continuous and events may reduce inocula, we also include an event-frequency factor (ϕ).

Sensitivity. The final stage is to compare the distribution of high sensitivity scores. A site scoring high in its sensitivity to, say, phosphorus indices, could be expected to benefit quickly from phosphorus load reduction, especially if it has little shallow sediment. One scoring low against phosphorus would consume a lot of investment before load reduction became effective, but if it is deep, management of the stratification looks a better bet. If ponds are insensitive to all four factors, they will be sufficiently small and shallow to permit ready biomanipulation of the trophic structure.

5. Modelling biological sensitivity to global change

Modelling sensitivity to global change combines all the above processes. Once it can be established what the altered weather pattern might be, we may model the water column response and seek from the modelled biotic tolerances the species most likely to succeed. In response to increasing radiation, rising air temperatures and weaker flushing, many lakes are more likely to be subject to growths of bloom-forming cyanobacteria, providing they are not acidic or oligotrophic. Cooler or windier summers might have an opposite response but there may be opportunities for filamentous cyanobacteria like *Planktothrix* to flourish (Reynolds in this volume, pp. 15–38). Wetter conditions might benefit water quality through enhanced dilution. There is a large number of possibile combinations and a large number of potential biotic responses, but we are confident that with the modelling appraoches outlined here, biotic responses can be accurately modelled and their magnitudes usefully predicted.

References

[1]. Imberger, J. & Hamblin, P. F. (1982). Dynamics of lakes, reservoirs and cooling ponds. *Annual Review of Fluid Mechanics* **14**, 153-187.

[2]. Imberger, J & Patterson, J. C. (1981). A dynamic reservoir simulation model - DYRESM 5. In: *Transport models for inland and coastal waters* (ed. H. B. Fisher). Academic Press, New York, pp. 310-361.

[3]. Reynolds, C. S. & Irish, A. E. (1997). Modelling phytoplankton dynamics in reservoirs: the problem of in situ growth rates. *Hydrobiologia* (in press).

[4]. Reynolds C. S. (1989). Physical determinants of phytoplankton succession. In: *Plankton ecology* (ed. U. Sommer). Brock-Springer, Madison, pp. 9-56.

[5]. Nielsen, S. L. & Sand-Jensen, K. (1990). Allometric scaling of maximal photosynthetic growth rate to surface to volume ratio. *Limnology and Oceanography* **35**, 177-181.

[6]. Sas, H. (1989). *Lake restoration by reduction of nutrient loading: expectations, experiences, extrapolations.* Academia Verlag Richarz, Skt Augustin.

[7]. Reynolds, C. S. (1992). Eutrophication and the management of planktonic algae: what Vollenweider couldn't tell us. In: *Eutrophication: research and application to water supply* (eds. D. W. Sutcliffe & J. G. Jones). Freshwater Biological Association, Ambleside, pp. 4-29.

[8]. Vollenweider, R. A. (1976). Advances in defining the the critical load levels of phosphorus in lake eutrophication. *Memorie dell' Istituto Italiano di Idrobiologia* **33**, 53-83.

WORKING GROUP TOPIC 2: DETECTING AND PREDICTING THE RESPONSES OF LAKES TO GLOBAL CLIMATE CHANGE

URSULA GAEDKE, RITA ADRIAN, H. BUCKA, L. HAVEL, J. KESKITALO, NATALIA MINEEVA AND I. WINFIELD

1. Analysing long-term datasets and interannual fluctuations

Meteorological records throughout Europe demonstrate that in recent decades there have been widespread changes in a number of individual weather variables, and in the frequency of various weather types. Some of these changes are clearly driven by cyclical events ([1]; also see Davies et al. in this volume, pp. 1–13, and George & Hewitt in this volume, pp. 223–244) but other trends appear to be influenced by anthropogenic factors. Further changes, for example in the frequency of various extreme events, are to be expected in the next decades, and will probably include higher air temperatures and a higher frequency of strong wind events in major areas. The Working Group agreed that most of these changes would have a major effect on lakes and concluded that much more attention should be given to the cross-correlation of meteorological and limnological datasets. Many of the larger lakes in Europe have been the subject of intensive study for several decades, but very few of these datasets have been processed in ways that highlight the effects of year-to-year changes in the weather. The Working Group recognised that the techniques used to analyse such datasets would have to be adapted to meet local conditions but suggested that the methods adopted should pay due regard to three general principles, enumerated below.

1.1. STANDARDISATION

The most effective way of identifying the key driving variables is to study the climatic response of individual lakes over long periods of time or compare the relative responses of lakes situated in different areas [2]. However, such comparisons can be done only in a meaningful way if the measurement techniques produce comparable results. This implies that data for a core set of variables should be recorded using similar techniques, and that the same, or at least comparable approaches, should be used to study different

D.G. George et al.(eds.), Management of Lakes and Reservoirs during Global Climate Change, 297–300.

lakes. These two demands may, in some instances, conflict with each other. For example, the harmonisation of measurement techniques used in the Czech Republic, with the methods used in the European Union, would imply that the new results could no longer be compared with the existing datasets, and this will complicate the detection of long-term changes. A similar and very common situation arises when the sampling method is changed or samples are collected from different locations in the same system. The Working Group acknowleged that no general solution exists to this problem but noted that such difficulties could be minimised by (1) organising comprehensive intercalibrations between old and new methods, (2) arranging frequent intercalibrations with other laboratories, (3) documenting any changes and archiving all original protocols, and (4) storing preserved samples or at least representative specimens or photographs.

1.2. DE-TRENDING

The long-term covariation of other potentially important factors has to be considered when analysing field observations for potential responses of lakes to changes in weather conditions. For example, the occurrence of a series of years with exceptionally mild winters (such as those recorded in Europe in the early 1990s) may be confounded with changes in the nutrient load or acidification.

1.3. EXTREME EVENTS AND SENSITIVE PROCESSES

The analysis should pay particular attention to time-periods and spatial regions (e.g. the uppermost part of the water column) in which the factor under consideration may indeed have an impact, rather than using indiscriminantly extended temporal mean values (e.g. yearly averages) over large water volumes. For example, small plankton organisms are controlled by severe grazing pressures during the clearwater phase, which suggests that this period should be omitted from analyses on the short-term impact of meteorological factors. Further aspects of this particular issue are discussed by George & Hewitt (in this volume, pp. 223–244) and Gaedke & Seifried (in this volume, pp. 39–55).

2. Predicting the responses of lakes to climatic changes

This section aims to provide a rough guideline to the potential sensitivity of various lake types to changes of different meteorological variables. Ideally, it implies a recommendation concerning which particular time-periods, spatial compartments, groups of organisms, and various abiotic factors, should be studied most carefully. We are fully aware that our actual knowledge enables us to make only rough and preliminary indications, the validity of which needs to be evaluated carefully in each case study. We assume that the quality of the predictions can be improved by modelling studies.

Air temperature, irradiance, wind and precipitation were regarded as the most influential meteorological variables affecting lake ecosystems; these are relatively well assessable by measurement methods that are widely available. In order to predict the sensitivity of lakes to unusual fluctuations of these four variables, the lakes were roughly classified in various respects. First, a simplification was introduced by distinguishing between lakes that are dominated by allochthonous impacts (type A) and those lakes where internal factors play the major role (type B). Examples for lakes of type A include (small) lakes and reservoirs with high exchange rates. For these, changes in atmospheric precipitation (rainfall) may rule out the effects of fluctuations of the other meteorological factors (for details see below). The amount of precipitation may not only exert a strong influence on water residence times and runoff, but also influence the concentrations of silt, humic or toxic substances, nutrients, and organic material, as well as the pH which, in turn, may be of great importance for the lake ecosystem. Since most lakes are influenced by both external factors and internal processes, the above-mentioned processes to some extent will also apply to lakes of type B. Similarly, the internal changes mentioned below may also influence allochthonously-dominated lakes of type A.

Regarding lakes of type B, a further distinction is required into non-stratified, shallow lakes (type B_1) and temporarily stratified waterbodies (type B_2). The biota in non-stratified, shallow lakes is directly susceptible to changes in the irradiance influencing primary production, and in the air temperature which influences all process rates. All changes of weather conditions may imply shifts in species composition (e.g. Reynolds in this volume, pp. 15–38) and food-web structure in the long term, which may partially compensate or enhance the above-mentioned effects. The frequency and strength of wind events may exert a pronounced indirect effect if the wind is sufficiently strong to provoke water movements which cause suspension of sediments and alter the internal nutrient and light regime. These effects may strongly influence the distribution of macrophytes and, thus, the entire food-web structure. Except for the last-mentioned case, lakes belonging to type B_1 generally appear to be less sensitive to changes of meteorological factors than stratified lakes (type B_2).

The biota of stratifying lakes is also influenced by fluctuations of the irradiance and air temperature. However, their influence may be overruled by the subsequent indirect effects of changes in water column stability during periods of weak stratification. Air temperature, irradiance and wind determine in concert the water column stability which, in turn, strongly influences the light history, and the nutrient, temperature, oxygen and pH regime which planktonic organisms experience. Especially around the onset and breakdown of stratification, water column stability and related variables depend critically on the actual meteorological conditions (e.g. Ollinger & Bäuerle in this volume, pp. 57–70). Thus, for example, individual storm events may postpone stratification for many days in spring. To conclude, plankton dynamics in temporally stratified lakes (type B_2) appear to be relatively sensitive to actual weather conditions when stratification is weak (commonly in spring and autumn).

Introducing a final discrimination between stratified lakes that are relatively large and deep, more exposed to wind and less eutrophic (type B_{2a}), and lakes that are smaller, more sheltered and more eutrophic (type B_{2b}), enables some speculations to be made on the potential effects of an increase in air temperature and strong wind events, as are postulated to occur in large areas of Europe during the next decades (e.g. Davies et al. in this volume, pp. 1–13). Lakes of type B_{2b} are likely to experience a longer period of stratification owing to higher temperatures. This may have various effects on chemical and biological parameters. For example, oxygen depletion in deep waters may become more severe, causing a reduction in the thickness of the aerobic surface water layer, and a greater release of phosphorous from the sediments, which may subsequently promote more pronounced autumn and winter algal blooms [3]. In contrast, in large, deep and more exposed lakes (type B_{2a}) an increase in air temperature and strong wind events counteract each other in a non-predictable way in respect to the duration and stability of stratification, which prevents further speculation on biological effects (Ollinger & Bäuerle, in this volume, pp. 57–70).

References

[1]. Rapp, J. & C. D. Schönwiese 1995. Atlas der Niederschlags- und Temperaturtrends in Deutschland 1891-1990. *Frankfurter Geowissenschaftliche Arbeiten. Serie B. Meteorologie und Geophysik, Band 5.*

[2]. Pace, 1993. Forecasting ecological responses to global change: the need for large-scale comparative studies. In: *Biotic interactions and global change* (eds. P. M. Kareiva, J. G. Kingsolver & R. B. Huey). Sinauer Associates Inc., Sunderland, Massachusetts.

[3]. Adrian, R., Deneke, R., Mischke, U., Stellmacher, R. & Lederer, P. (1995). A long-term study of the Heiligensee (1975-1992). Evidence for effects of climate change on the dynamics of eutrophied lake ecosystems. *Arch. Hydrobiol.* **133**, 315-337.

WORKING GROUP TOPIC 3: MANAGING WATER QUALITY IN A CHANGING WORLD

D. G. GEORGE, T. L. CONSTANTINESCU,
J. DURAS, D. GERDEAUX, S. HORICKA AND T. OZIMEK

1. Introduction

In recent years, it has become clear that day-to-day changes in the weather can have a major effect on the quality of the water that we abstract from our lakes and reservoirs. For example, blooms of toxic algae that were once regarded as an inevitable consequence of eutrophication are also influenced by climatic factors, such as high winter temperatures and low summer wind speeds [1, 2]. Long-term investigations in a number of lake districts have also shown that some water quality problems originally regarded as local phenomena are actually influenced by changes in the weather that operate on regional and even global scales. A great deal of attention is currently being paid to the physical effects of short-term changes in the weather on the seasonal succession of phytoplankton [3, 4], but much less is known about the effects of longer-term changes in the weather. In the lakes of the English Lake District, some of these long-term variations are now known to be influenced by quasi-cyclical variations in the position of the Gulf Stream in the Atlantic Ocean (see [5], and George & Hewitt, in this volume, pp 223–244). The implications of such large-scale patterns of change cannot be appreciated by researchers operating on a regional level but require the concerted effort of meteorologists and limnologists working on a European scale.

This workshop session was planned as a forum where specialists from several different countries would be able to discuss long-term trends and compare the climatic responses of different lake systems. It became clear at the outset that the participants had very different perspectives on the impact of climate change. Several were concerned with the practical problems of managing their particular systems and were not aware of the long-term trends that have recently been identified in many regional datasets. Therefore some time was spent on discussing four rather general topics, enumerated below:-

1. The methods used to develop plausible scenarios of future changes in the weather.

D.G. George et al.(eds.), Management of Lakes and Reservoirs during Global Climate Change, 301–306.

2. The ways in which these scenarios can be used to develop "underwater weather" scenarios for different types of lakes.
3. The methods used to analyse the climatic response of lakes that are also subject to local anthropogenic influences.
4. The methods used to assess the sensitivity of different lakes to defined patterns of climatic change.

2. Developing climate change scenarios

In cases where it is difficult to predict future developments, it has become customary to use scenario analyses. Climate scenarios are not forecasts of forthcoming climatic events but rather they provide a set of internally consistent descriptions of what could occur given a defined change in the driving variables. At present there are two main ways of constructing scenarios of climate change in a warmer world. The first is a physical method that uses various numerical models of the atmospheric circulation to simulate the regional and seasonal patterns of change. These General Circulation Models (GCMs) basically can be regarded as weather forecasting models that have been adapted for running over long periods of simulated time, and they range in complexity according to the number of driving variables that are included [6]. The second is the analogue method [7] in which the regional and seasonal contrasts between past warm and cold periods are used to construct maps of the relative changes likely to appear in a warmer world. A critical consideration in all such scenarios is the selection of years included in the comparison between warm and cold periods. Warm years can either be selected individually, thus maximimising the warm anomaly, or contrasts are made between contiguous blocks of exceptionally cold and warm years.

The two methods have different strengths and weaknesses and each may be more or less appropriate depending on the area of interest and the predictions demanded of a particular scenario. The main advantage of the (physical) numerical methods is the range of their predictive capabilities. They can be used to describe not only past and present climates, but also the climate that might possibly develop under future extremes. Their main disadvantage is a lack of spatial resolution and a failure to simulate many critical features of our present climate. The lack of resolution is particularly important for small countries like the UK, as the current generation of GCMs cannot resolve the cyclonic features that have such an important effect on weather in the UK. The main advantage of the analogue methods is that they are based on warm climates that have actually existed. A major disadvantage of these methods is the limited range of "extreme year" contrasts that are currently available in the instrumental records.

3. Developing "underwater weather" scenarios for different lake types

The "climate change" scenarios that currently are being produced by atmospheric modellers will have to be refined in a number of important respects before they can be used for impact studies on lakes. In the terrestrial environment, the key climatic driving variables are the changes in air temperature, rainfall, relative humidity and the concentration of atmospheric CO_2. Changes in rainfall may have an important effect on the seasonal dynamics of shallow lakes but deep, thermally stratified lakes are more sensitive to changes in solar radiation and wind speed. The thermal characteristics of deep, thermally stratified lakes are determined by the complex interaction of fixed topographic and variable climatic factors. When a lake becomes thermally stratified its surface temperature is partly determined by the level of incoming solar radiation, and partly by the intensity of turbulent mixing. Most existing climate change scenarios include projections of the regional change in air temperature, rainfall and relative humidity, but very few include any indication of the likely change in wind speed. At present, the most reliable method of generating high-resolution wind field maps is to devise an objective classification of the circulation patterns that exist in the different regions. Jones et al. [8] have devised a system that utilises the daily grid-point sea-level pressure data that are available on a 5° × 10° grid for the UK. Different indicators of wind flow can then be derived from the gridded pressure data and the resultant historical time-series used to assess the likely impact of extreme changes in the weather.

4. Analysing long-term patterns of change

The Working Group spent some time discussing the methods used to analyse long-term datasets and assessing the significance of interannual variations. Particular attention was paid to the problem of distinguishing episodic patterns of change, that could be related to the weather, from the more progressive patterns of change associated with enrichment of waterbodies. The group discussed the availability of meteorological records in Europe and commented that more emphasis should be placed on identifying weather patterns rather than comparing "single-variable" averages. Some time was also spent assessing the sensitivity of different lake systems to changes in the weather. A simple check-list scheme was devised to highlight the key driving variables and the "optimum" and "acceptable" sampling frequencies for a number of water quality variables that were discussed. Problems often arise when the methods used to collect samples have changed, particularly where the efficiencies of the two methods have not been assessed by synoptic measurements. The Working Group strongly advocates the use of simple methods that could provide useful insights into regional trends, e.g. Secchi-disc measurements of water transparency and settled-volume estimates of zooplankton biomass. In some situations, daily measurements of maximum/minimum surface temperatures could prove to be invaluable as long as they were taken at a fixed site using a calibrated thermometer.

The Working Group also recommended that participants should explore the possible use of "low cost" ways of assembling long time-series of data that could include records collated by other organisations; e.g., ice-cover records by local residents, or fish-catch returns from established angling clubs. Angling returns, although selective, can provide very useful information where it is possible to quantify the catch-per-unit effort. Some consideration should also be given to measurements that record the integrated responses of a lake to changes in the weather and the catchment. A good example of such an integrated measurement is the hypolimnetic oxygen deficit, which can be calculated by comparing the vertical distribution of oxygen in a thermally stratified lake at the beginning and end of the growing season. The Working Group also highlighted the value of high-resolution time-series where appropriate instruments are available. Automatic water quality monitoring systems are now much more reliable and can download data via telephone links or satellite networks. Details of a new automatic water quality monitoring station, designed by the Institute of Freshwater Ecology (Windermere, UK), were presented to the group, together with a brief account of a demonstration programme that is currently being supported by the European Union. This system includes a new sensor that can be used to estimate the concentration of phytoplankton and identify some important functional groups, e.g. blue-green algae (cyanobacteria). The data currently being recorded by the system are used to support a number of scientific projects but, in future, similar systems could be installed in reservoirs and used for operational tasks.

5. Impact assessments

The Working Group then discussed ways in which long-term datasets could be used to quantify the sensitivity of different lake types to changes in the weather. The group recognised that different lakes would respond in different ways and highlighted the need to identify the critical driving variables on a site-by-site basis. In most instances, these sensitivity analyses can be attempted only on a subjective basis, but usually a clear distinction can be drawn between isothermal lakes that are strongly influenced by air temperature and rainfall, and thermally stratified lakes that are also influenced by the intensity of wind-induced mixing. Various examples were discussed and a "sensitivity matrix" was devised for a notional lake subjected to a variety of catchment and climatic influences. Table 1 shows how a check-list of key attributes can be used to assess the sensitivity of a lake to a progressive increase in temperature. In this example, the lake is a large, productive, thermally stratified lake that typically remains frozen for several weeks in winter. The number of crosses in the matrix represent the impact of a particular weather variable on the four listed measures of water quality. In a warmer world, we would expect the lake to remain free of ice for most of the winter but be exposed to more prolonged "flushing" periods which could have a significant effect on the seasonal distribution of nutrients. The most important summer effects are those related to the timing and intensity of thermal stratification. A period of warming in early summer almost certainly would promote the earlier development of algal blooms, which could lead to increased deep water anoxia if the seasonal thermocline was relatively stable.

6. Conclusions

In recent years it has become clear that natural systems have to be monitored for some considerable time if we are to identify any persistent patterns of change. This is particularly so in the area of climate research where the inherent variablity of the weather greatly complicates any analysis of the secular trend. European limnologists are exceptionally fortunate to have inherited long-term physical, chemical and biological records from a number of lakes that cover a range of latitudes and altitudes. So far,

TABLE 1. A simple subjective method of assessing the sensitivity of a particular lake to a progressive increase in temperature.

Variables	Nutrient loading	Deep water anoxia	Allochthonous inputs	Algal blooms
Ice cover		+		+
Winter temperature	+		+ +	+
Winter rainfall	+ + +		+ +	+
Winter wind speed			+	
Summer temperature	+	+ + +		+ +
Summer rainfall	+		+	+
Summer wind speed	+	+ + +		+ + +

only a few of these records have been analysed systematically and even fewer processed in ways that make them accessible to climatologists and environmental physicists.

In the early 1970s the International Biological Programme [9] served as an effective focus for comparative research in eastern and western Europe. The 1995 workshop in Prague has provided a unique opportunity to renew these contacts and plan new integrated studies for the future. In this era of global environmental research, freshwater ecologists will also need to strengthen their contacts with observers and modellers who report on long-term changes in the weather. If we are to plan research programmes that address the right questions we must know what climatic changes are likely to occur and where these changes are likely to be most severe. Several participants in the Working Group were clearly unaware of recent developments in the field of climate research. In the UK, the Climatic Research Unit at East Anglia (represented at the meeting by Professor Davies) provides regular updates to the scientific community via the UK LINK programme and the European ECLAT project [10]. These programmes could serve as a model for other countries and demonstrate what can be achieved when a single organisation is charged with disseminating information and providing advice on the development and interpretation of climate change scenarios.

The Working Group concluded by considering some of the likely economic consequences of changes in the weather. In practical terms, very little can be done to ameliorate the effects of extreme variations in the weather, but water quality managers still need to be able to evaluate the cost of implementing new water treatment procedures in relation to the perceived risk. It is also important to recognise when the factors responsible for a sudden deterioration in water quality are outside the control of the regulatory authorities. For example, phosphate stripping plants are now being installed on the inflows to many lakes and reservoirs. The public expect these measures to produce immediate results, but undesirable algal growths may still appear when mild winters are followed by calm summers.

References

[1]. Steinberg, E. W. & Hartmann, H. M. (1988). Planktonic bloom-forming Cyanobacteria and the eutrophication of lakes and rivers. *Freshwater Biology* **20**, 297-287.

[2]. George, D. G., Hewitt, D. P., Lund, J. W. G. & Smyly, W. J. P. (1990). The relative effects of enrichment and climate change on the long-term dynamics of *Daphnia* in Esthwaite Water, Cumbria. *Freshwater Biology* **23**, 55-70.

[3]. Reynolds, C. S. (1993). Scales of disturbance and their importance in plankton ecology. *Hydrobiologia* **249**, 157-171.

[4]. Reynolds, C. S. (1994). The role of fluid motion in the dynamics of phytoplankton. In: *Aquatic ecology: scale, pattern and process* (eds. P. S. Giller, A. G. Hildrew & D. Rafaelli). Blackwell Scientific Publications, Oxford, pp. 141-187.

[5]. George, D. G. & Taylor, A. H. (1995). UK lake plankton and the Gulf Stream. *Nature* **378**, 139.

[6]. Gates, W. L., Mitchell, J. F. B, Boer, G. J., Cubasch, U. & Maleshko, V. P. (1992). Climate modelling, climate prediction and model validation. In: *Climate change: the supplementary report to the IPCC Scientific Assessment* (eds. J. T. Houghton, B. A. Callander & S. K.Varney). Cambridge University Press, pp. 97-134.

[7]. Palutikof, J. P., Wigley, T. M. L. & Lough, J. M. (1984). Seasonal climate scenarios for Europe and North America in a high-CO_2, warmer world. Report prepared for the US Department of Energy, DOE / EV / 10098-5, 70 pp.

[8]. Jones, P. D., Hulme, M. & Briffa, K. R. (1993). A comparison of Lamb circulation types with an objective classification scheme. *International Journal of Climatology* **13**, 655-663.

[9]. Le Cren. E. D. & McConnell, R. H. (1980). *The functioning of freshwater ecosystems.* International Biological Programme, Volume 22, Cambridge University Press, 588 pp.

[10]. The ECLAT Project (1996). *Climate change scenarios for European climate change impacts assessments: network for the dissemination of climate data, climate change scenarios and scientific advice.* Conclusions and recommendations of the ECLAT Workshop.

WORKING GROUP TOPIC 4: INTERNATIONAL NETWORK OF LAKE SITES

T. K. KRATZ, B. BOJANOVSKY,
V. DRABKOVA AND V. STRAŠKRABOVÁ

1. Introduction

The responses of lakes to year-to-year fluctuations in the weather can give important clues concerning how lakes might respond to future climatic changes. Lakes with long-term records of physical, chemical, and biological data can be especially useful in these analyses. A number of lakes in Europe and North America have now been studied intensively for at least twenty years. More recently, some of these datasets have been assembled into regional and national networks that, very often, were established to resolve specific water quality problems. The Working Group discussed the rationale behind a number of existing networks and highlighted the need to sustain such networks over a long period of time. The structure of two (the US Long Term Ecological Research Network and the UK Environmental Change Network) was discussed in greater detail and compared with more specialised networks that are still operating in Europe. At the end of the session, the group agreed to produce a database of all known long-term lake sites in Europe and North America, and devised a questionaire that could be circulated to all participants. In the following Appendix Table we have assembled some summary data from a number of lakes in Europe and North America that are regularly monitored using standard methods. This list of lakes and reservoirs is not exhaustive but includes examples of waterbodies from all of the countries represented at the NATO workshop. The group expressed the hope that this information would provide a useful baseline for investigators planning collaborative research on environmental change, and recommended that participants should explore ways of expanding the network to include a greater variety of lake-types.

D.G. George et al.(eds.), Management of Lakes and Reservoirs during Global Climate Change, 307–320.

Appendix Table. Characteristics of selected lakes having long-term data.

Lake Name	Mondsee	Neusiedler See	8 lakes near Dorset Research Centre	12 lakes of Experimental Lakes Area	Slapy Reservoir
Country	Austria	Austria	Canada	Canada	Czech Republic
Latitude	47°50'N	47°47'N	44°45'N	49°40'N	49°37'N
Longitude	13°23'E	16°46'E	78°35'W	93°43'W	14°20'E
Elevation (masl)	481	113	327-379	388-424	271
Surface Area (ha)	1420	32000	21-94	5-56.1	1310
Volume (10^6 m^3)	510	200	0.7-9.5	0.2-7 9	270
Max Depth (m)	68.3	1.8	5.8-38	5.7-32.7	53
Shoreline Length (km)	28		2.7-8.2		150
Catchment Area (km^2)	247	1200	0.9-6	0.3-7.2	12900
Catchment Type	forested, grassland	agricultural	forested	forested	agricultural, forested
Retention Time (yr)	1.7	>3	1.2-4.5	1->3	<1
Trophic Status	oligo-mesotrophic	mesotrophic	oligotrophic	oligotrophic, mesotrophic	eutrophic
Contact	1	1	2	3	4
Summary Reference	1	2	-	3,4	5
Available Data	physical, chemical, biological weekly to monthly since 1981	physical, chemical, biological biweekly to monthly since 1968	physical, chemical, biological sampled 6-12 times per year, 1976-present	physical, chemical, biological daily to seasonally for 12-28 years.	physical, chemical, and biological data every 3 weeks since 1979.

Appendix Table (continued)

Lake Name	Rimov Reservoir	Cerne Lake	Paijanne	Paajarvi	Lentua
Country	Czech Republic	Czech Republic	Finland	Finland	Finland
Latitude	48°52'N	49°11'N	61°45'N	61°04'N	64°15'N
Longitude	14°35'E	13°11'E	25°18'E	25°08'E	29°32'E
Elevation (masl)	470	1008	78	103	168
Surface Area (ha)	206	18.4	110000	1350	9100
Volume (10^6 m^3)	34	2.6	17800	206	600
Max Depth (m)	43	39.5	98	87	52
Shoreline Length (km)	24	24.5	1149	33	136
Catchment Area (km^2)	488	1.3	25400	255	2065
Catchment Type	forested, agricultural	forested	forested, wetland	forested	forested, wetland
Retention Time (yr)	<1	3	2.7	3.5	0.75
Trophic Status	eutrophic	oligotrophic	eutrophic	oligo-mesotrophic	oligotrophic
Contact	4	5	6	6	6
Summary Reference	6	7	-	-	-
Available Data	physical, chemical, and biological data every 3 weeks since 1960.	physical, chemical, biological data 1-3 times/year since 1987.	physical, chemical, biological: daily water temp since 1980; 1-9 samples/year since early 1960's	physical, chemical, biological: 1-11 samples/year since early 1960's	physical, chemical, biological: 1-7 samples/year since early 1960's

Appendix Table (continued)

Lake Name	Pielinen	Inari	Lappajarvi	Pyhajarvi	Pihlajavesi
Country	Finland	Finland	Finland	Finland	Finland
Latitude	62°40'N	69°00'N	63°08'N	60°59'N	62°21'N
Longitude	29°30'E	27°45'E	23°39'E	22°19'E	24°17'E
Elevation (masl)	94	119	69	45	139
Surface Area (ha)	86700	105000	14200	15400	2000
Volume (10^6 m^3)	8500	15100	1120	779	103
Max Depth (m)	60	96	38	26	18
Shoreline Length (km)	610	2776	107	80	67
Catchment Area (km^2)	12823	13400	1526	617	370
Catchment Type	forested, wetland	forested, wetland	forested, wetland	forested, agricultural	forested, wetland
Retention Time (yr)	1.9	3.4	2.8	3	1
Trophic Status	mesotrophic	oligotrophic	eutrophic	mesotrophic	mesotrophic
Contact	6	6	6	6	6
Summary Reference	-	-	8	9	-
Available Data	physical, chemical, biological: daily water temp; 1-9 samples/year since early 1960's (physical data since 1944)	physical, chemical, biological: daily water temp.; 1-4 samples/year since early 1960's (physical data since 1950)	physical, chemical, biological: 2-4 samples/year since early 1960's; biological data since 1972	physical, chemical, biological: 1-17 samples/year since early 1960's	physical, chemical, biological: 1-9 samples/year since 1972, biological since 1963

Appendix Table (continued)

Lake Name	Tuusulanjarvi	Arendsee Reservoir	Barleber See	Stechlinsee	Heiligensee
Country	Finland	Germany	Germany	Germany	Germany
Latitude	60°26'N	52°53'N	52°13'N	53°10'N	52°36'N
Longitude	25°03'E	11°29'E	11°39'E	13°02'E	13°13'E
Elevation (masl)	37	23	42	60	31
Surface Area (ha)	600	514	103	430	30
Volume (10^6 m^3)	19	147	6.9	97	1.9
Max Depth (m)	10	48.7	11	68	9.5
Shoreline Length (km)	22	11	4.1	16.1	2.3
Catchment Area (km^2)	92	29.8	-	12.4	-
Catchment Type	forested, agricultural	agricultural, forested	-	forested	urban, agriculltural
Retention Time (yr)	0.7	114	-	>3	-
Trophic Status	hypereutrophic	hypereutrophic	mesotrophic	oligotrophic	hypereutrophic
Contact	6	7	7	8	9
Summary Reference	10	11	12	13	14
Available Data	physical, chemical, biological: 1-11 samples/year since early 1960's, daily water temp. since 1985	physical, chemical, biological data every two weeks since 1977	physical and chemical every two weeks since 1986, biological data every two weeks since 1994	physical, chemical, biological data since1959.	physical since 1975, chemical since 1980, biological since 1986, weekly to monthly sampling

Appendix Table (continued)

Lake Name	Mugglesee	Bodensee	Mikolajskie Lake	Sniardwy	Talty-Rynskie Lake
Country	Germany	Germany, Switzerland, Austria	Poland	Poland	Poland
Latitude	52°26'N	47°30'N	53°46.5'N	53°44.5'N	53°51.5N
Longitude	13°39'E	9°E	21°35.5'E	21°45'E	21°32'E
Elevation (masl)	32	397	116	115.7	116
Surface Area (ha)	730	47600	460	11340	1831
Volume (10^6 m^3)	36	47700	56	660	248
Max Depth (m)	8	252	25.9	23.4	50.8
Shoreline Length (km)	11.5	165	15.1	97.1	58.7
Catchment Area (km^2)	7000	10819	16.9	223	52
Catchment Type	agricultural, forested, urban	forested, agricultural, urban	forested, agricultural, urban	forested, agricultural, urban	agricultural, urban
Retention Time (yr)	0.1	>3	0.16	1.4	1.7
Trophic Status	hypereutrophic	mestrophic	hypereutrophic	eutrophic	polytrophy
Contact	10	11	12	12	12
Summary Reference	15	16	-	-	-
Available Data	physical since 1973, chemical since 1980, biological since 1970's, weekly to biweekly sampling	physical, chemical, biological every 2-4 weeks since 1963.	physical and chemical: 2 samples/year; biological: 1 sample per year; annually since 1985 (1987 and 1989 missing).	physical and chemical: 2 samples/year; biological: 1 sample per year; annually since 1985 (1987 and 1989 missing).	physical and chemical: 2 samples/year; biological: 1 sample per year; annually since 1985 (1987 and 1989 missing).

Appendix Table (continued)

Lake Name	Niegocin	Beldany	Kisajno	Dargin	Izvorul Muntelui Reservoir
Country	Poland	Poland	Poland	Poland	Romania
Latitude	54°00'N	53°43'N	54°04'N	54°07.5'N	47°N
Longitude	21°47'E	21°35'E	21°42.5'E	21°41.2'E	26°E
Elevation (masl)	116	117	116	116	500
Surface Area (ha)	2600	941	1896	3030	3200
Volume (10^6 m^3)	259	95	160	322	1230
Max Depth (m)	39.7	46	25	37.6	90
Shoreline Length (km)	35.4	34	50	33	78
Catchment Area (km^2)	47	980	-	-	-
Catchment Type	agricultural, urban	forested, agricultural	agricultural, forested	agricultural, forested	forested
Retention Time (yr)	4	0.5	14	14	0.33
Trophic Status	hypereutrophic	eutrophic	mesotrophic	mesotrophic	oligo-mesotrophic
Contact	12	12	12	12	13
Summary Reference	-	-	-	-	-
Available Data	physical and chemical: 2 samples/year; biological: 1 sample per year; annually since 1985 (1987 and 1989 missing).	physical and chemical: 2 samples/year; biological: 1 sample per year; annually since 1985 (1987 and 1989 missing).	physical and chemical: 2 samples/year; biological: 1 sample per year; in 1984, '85, '93, '94, and '95	physical and chemical: 2 samples/year; biological: 1 sample per year; in 1984, '85, '93, '94, and '96	physical,chemical and biological; 3 samples/yr for >10years

Appendix Table (continued)

Lake Name	**Stanca-Costesti Reservoir**	**Porti de Fier Reservoir**	**Razelm**	**Pleshcheevo**	**Darwin National Parks Lakes (7 lakes)**
Country	Romania	Romania	Romania	Russia	Russia
Latitude	48°N	44°50'N	45°E	56°50'N	58°36'N
Longitude	27°E	22°E	29°E	38°45'E	37°32'E
Elevation (masl)	90	69	0	137	103-105
Surface Area (ha)	9000	17200	60000	5150	.5-200
Volume (10^6 m^3)	1400	2900	909	582	
Max Depth (m)	35	28	5	24.3	1.5-4
Shoreline Length (km)	140	250	120		0.3-5
Catchment Area (km^2)	-	-	-	406	0.15-7.6
Catchment Type	forested, agricultural	forested	delta	agricultural, forested, urban	forested, wetland
Retention Time (yr)	<1	<2	<3	>3	1-3
Trophic Status	oligotrophic	oligotrophic	oligotrophic	mesotrophic	oligtrophic, mesotrophic, eutrophic, dystrophic
Contact	13	13	13	14	15
Summary Reference	-	-	-	17	18
Available Data	physical,chemical and biological; 4 samples/yr for >10years	physical,chemical and biological; 6 samples/yr for >10years	physical,chemical and biological; 3 samples/yr for >10years	physical and chemical: 1-6 samples/year; continuous since 1984, scattered earlier records. biological: 1-10 samples/year; scattered years ending in 1986	physical, chemical, biological: 1-9 samples/year; since 1984

Appendix Table (continued)

Lake Name	Rybinsk Reservoir	Kiev Reservoir	Kremenchug Reservoir	Dnieprodzerzhinsk Reservoir	Zaporozhye Reservoir
Country	Russia	Ukraine	Ukraine	Ukraine	Ukraine
Latitude	58°30'N	-	-	-	-
Longitude	38°20'E	-	-	-	-
Elevation (masl)	102	-	-	-	-
Surface Area (ha)	455000	32200	225000	56700	41000
Volume (10^6 m^3)	25420	3700	13500	2450	3300
Max Depth (m)	30	14.5	20	16	45
Shoreline Length (km)	2150	-	-	-	-
Catchment Area (km^2)	150500	-	-	-	-
Catchment Type	forested, agricultural	agricultural, urban	agricultural, urban	agricultural, urban	agricultural, urban
Retention Time (yr)	<1	10.5	3	19	13
Trophic Status	eutrophic	eutrophic	eutrophic	eutrophic	eutrophic
Contact	16	17	17	17	17
Summary Reference	19	-	-	-	-
Available Data	physical, chemical, and biological data weeklly to monthly for >30 years	physical, chemical, and biological data for >30 years	physical, chemical, and biological data for >30 years	physical, chemical, and biological data for >30 years	physical, chemical, and biological data for >30 years

Appendix Table (continued)

Lake Name	Kakhovka Reservoir	Lake Washington	Castle Lake	Lake Tahoe	7 Wisconsin Long-Term Ecological Research Lakes
Country	Ukraine	United States	United States	United States	United States
Latitude	-	47°38'N	41°13'N	39°10'N	46°01'N
Longitude	-	122°15W	122°22'W	120°08'W	89°40'W
Elevation (masl)	-	9	1657	1898	492-502
Surface Area (ha)	215000	8762	20	49900	0.5-1608
Volume (10^6 m^3)	18200	2885	2.3	156000	0.009-132
Max Depth (m)	32	65.2	35	501	2.5-35.7
Shoreline Length (km)	-	115	2150	113	0.2-31
Catchment Area (km^2)	-	1588	0.8	816	0.3-104
Catchment Type	agricultural, urban	forested, agricultural, urban	forested	forested	forested
Retention Time (yr)	2.5	1 to 3	>3	>700	<1 - >3
Trophic Status	eutrophic	mesotrophic	meso-oligotrophic	ultra-oligotrophic	oligotrophic, mesotrophic, dystrophic
Contact	17	-	18	18	19
Summary Reference	-	20	21	22	23
Available Data	physical, chemical, and biological data for >30 years	physical, chemical, biological weekly to monthly since 1956	physical, chemical, biological weekly to monthly since 1959	physical, chemical, biological at 10-30 day intervals since 1962	physical, chemical, biological data biweekly to annually since 1981

Appendix Table (continued)

Lake Name	Lake Mendota	Windermere North Basin	Windermere South Basin	Esthwaite Water
Country	United States	United Kingdom	United Kingdom	United Kingdom
Latitude	43° 06'N	54° 23'N	54° 20'N	54° 21'N
Longitude	89° 25'W	2° 56'E	2° 56'E	2° 59'E
Elevation (masl)	259	39	39	65
Surface Area (ha)	3985	8.04	6.71	1.00
Volume (10^6 m^3)	506	201.8	112.7	6.4
Max Depth (m)	25.3	64	42	15
Shoreline length (km)	35.2	---	---	---
Catchment Area (km^2)	602	230 (whole lake)	230 (whole lake)	17.1
Catchment Type	agricultural, urban	grassland, woodland	grassland, woodland	grassland,woodland
Retention Time (yr)	6.5	0.51	0.27	0.24
Trophic Status	eutrophic	mesotrophic	mesotrophic	eutrophic
Contact	20	21	21	21
Summary Reference	24	25, 26, 27	25, 26, 27	25, 26, 27
Available Data	physical, chemical, biological, weekly to monthly since 1975	physical, chemical and biological, weekly from 1946 to 1982, weekly in summer and fortnightly in winter from 1983 to present		

Contacts

[1]Dr. M. Dokulil, Inst. fur Limnologie, Gaisberg 116, a-5310 Mondsee, Austria
[2]Dr. P. Dillon, Ontario Ministry of Environment and Energy, Dorset Research Centre, Bellwood Acres Road, P.O. Box 39, Dorset, Ontario, P0A 1E0
[3]Dr. John Shearer, Freshwater Institute, Winnipeg, Canada
[4]Hydrobiological Institute, Academy of Sciences of the Czech Republic, Na sadkach 7, 37005 C. Budejovice, Czech Republic
[5]Dr. Jan Fott, Department of Hydrobiology, Faculty of Sciences, Charles University, Vinicna 7, 12844 Prague, Czech Republic
[6]Mrs. Sari Antikainen, Finnish Environment Agency, P.O. Box 140, FIN-00251, Finland
7Dr. H. Ronicke, Environmental Research Centre Leipzig-Halle, Department for Inland Water Research, Magdeburg, Germany
[8]Dr R. Koschel, Institut fur Gewasserokologie und Binnenfischerei, Alte Fischerhutte 2, D-16775 Neuglobsow, Germany
[9]Dr.Rita Adrian, Institute fur Gewasserokologie und Binnenfischerei, Muggelseedamm 260, D-12587 Berlin, Germany
[10]Dr. Norbert Walz, Institut fur Gewasserokologie und Binnenfisherei, Muggelseedamm 260, D 12587 Berlin, Germany
[11]Dr. Ursula Gaedke, Limnologisches Institut, Mainaustr 212, D-78434 Konstanz, Germany
[12]Dr. Jo'anta Ejsmont-Karabin, Institute of Ecology PAS, Hydrobiological Station, ul Lesna 13, 11-730 Mikolajki, Poland
[13] Romanian Waters Authority
[14]Dr. Ernst S. Bikbulatov, Institute for Biology of Inland Waters, Russian Academy of Sciences, 152742, Borok, Yaroslavl, Russia
[15]Dr. Victor T. Komov, Institute for Biology of Inland Waters, Russian Academy of Science, 152742, Borok, Yaroslavl, Russia
[16]Dr. Alexander I. Kopylov, Institute for Biology of Inland Waters, Russian Academy of Science, 152742, Borok, Yaroslavl, Russia
[17]Dr. L Mikhaylenko, Institute of Hydrobiology of NAS of Ukraine, 12 Geroyev Stalingrada, Kiev-210, 254655, Ukraine
[18]Dr. Charles R. Goldman, Division of Environmental Studies, University of California, Davis, California, 95616 USA
[19]Dr. Tim Kratz, University of Wisconsin Trout Lake Station, 10810 County Highway N, Boulder Junction, Wisconsin, 54512, USA

[20]Dr. Richard C. Lathrop, Wisconsin Department of Natural Resources, 1350 Femrite Drive, Monona, Wisconsin 53716, USA
[21]Dr. Glen George, Institute of Freshwater Ecology, Windermere Laboratory, Far Sawrey, Ambleside, Cumbria LA22 0LP, UK

References

[1]Dokulil, M. and C. Skolaut. 1986. Succession of phytoplankton in a deep stratifying lake: Mondsee, Austria. Hydrobiologia 138:9-24

[2]Loffler, H. (ed.) 1979. Neusiedlersee: the limnology of a shallow lake in central Europe. Monographiae Biologicae 37. Junk Publisher, The Hague

[3]Brunskill, G.J., and D.W. Schindler. 1971. J. Fish. Res. Board. Can. 28:139-155.

[4]Beaty, K.G. 1981. Can. Data Rep. Fish. Aquat. Sci. 285: 367p.

[5]Straskraba, M. 1991. Lake Slapy. pp. 2-20 in Lake Biwa Research Institute (eds.)Data book of world lake environments - a survey of the state of world lakes.

[6]Hejzlar, J., and M. Straskraba. 1989. On the horizontal distribution of limnological variables in Rimov and other stratified Czechoslovak reservoirs. Arch. Hydrobiol. Beih., Ergebn. Limnol. 33:41-55.

[7]Fott, J., Prazakova, M., Stuchlik, E., Stuchlikova, Z., 1994. Acidification of lakes in Sumava (Bohemia) and in the high Tatra Mountains (Slovakia). Hydrobiologia 274:1-11.

[8]Huttula, T. 1992. Modelling resuspension and settling in lakes using a one dimensional vertical model. Aqua Fennica 22:23-34.

[9]Sarvala, J. and K. Jumppanen. 1988. Nutrients and planktivorous fish as regulators of productivity in Lake Pyhajarvi, SW Finland. Aqua Fennica 18:137-155.

[10]Pekkarinen, M. 1990. Comprehensive survey of the hypertrophic Lake Tuusulanjarvi, nutrient loading, water quality and prospects of restoration. Aqua Fennica 20:13-25.

[11]Ronicke, H., H. Klapper, J. Tittel, B. Zippel, and M. Beyer. 1995. Control of phosphorus and plankton by calcite flushing in Lake Arendsee: enclosure-experiments. XXVI Congress of Int. Assn. of Theor. and App. Limn., Sao Paulo, Brasil. Proceedings.

[12]Ronicke, H., H. Klapper, and M. Beyer. 1993. Control of phosphorus and blue greens by nutrient precipitation. 5th International Conference on the Conservation and Management of Lakes, Stresa, Italy, pp. 177-179

[13]Casper, S.J. (ed.). 1985. Lake Stechlin: a temperate oligotrophic lake. Monographien Biol. 58. Junk Publishers, Dordrecht.
[14]Adrian, R., R. Deneke, U. Mischke, R. Stellmacher, and P. Lederer. 1995. A long-term study of the Heiligensee (1975-1992): evidence for effects of climate change on the dynamics of eutrophied lake ecosystems. Arch Hydrobiol. 133:315-337.
[15]Nixdorf, B. and S. Hoeg. 1993. Phytoplankton-community structure, succession and cholorphyll content in Lake Muggelsee from 1979 to 1990. Int. Revue ges. Hydrobiol. 78:359-377.
[16]Gaedke, U. and A. Schweizer. 1993. The first decade of oligotrophication in Lake Constance. I. The response of phytoplankton biomass and cell size. Oecologia 93:268-275.
[17]Butorin, N.V. ed. 1989. Ecosystem of Lake Pleshceevo. Nauka Press, Leningrad. 264p. (in Russian)
[18]Komov, V.T. (ed.) 1994. The Structure and Functioning of Acid Lake Ecosystems. Nauka Press, Saint Petersburg, 248pp. (in Russian)
[19]Hayka, A. 1972. Rybinsk Reservoir and its Life. Nauka Press, Leningrad. 364pp (in Russian)
[20]Edmondson, W. T. 1994. Sixty years of Lake Washington: a curriculum vitae. Lake and Reservoir Management 10:75-84.
[21]Goldman, C.R., A.D. Jassby, and T.M. Powell. 1989. Interannual fluctuations in primary production: meteorological forcing at two subalpine lakes. Limnology and Oceanography 34:310-323.
[22]Goldman, C.R. 1990. The importance of long-term limnological research with emphasis on Lake Tahoe and Castle Lake. pp 221-231 In R. de Bernadi et al. (eds.), Scientific Perspectives in Theoretical and Applied Limnology. C.N.R., Instituto Italia di Idrobiologia, Pallanza, Italy.
[23]Magnuson, J.J., and C.J. Bowser. 1990. A network for long-term ecological research in the United States. Freshwater Biology 23:137-143.
[24]Kitchell, J.F. (ed.) 1992. Food Web Management: a case study of Lake Mendota. Springer-Verlag, New York.
[25]Kadiri, M.O., and C.S. Reynolds. 1993. Long-term monitoring of the conditions of lakes: the example of the English Lake District. Arch. Hydrobiol. 129:157-178.
[26]Talling, J.F. 1993. Comparative seasonal changes, and inter-annual variability and stability, in a 26-year record of total phytoplankton biomass in four English lake basins. Hydrobiologia 268:65-98.
[27]Heaney, S.I., J.E. Parker, C. Butterwick, and K.J. Clarke. 1996. Interannual variability of algal populations and their influence on lake metabolism. Freshwater Biology 35:561-577.

INDEX

9 780792 350552

MIX
Papier aus verantwortungsvollen Quellen
Paper from responsible sources
FSC® C105338

If you have any concerns about our products,
you can contact us on
ProductSafety@springernature.com

In case Publisher is established outside the EU,
the EU authorized representative is:
Springer Nature Customer Service Center GmbH
Europaplatz 3, 69115 Heidelberg, Germany

Printed by Libri Plureos GmbH
in Hamburg, Germany